Functional Materials
Electrical, Dielectric, Electromagnetic, Optical and Magnetic Applications

Second Edition

Engineering Materials for Technological Needs

Series Editor: Deborah D. L. Chung
(University at Buffalo, The State University of New York)

Vol. 4:   *Functional Materials: Electrical, Dielectric, Electromagnetic, Optical and Magnetic Applications*
Second Edition
by Deborah D. L. Chung (University at Buffalo, The State University of New York)

Vol. 3:   *Carbon Materials: Science and Applications*
by Deborah D. L. Chung (University at Buffalo, The State University of New York)

Vol. 2:   *Functional Materials: Electrical, Dielectric, Electromagnetic, Optical and Magnetic Applications*
(With Companion Solution Manual)
by Deborah D. L. Chung (University at Buffalo, The State University of New York)

Vol. 1:   *High-Performance Construction Materials: Science and Applications*
edited by Caijun Shi (Hunan University, China) &
Yilung Mo (University of Houston, USA)

Functional Materials

Electrical, Dielectric, Electromagnetic, Optical and Magnetic Applications

Second Edition

Vol. 4

ENGINEERING MATERIALS FOR TECHNOLOGICAL NEEDS

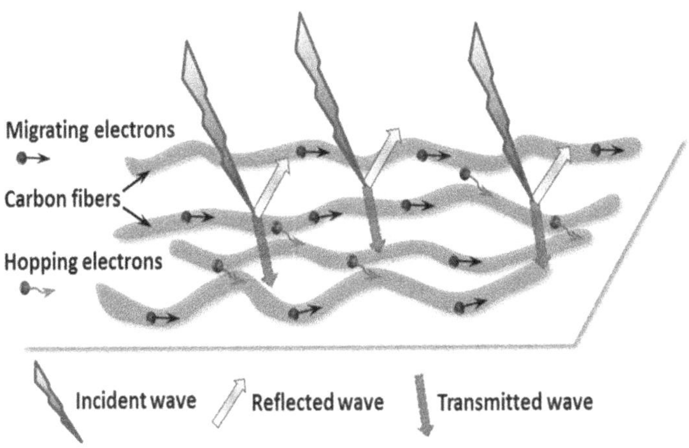

Deborah D. L. Chung

University at Buffalo, The State University of New York, USA

World Scientific

NEW JERSEY · LONDON · SINGAPORE · BEIJING · SHANGHAI · HONG KONG · TAIPEI · CHENNAI · TOKYO

Published by

World Scientific Publishing Co. Pte. Ltd.
5 Toh Tuck Link, Singapore 596224
USA office: 27 Warren Street, Suite 401-402, Hackensack, NJ 07601
UK office: 57 Shelton Street, Covent Garden, London WC2H 9HE

British Library Cataloguing-in-Publication Data
A catalogue record for this book is available from the British Library.

First published 2022 (hardcover)
Reprinted 2024 (in paperback edition)
ISBN 978-981-12-9033-6 (pbk)

Engineering Materials for Technological Needs — Vol. 4
FUNCTIONAL MATERIALS
Electrical, Dielectric, Electromagnetic, Optical and Magnetic Applications
Second Edition

ISBN 978-981-123-883-3 (hardcover)
ISBN 978-981-123-884-0 (ebook for institutions)
ISBN 978-981-123-885-7 (ebook for individuals)

For any available supplementary material, please visit
https://www.worldscientific.com/worldscibooks/10.1142/12331#t=suppl

In memory of my maternal grandmother, Dr. Lee Sun Chau 周理信, M.D. (1890–1979).

Photo taken in ~1915
Graduation from Hackett Medical College for Women,
Guangzhou, China.
https://en.wikipedia.org/wiki/Lee_Sun_Chau

Preface to 2nd Edition

Functional materials refer to materials that have been tailored to exhibit certain desirable functional properties that are useful for applications. Functional properties are to be distinguished from structural properties, which have dominated the field of materials for decades.

The field of functional materials has grown considerably since the publication of the 1st Edition of this book in 2010. The growth is fueled by the widening of the discipline of materials science and engineering, and the increase in cross-disciplinary work. In particular, the growth involves the broadening of the scope of multifunctional structural materials to include functionalities other than the basic function of sensing, the increased attention to electromagnetic functionalities (due to the importance of telecommunication, wireless communication and 5G), and the intensive research on electrical and dielectric functionalities (due to the need for energy storage and energy conversion).

As in the 1st Edition, the 2nd Edition addresses the science and applications in a tutorial fashion, covering the basic concepts on the electrical, dielectric, electromagnetic, optical and magnetic behavior of materials. The 2nd Edition provides enhancement of all the chapters in both substance and pedagogy, particularly concerning the dielectric and electromagnetic behavior. Moreover, the introductory chapter has been strengthened to provide an overview. In order to facilitate learning, numerous cross references and example problems have been added. The 2nd Edition is longer than the 1st Edition by 46%.

I hope that this book will catalyze further growth of the field, in addition to widening the workforce that is necessary for the growth. My thanks go to my numerous students and scholars who have worked with me on this subject over the years. I am grateful to God for enabling me to pour out the science from my heart through this book.

Deborah D.L. Chung
Buffalo, NY, U.S.A., August 21, 2021
https://en.wikipedia.org/wiki/Deborah_Chung

Preface to 1st Edition

The field of materials science and engineering started decades ago with emphasis on structural materials. As the development of structural materials matures and the technological need for functional materials intensifies, the field has moved from an emphasis on structural materials to one on functional materials. The development of functional materials is at the heart of current technological needs and is at the forefront of current materials research. Knowledge of the competing methods for providing a function or a multiplicity of functions is necessary for today's design engineers.

This book is a textbook for use in a course on functional materials, which refer to engineering materials that are for functional applications, such as electrical, dielectric, electromagnetic, optical, and magnetic applications. The functional materials include all types of materials, including metals, polymers, ceramics, composites, carbons and semiconductors. Smart materials and electronic materials are subsets of functional materials covered in this book.

The functions mentioned above are relevant to smart structures, smart fluids, heating, deicing, microelectronics, capacitors, electrical insulation, batteries, computer memories, optics, lighting, lasers, light detectors, photocopying, communication, sensors, structural health monitoring, nondestructive evaluation, electromagnetic interference shielding, low observability (Stealth technology), radio frequency identification (RFID), magnetic recording, actuators, energy harvesting, motors, etc.

Textbooks that are relevant to the subject of functional materials have not kept pace with the technological needs and the associated scientific advances. Textbooks on introductory materials science give rather cursory coverage of functional materials. Limited textbooks on electronic materials are focused on semiconductor materials in relation to electrical and optical functions. Books on smart materials and structures are focused on a relatively small number of specific kinds of dielectric, magnetic and optical properties that are typically used to provide multifunctionality to structures. Furthermore, the coverage of applications is limited in textbooks that are relevant to functional materials, in spite of the importance of linking functional properties to applications in the learning experience. This book provides a comprehensive treatment of the science and applications of functional materials.

This book assumes that the readers have had a one-semester introductory undergraduate course on materials science, so that they have the basic background on crystal structures, imperfections and phase diagrams. In this book, the coverage on functional materials is much broader and deeper than that in an introductory materials science course. Due to the limited time for functional materials coverage in an introductory materials science course, this book covers the functional materials topics without assuming that the readers have acquired basic background in any of the functional properties. Thus, this book introduces the scientific concepts behind each functional property from scratch. The introduction of the basic concepts does not involve quantum mechanics, so the physics background required of the readers is basic. The book does not involve complicated organic or inorganic chemistry, so the chemistry background required of the readers is just introductory chemistry. In addition, this book applies the concepts to a large variety of applications, which are relevant to the electronic, communication, information, energy, aerospace, automotive, transportation, homeland security, medical, construction and control industries. All concepts covered are strongly linked to applications.

In order to help the readers learn thoroughly, most chapters include sections on example problems, review questions and supplementary reading. The book features hundreds of illustrations for helping the readers grasp the concepts involved. The illustrations include numerous graphs that contain real data obtained on real materials. The data help solidify the learning experience, in addition to making the book up-to-date.

This book is suitable for use as a textbook in undergraduate and graduate courses. Relevant academic disciplines include materials, chemical, mechanical, electrical, aerospace, civil and industrial engineering.

This book is also suitable for use as a reference book for students and professionals that are interested in the use of functional materials and the development of applications that involve functional material. Relevant industries include electronic, computer, communication, aerospace, automotive, transportation, construction, energy and control industries.

Deborah D.L. Chung
Buffalo, NY
Oct. 1, 2009
http://alum.mit.edu/www/ddlchung

Contents

Chapter 1

Introduction to Functional Materials and their Applications

1.1 Types of materials

Engineering materials constitute the foundation of technology, whether the technology pertains to structural, electronic, thermal, electro-chemical, environmental, biomedical or other applications. The history of human civilization evolved from the Stone Age to the Bronze Age, the Iron Age, the Steel Age and to the Space Age (simultaneously the Electronic Age). Each age is marked by the advent of certain materials. The Iron Age brought tools and utensils. The Steel Age brought rails and the industrial revolution. The Space Age was brought by structural materials (e.g., composite materials) that are both strong and lightweight. The Electronic Age was brought by semiconductors. Materials include metals, polymers, ceramics, semiconductors and composite materials.

Metals (including alloys) consist of atoms and are characterized by metallic bonding (i.e., the valence electrons of each atom being delocalized and shared among all the atoms). Most of the elements in the Periodic Table are metals. Examples of alloys are Cu-Zn (brass), Fe-C (steel) and Sn-Pb (solder). Alloys are classified according to the majority element present. The main classes of alloys are iron-based alloys (for structures), copper-based alloys (for piping, utensils, thermal conduction, electrical conduction, etc.) and aluminum-based alloys (for lightweight structures and for metal-matrix composites). Alloys are almost always in the polycrystalline form.

Ceramics are inorganic compounds such as Al_2O_3 (for spark plugs and for substrates for microelectronics), SiO_2 (for electrical insulation in microelectronics), Fe_3O_4 (ferrite for magnetic memories used in computers), silicates (e.g., clay, cement, glass, etc.), SiC (an abrasive), etc. The main classes of ceramics are oxides, carbides, nitrides and silicates. Ceramics are typically partly crystalline and partly amorphous. They consist of ions (often atoms as well) and are characterized by ionic bonding (often covalent bonding as well).

1

Polymers in the form of thermoplastics (e.g., nylon, polyethylene, polyvinyl chloride, rubber, etc.) consist of molecules which have covalent bonding within each molecule and van der Waals forces between the molecules. Polymers in the form of thermosets (e.g., epoxy, phenolics, etc.) consist of a network of covalent bonds. Polymers are amorphous, except for a minority of thermoplastics. Due to the bonding, polymers are typically electrical and thermal insulators. However, conducting polymers can be obtained by doping and conducting polymer-matrix composites can be obtained by the use of conducting fillers.

Semiconductors are characterized by having the highest occupied energy band (the valence band, where the valence electrons reside energetically) being full, such that the energy gap between the top of the valence band and the bottom of the empty energy band above (called the conduction band) is small enough for some fraction of the valence electrons to be excited from the valence band to the conduction band by thermal, optical or other forms of energy. Conventional semiconductors, such as silicon, germanium and gallium arsenide (GaAs, a compound semiconductor), are covalent network solids. They are usually doped in order to enhance the electrical conductivity. They are typically used in the form of single crystals without dislocations, as grain boundaries and dislocations would degrade the electrical behavior.

Composite materials are multi-phase materials obtained by artificial combination of different materials, so as to attain properties that the individual components by themselves cannot attain. They are not multiphase materials in which the different phases are formed naturally by reactions, phase transformations, or other phenomena. An example is carbon fiber reinforced polymer. Composite materials are to be distinguished from alloys which may comprise two or more components but are formed naturally through processes such as casting. Composite materials can be tailored for various properties through choice of the components, their proportions, their distributions, their morphologies, their degrees of crystallinity, their crystallographic textures, as well as choice of the structure and composition of the interface between one component and another. Due to the strong tailorability, composite materials can be designed to satisfy the needs of technologies, which relate to aerospace, automobile, electronics, construction, energy, bio-medical and other industries. As a result, composite materials constitute the majority of commercial engineering materials.

1.2 Composite materials

An example of a composite material is a lightweight structural composite that is obtained by embedding continuous carbon fibers in one or more orientations in a polymer matrix [1]. The fibers provide the strength and stiffness, while the polymer serves as the binder. In particular, carbon fiber polymer-matrix composites have the following attractive properties:

- low density (lower than aluminum)
- high strength (as strong as high-strength steels)
- high stiffness (stiffer than titanium, yet much lower in density)
- good fatigue resistance
- good creep resistance
- low friction coefficient and good wear resistance
- toughness and damage tolerance (as enabled by using appropriate fiber orientations)
- chemical resistance (chemical resistance controlled by the polymer matrix)
- corrosion resistance
- dimensional stability (can be designed for zero CTE)
- vibration damping ability
- low electrical resistivity
- high electromagnetic interference (EMI) shielding effectiveness
- high thermal conductivity

Another example of a composite is concrete, which is a structural composite obtained by combining (through mixing) cement (the matrix, i.e., the binder, obtained by a reaction, known as hydration, between cement and water), sand (fine aggregate), gravel (coarse aggregate) and optionally other ingredients that are known as admixtures. Short fibers and silica fume (a fine SiO_2 particulate) are examples of admixtures. In general, composites are classified according to their matrix material. The main classes of composites are polymer-matrix, cement-matrix, metal-matrix, carbon-matrix and ceramic-matrix composites.

Polymer-matrix and cement-matrix composites are the most common, due to the low cost of fabrication. Polymer-matrix composites are used for lightweight structures (aircraft, sporting goods, wheel chairs, etc.), in addition to vibration damping, electronic enclosures, asphalt (composite with pitch, a polymer, as the matrix), solder replacement, etc. Cement-matrix composites in the form of concrete (with fine and coarse aggregates), steel reinforced concrete, mortar (with fine aggregate, but no coarse aggregate) or cement paste (without any aggregate) are used for civil structures, prefabricated housing, architectural precasts, masonry, landfill cover, thermal insulation, sound absorption, etc.

Carbon-matrix composites are important for lightweight structures (e.g., Space Shuttle) and components (e.g., aircraft brakes) that need to withstand high temperatures, but they are relatively expensive due to the high cost of fabrication. Carbon-matrix composites suffer from their tendency to be oxidized ($2C + O_2 \rightarrow 2CO$), thereby becoming vapor.

Carbon fiber carbon-matrix composites, also called carbon-carbon composites, are the most advanced form of carbon, as the carbon fiber reinforcement makes them stronger, tougher, and more resistant to thermal shock than conventional graphite. With the low density of carbon, the specific strength (strength/density), specific modulus (modulus/density) and specific thermal conductivity (thermal conductivity/ density) of carbon-carbon composites are the highest among composites. Furthermore, the coefficient of thermal expansion is near zero.

Ceramic-matrix composites are superior to carbon-matrix composites in the oxidation resistance, but they are not as well developed as carbon-matrix composites. Metal-matrix composites with aluminum as the matrix are used for lightweight structures and low-thermal-expansion electronic enclosures, but their applications are limited by the high cost of fabrication and by galvanic corrosion.

Metal-matrix composites are gaining importance because the reinforcement serves to reduce the coefficient of thermal expansion (CTE) and increase the strength and modulus. If a relatively graphitic kind of carbon fiber is used, the thermal conductivity can be enhanced also. The combination of low CTE and high thermal conductivity makes them very attractive for electronic packaging (e.g., heat sinks). Besides good thermal properties, their low density makes them particularly desirable for aerospace electronics and orbiting space structures; orbiters are thermally cycled by moving through the earth's shadow.

Compared to the metal itself, a carbon fiber metal-matrix composite is characterized by a higher strength-to-density ratio (i.e., specific strength), a higher modulus-to-density ratio (i.e., specific modulus), better fatigue resistance, better high temperature mechanical properties (a higher strength and a lower creep rate), a lower CTE, and better wear resistance.

Compared to carbon fiber polymer-matrix composites, a carbon fiber metal-matrix composite is characterized by higher temperature resistance, higher fire resistance, higher transverse strength and modulus, the lack of moisture absorption, a higher thermal conductivity, a lower electrical resistivity, better radiation resistance, and absence of outgassing.

On the other hand, a metal-matrix composite has the following disadvantages compared to the metal itself and the corresponding polymer-matrix composite: higher fabrication cost and limited service experience.

Fibers used for load-bearing metal-matrix composites are mostly in the form of continuous fibers, but short fibers are also used. The matrices used include aluminum, magnesium, copper, nickel, tin alloys, silver-copper, and lead alloys. Aluminum is by far the most widely used matrix metal because of its low density, low melting temperature (which makes composite fabrication and joining relatively convenient), low cost, and good mechinability. Magnesium is comparably low in melting temperature, but its density is even lower than aluminum. Applications include structures (aluminum, magnesium), electronic heat sinks and substrates (aluminum, copper), soldering and bearings (tin alloys), brazing (silver-copper), and high-temperature applications (nickel).

Although cement is a ceramic material, ceramic-matrix composites usually refer to those with silicon carbide, silicon nitride, alumina, mullite, glasses and other ceramic matrices that are not cement.

Ceramic-matrix fiber composites are gaining increasing attention because the good oxidation resistance of the ceramic matrix (compared to a carbon matrix) makes the composites attractive for high-temperature applications (e.g., aerospace and engine components). The fibers serve mainly to increase toughness and strength (tensile and flexural) of the composite due to their tendency to be partially pulled out during the deformation. This pullout absorbs energy, thereby toughening the composite. Although the fiber pullout has its advantage, the bonding between the fibers and the matrix must still be sufficiently good in order for the fibers to strengthen the composite effectively. Therefore, the control of the bonding between the fibers and the matrix is important for the development of these composites.

In case of the reinforcement being carbon fibers, the reinforcement has a second function, which is to increase the thermal conductivity of the composite, as the ceramic is mostly thermally insulating whereas carbon fibers are thermally conductive. In electronic, aerospace, and engine components, the enhanced thermal conductivity is attractive for heat dissipation.

A third function of the reinforcement is to decrease the drying shrinkage in the case of ceramic matrices prepared by using slurries or slips. In general, the drying shrinkage decreases with increasing solid content in the slurry. Fibers are more effective than particles for decreasing the drying shrinkage. This function is attractive for the dimensional control of parts made from the composites.

Fiber reinforced glasses are useful for space structural applications, such as mirror back structures and supports, booms and antenna structures. In low earth orbit, these structures experience a temperature range from -100 to $80°C$, so they need an improved thermal conductivity and a reduced coefficient of thermal expansion. In addition, increased toughness, strength and modulus are desirable. Due to the environment degradation resistance of carbon fiber reinforced glasses, they are also potentially useful for gas turbine engine components. Additional attractions are low friction, high wear resistance, and low density.

The glass matrices used for fiber reinforced glasses include borosilicate glasses (e.g., Pyrex), aluminosilicate glasses, soda lime glasses and fused quartz. Moreover, a lithia aluminosilicate glass-ceramic and a $CaO\text{-}MgO\text{-}Al_2O_3\text{-}SiO_2$ glass-ceramic have been used.

1.3 Carbon materials

Not included in the five material categories mentioned in Sec. 1.1 is carbon, which can be in the form of graphite (most common form, Example Problems 2 and 3 at the end of the chapter), diamond (Example Problems 1 and 33) and fullerenes (a relatively new form, Example Problem 4). They are not ceramics, because they are not compounds.

Graphite (a semimetal) consists of carbon atom layers stacked in the AB sequence, such that the bonding is covalent (due to sp^2 hybridization) and metallic (two-dimensionally delocalized $2p_z$ electrons) within a layer and is van der Waals between the layers. This bonding makes graphite very anisotropic, so that it is a good lubricant (due to the ease of sliding of the layers with respect to one another). Graphite is also used for pencils because of this property. Moreover, graphite is an electrical and thermal conductor within the layers, but is an insulator in the direction perpendicular to the layers. The electrical conductivity is valuable for the use of graphite for electrochemical electrodes. Graphite is chemically quite inert. However, due to the anisotropy, graphite can undergo a reaction (known as intercalation) in which foreign species (called the intercalate) is inserted between the carbon layers. Disordered carbon (called turbostratic carbon) also has a layered structure, but, unlike graphite, it does not have the AB stacking order and the layers are bent. Upon heating, disordered carbon becomes more ordered, as the ordered form (graphite) has the lowest energy. Graphitization refers to the ordering process that leads to graphite. Conventional carbon fibers are mostly disordered carbon, such that the carbon layers are preferentially along the fiber axis. A form of graphite called flexible graphite is formed by compressing a collection of intercalated graphite flakes that have been exfoliated (i.e., allowed to expand by over 100 times along the direction perpendicular to the layers, typically through heating after intercalation). The exfoliated flakes are held together by mechanical interlocking, as there is no binder. Flexible graphite is typically in the form of sheets, which are resilient in the direction perpendicular to the sheet. The resilience allows flexible graphite to be used as a gasket for fluid sealing.

Diamond is a covalent network solid exhibiting the diamond crystal structure due to sp^3 hybridization (akin to silicon). It is used as an abrasive and as a thermal conductor. The thermal conductivity is the highest among all materials. However, it is an electrical insulator. Due to the high material cost, diamond is typically used in the form of powder or thin-film coating. Diamond is to be distinguished from diamond-like carbon (DLC), which is amorphous carbon having carbon that is sp^3-hybridized. Diamond-like carbon is mechanically weaker than diamond, but it is less expensive than diamond.

Fullerenes are molecules (C_{60}) with covalent bonding within each molecule. Adjacent molecules are held by van der Waals forces. However, fullerenes are not polymers. Carbon nanotubes are a derivative of the fullerenes, as they are essentially fullerenes with extra carbon atoms at the equator. The extra atoms cause the fullerene to be longer. For example, ten extra atoms (i.e., one equatorial band of atoms) exist in the molecule C_{70}. Carbon nanotubes can be single-wall or multi-wall nanotubes, depending on the number of carbon layers in the wall of the nanotube.

1.4 Smart structures

Smart structures refer to structures exhibiting certain functional behavior, so that they can respond to stimuli in a desirable fashion. The most basic functional property is sensing. In response to the information sensed, the structure may respond. For example, in response to the strain sensed, the structure may activate a mechanism to reduce the strain. In another example, in response to the damage sensed, the structure may activate a mechanism to repair or heal. It is important for the functional properties to be rendered without adversely affecting the structural behavior, which is critical to a structure.

Smart structures are important because of their relevance to hazard mitigation, structural vibration control, structural health monitoring, transportation engineering, operation control, security and energy management. Sensing is the most fundamental aspect of a smart structure. A structural material which is itself a sensor is multifunctional.

Whether a structural material is load bearing or not in a structure, its strength and stiffness are important. Although purely structural applications dominate, combined structural and nonstructural applications are increasingly important as smart structures and electronics become more common. Such combined applications are facilitated by smart materials, which include multifunctional structural materials and nonstructural functional materials. The functions include sensing, vibration damping, thermal insulation, electrical grounding, resistance heating (e.g., for deicing), controlled electrical conduction, electromagnetic interference shielding, lateral guidance of vehicles, energy generation and energy storage.

In addition to mechanical properties, a structural material may be required to have other properties, such as low density (light weight) for fuel saving in the case of aircraft and automobiles, for high speed in the case of race bicycles, and for handleability in the case of wheelchairs and armor. Another property that is often required is corrosion resistance, which is desirable for the durability of all structures, particularly automobiles and bridges. Yet another property that may be required is the ability to withstand high temperatures and/or thermal cycling as heat may be encountered by the structure during operation, maintenance or repair.

A relatively new trend is for a structural material to be able to serve functions other than the structural function, so that the material becomes multifunctional (akin to killing two or more birds with one stone, thereby saving cost and simplifying design). An example of a non-structural function is the sensing of damage. Such sensing, also called structural health monitoring, is valuable for the prevention of hazards. It is particularly important to aging aircraft and bridges. The sensing function can be attained by embedding sensors (such as optical fibers, the damage or strain of which affects the light throughput) in the structure. However, the embedding usually causes degradation of the mechanical properties and the embedded devices are costly and poor in durability compared to the structural material. Another way to attain the sensing function is to detect the change in property (e.g., the electrical resistivity) of the structural material due to damage. In this way, the structural material serves as its own sensor and is said to be self-sensing. Such multifunctional structural materials are also referred to as intrinsically smart materials. Intrinsic smartness is to be distinguished from extrinsic smartness, which relies on embedded or attached devices rather than the structural materials themselves in order to attain a certain non-structural function.

1.5 Intrinsic smartness

A multifunctional structural material is an intrinsically smart material. It is a structural material that is capable of both the structural function and a certain non-structural function without the need for embedded or attached devices. Nonstructural functions include the following.

(i) Sensing (sensing of strain, stress, damage, temperature, etc., with the multifunctionality based on piezoresistivity, direct piezo-electricity and thermoelectricity).

(ii) Electromagnetic interference (EMI) shielding (as needed to protect electronics and shield radio wave sources, with the multi-functionality based on electromagnetic radiation absorption and reflection).

(iii) Heating (deicing and anti-icing, with the multifunctionality based on the Joule effect).

(iv) Energy generation (generation of electricity from other forms of energy, with the multifunctionality based on thermoelectricity and direct piezoelectricity).

(v)   Energy storage (structural batteries and structural capacitors for storing energy that is intermittently generated by photovoltaics and wind turbines, and for managing the electric grid, with the multifunctionality provided by use of the reinforcing carbon fibers as the electrodes and suitable modification of the structural composite).

(vi) Actuation (providing strain or movement, with the multifunctionality based on the converse piezoelectric effect).

(vii) Self-healing (automatic repair when damage occurs, with the multifunctionality provided by encapsulated healing agents, remendable polymers, thermoplastic polymers and solvents).

1.5.1 *Self-sensing*

Self-sensing refers to the ability of a structural material to sense itself without sensor incorporation [3]. This means that the structural material is multifunctional. Compared to the use of embedded or attached sensors, self-sensing is advantageous in the low cost, high durability, large sensing volume and the absence of mechanical property loss. Decrease in strength and modulus occurs in case of embedded sensors, particularly if the sensors are not microscopic in size. In spite of these advantages, the use of embedded or attached sensors is much more common than self-sensing. This is partly due to the fact that the self-sensing structural materials tend to differ in composition from those of the commonly used structural materials. The requirement for the composition to be special limits the applications to new structures. In contrast, the attachment of sensors is applicable to existing structures. The applicability to both existing and new structures greatly widens the scope of applications.

There are two approaches to achieve self-sensing. One approach involves electrical resistance measurement. The resistance relates to the resistivity, which pertains to the conduction behavior (Ch. 2). The other approach involves capacitance measurement. The capacitance relates to the electric permittivity, which pertains to the dielectric behavior (Ch. 3).

By using an array of electrical contacts (electrodes) and measuring the resistance, capacitance or impedance between various electrodes in the array, spatially resolved sensing (one-dimensional, two-dimensional or three-dimensional) can be achieved. This technique is known as tomography.

Resistance-based self-sensing is the original self-sensing approach and has been extensively reported in cement-based materials containing various electrically conductive admixtures. The requirement of admixtures limits the application to new structures. Resistance-based self-sensing has also been extensively reported in continuous carbon fiber polymer-matrix composites, which do not need to be modified.

Capacitance-based self-sensing has received much less attention than resistance-based self-sensing, but it is attractive. In case of cement-based materials, capacitance-based self-sensing does not require any particular admixture and thus it is applicable to both existing and new structures. However, a disadvantage of capacitance-based sensing is the sensitivity to moisture, the level of which in a cement-based material changes when it rains. However, for indoor building applications, this is not an issue.

Self-sensing is valuable for building facility management through weight-based monitoring of the occupany of each HVAC (abbreviation for heating, ventilation, and air conditioning) serving area. In this way, the heating, cooling and ventilation of each HVAC serving area can be controlled in real time in accordance with the occupancy, thereby saving energy. The weight-based occupancy monitoring is advantageous to and more accurate than the use of cameras (which usurp privacy), motion detectors or carbon dioxide detectors for occupancy monitoring.

Tesistance-based sensing is based on the effect of strain on the resistivity. This phenomenon is known as piezoresistivity, which is unrelated to piezoelectricity. However, the temperature and damage also affect the resistivity. Damage increases the resistivity irreversibly, whereas elastic strain affects the resistivity reversibly. The irreversible resistivity change correlates with the irreversible strain, whereas the reversible resistivity change correlates with the reversible strain. Thus, by distinguishing between reversible and irreversible effects, elastic strain, and damage effects can be distinguished.

Increase in temperature decreases the resistivity, due to the activation energy associated with the electrons jumping across certain interfaces in the material. In the case of fiber-reinforced cement, the fiber–matrix interface and fiber-fiber interface are relevant. In case of unreinforced plain cement paste, diffuse interfaces exist. The temperature effect means that the material serves as a thermistor, which refers to a temperature-measuring device that is based on the effect of temperature on the resistance.

Both elastic strain and temperature give reversible effects on the resistivity. As a result, their decoupling needs an additional means, such as an additional mean of temperature measurement.

1.5.1.1 *Self-sensing cement-matrix composites*

The electrical resistance of strain sensing concrete (without embedded or attached sensors) changes reversibly with strain, such that the gage factor (fractional change in resistance per unit strain) is up to 700 under compression or tension. The resistance (DC/AC) increases reversibly upon tension and decreases reversibly upon compression, owing to fiber pull-out upon microcrack opening (<1 μm) and the consequent increase in fiber-matrix contact resistivity [4]. The concrete contains as low as 0.2 vol.% short carbon fibers, which are preferably those that have been surface treated. The fibers do not need to touch one another in the composite. The treatment improves the wettability with water. The presence of a large aggregate or of damage decreases the gage factor, but the strain sensing ability remains sufficient for practical use. Strain sensing concrete works even when data acquisition is wireless. The applications include structural vibration control and traffic monitoring.

Figure 1.1 shows the fractional changes in the longitudinal resistance for carbon fiber (0.5 vol.%) silica-fume cement paste at 28 days of curing during repeated uniaxial compressive loading at a series of various strain amplitudes [4]. The strain essentially returns to zero at the end of each cycle, indicating elastic deformation. The resistance decreases reversibly upon compression. Figure 1.2 [4] shows that the resistivity at the peak strain varies essentially linearly with the peak strain, thereby allowing strain sensing by resistance measurement. Figure 1.3 [4] shows that the gage factor (fractional change in resistance per unit strain) approaches 300, which is high compared to the value of 2 that is typical of metallic strain gages. However, the gage factor decreases with increasing strain amplitude, as shown in Fig. 1.3 [4]. Upon uniaxial tension, the resistivity increases, as shown in Fig. 1.4 [5] and Fig. 1.5 [5] for the longitudinal and transverse directions respectively. In the absence of the fibers, the resistance varies much less upon loading and the effect is much less reversible.

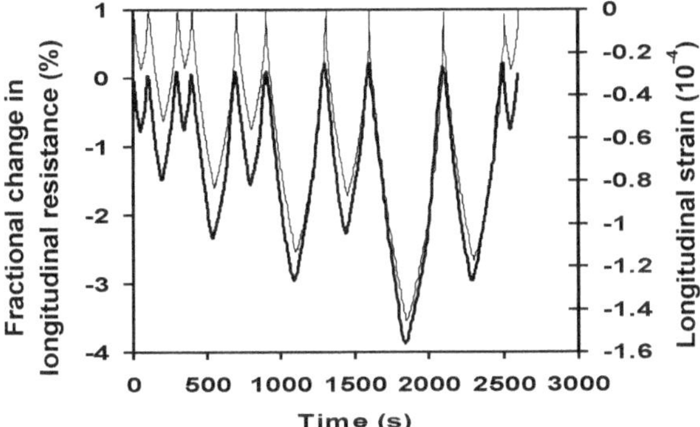

Fig. 1.1 Curves of fractional change in longitudinal resistance (thick curve) vs. time and of longitudinal strain (thin curve) vs. time during repeated compression at various strain amplitudes for a carbon fiber reinforced cement [4].

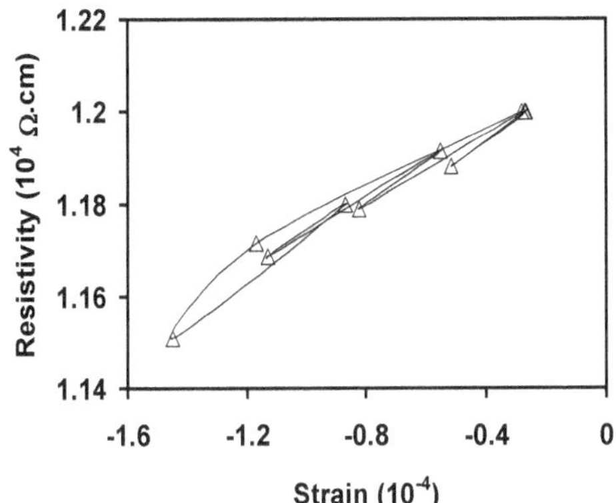

Fig. 1.2 Variation of the resistivity at the peak strain under compression vs. the strain amplitude for carbon fiber reinforced cement [4]. The data points are connected with a line drawn to indicate the order in which the various strain amplitudes are imposed. The order and data correspond to those of Fig. 1.1.

Moisture in the form of free water increases the electrical conductivity and decreases the gage factor of carbon fiber reinforced cement [6]. In addition, it increases the variability of the gage factor with the strain amplitude and with the strain history. Piezoresistivity involves electrical conduction across the interface between the fiber and the cement matrix [2]. The diminished piezoresistivity due to excessive moisture is because of the water at the fiber-matrix interface interfering the electronic conduction across the interface. In spite of the lower piezoresistive performance, the piezoresistivity remains strong and the relationship between resistivity and strain remains quite linear [6].

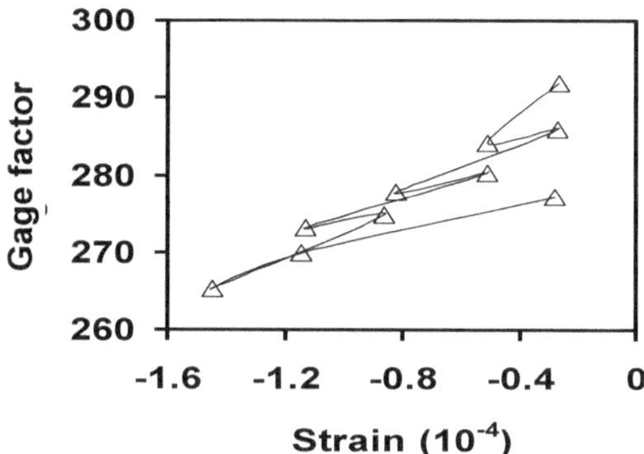

Fig. 1.3 Effect of strain amplitude on the gage factor for carbon fiber reinforced cement [4]. The data points are connected with a line drawn to indicate the order in which the various strain amplitudes are imposed. The order and data correspond to those of Fig. 1.1.

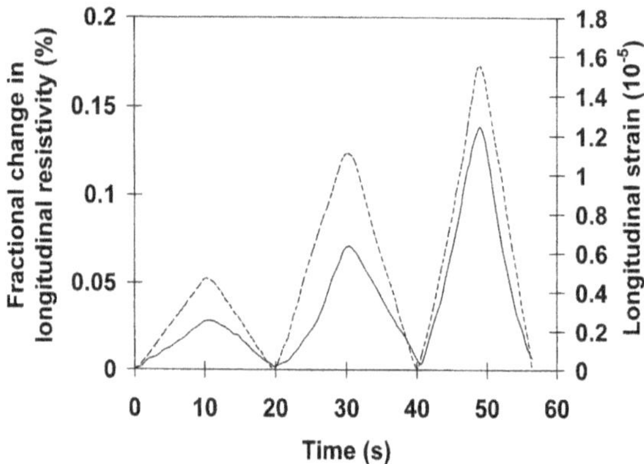

Fig. 1.4 Variation of the fractional change in longitudinal electrical resistivity with time (solid curve) and of the longitudinal strain with time (dashed curve) during dynamic uniaxial tensile loading at increasing stress amplitudes within the elastic regime for carbon fiber reinforced cement [5].

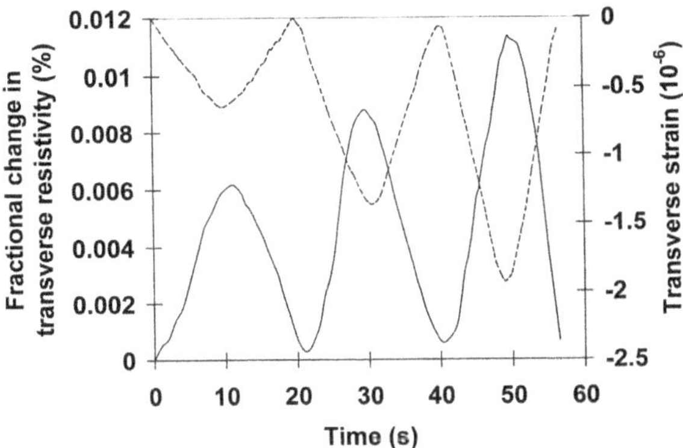

Fig. 1.5 Variation of the fractional change in transverse electrical resistivity with time (solid curve) and of the transverse strain with time (dashed curve) during dynamic uniaxial tensile loading at increasing stress amplitudes within the elastic regime for carbon fiber reinforced cement [5].

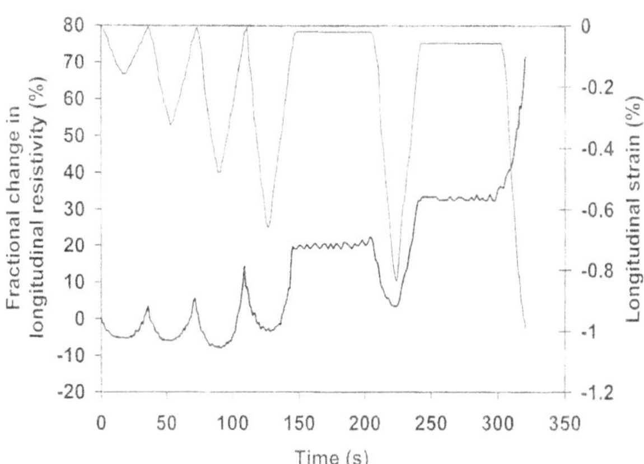

Fig. 1.6 Variation of the fractional change in longitudinal resistivity (thick curve) with time and of the longitudinal strain (thin curve) with time during uniaxial compression at progressively increasing stress amplitudes for carbon fiber reinforced cement [7].

Carbon fiber reinforced cement is also capable of sensing damage, as damage causes the resistance to increase irreversibly, as shown in Fig. 1.6 [7], which is obtained during uniaxial compression at progressively increasing strain amplitudes. The higher is the strain

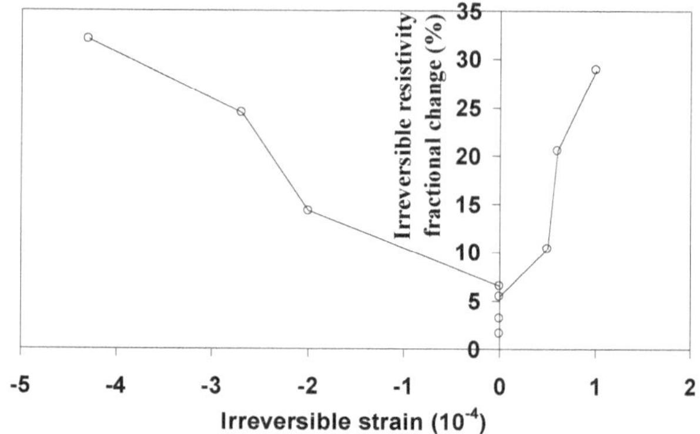

Fig. 1.7 Relationship of the irreversible resistivity fractional change with the irreversible strain for carbon fiber reinforced cement under uniaxial compression [7]. Negative strain is in the longitudinal direction; positive strain is in the transverse direction.

amplitude, the more is the damage, and the more is the irreversible resistance increase at the end of a loading cycle. The irreversible resistance increase correlates with the irreversible strain, as shown in Fig. 1.7 [7].

That both strain and damage can be sensed simultaneously through resistance measurement means that the strain/stress condition (during dynamic loading) under which damage occurs can be obtained, thus facilitating damage origin identification.

1.5.1.2 *Self-sensing polymer-matrix composites*

Polymer-matrix composites for structural applications typically contain continuous fibers such as carbon, polymer and glass fibers, as continuous fibers tend to be more effective than short fibers as a reinforcement. Polymer-matrix composites with continuous carbon fibers are used for aerospace, automobile and civil structures. (In contrast, continuous fibers are too expensive for reinforcing concrete.) Due to the fact that carbon fibers are electrically conducting, whereas polymer and glass fibers are not, carbon fiber composites are predominant among polymer-matrix composites that are intrinsically smart.

Polymer-matrix composites containing continuous carbon fibers are important structural materials owing to their high tensile strength, high tensile modulus and low density. They are used for lightweight structures such as satellites, aircraft, automobiles, bicycles, ships, submarines, sporting goods, wheelchairs, armor and rotating machinery (such as turbine blades and helicopter rotors).

Carbon fibers are electrically conductive, while the polymer matrix is electrically insulating (except for the uncommon situation in which the polymer is an electrically conducting one). The continuous fibers in a composite laminate are in form of layers called laminae. Each lamina comprises many bundles (called tows) of fibers in a polymer matrix. Each tow consists of thousands of fibers. There may or may not be twist in a tow. Each fiber has a diameter typically ranging from 7 to 12 μm. The tows within a lamina are typically oriented in the same direction, but tows in different laminae may or may not be in the same direction. A laminate with tows in all the laminae oriented in the same direction is said to be unidirectional. A laminate with tows in adjacent laminae oriented at an angle of 90° is said to be crossply. In general, an angle of 45° and other angles may also be involved for the various laminae, as desired for attaining the mechanical properties required for the laminate in various directions in the plane of the laminate.

Within a lamina with tows in the same direction, the electrical conductivity is highest in the fiber direction. In the transverse direction in the plane of the lamina, the conductivity is not zero, even though the polymer matrix is insulating. This is because there are contacts between fibers of adjacent tows. In other words, a fraction of the fibers of one tow touch a fraction of the fibers of an adjacent tow here and there along the length of the fibers. These contacts result from the fact that fibers are not perfectly straight or parallel (even though the lamina is said to be unidirectional), and that the flow of the polymer matrix (or resin) during composite fabrication can cause a fiber to be not completely covered by the polymer or resin (even though, prior to composite fabrication, each fiber may be completely covered by the polymer or resin, as in the case of a prepreg, i.e., a fiber sheet impregnated with the polymer or resin). Fiber waviness is known as marcelling. Thus, the transverse conductivity gives information on the number of fiber-fiber contacts in the plane of the lamina.

For similar reasons, the contacts between fibers of adjacent laminae cause the conductivity in the through-thickness direction (direction perpendicular to the plane of the laminate) to be non-zero. Thus, the through-thickness conductivity gives information on the number of fiber-fiber contacts between adjacent laminae.

Matrix cracking between the tows of a lamina decreases the number of fiber-fiber contacts in the plane of the lamina, thus decreasing the transverse conductivity. Similarly, matrix cracking between adjacent laminae (as in delamination) decreases the number of fiber-fiber contacts between adjacent laminae, thus decreasing the through-thickness conductivity. This means that the transverse and through-thickness conductivities can indicate damage in the form of matrix cracking.

Fiber damage (as distinct from fiber fracture) decreases the conductivity of a fiber, thereby decreasing the longitudinal conductivity (conductivity in the fiber direction). However, owing to the brittleness of carbon fibers, the decrease in conductivity due to fiber damage prior to fiber fracture is rather small.

Fiber fracture causes a much larger decrease in the longitudinal conductivity of a lamina than fiber damage. If there is only one fiber, a broken fiber results in an open circuit, i.e., zero conducitivity. However, a lamina has a large number of fibers and adjacent fibers can make contact here and there. Therefore, the portions of a broken fiber still contribute to the longitudinal conductivity of the lamina. As a result, the decrease in conductivity due to fiber fracture is less than what it would be if a broken fiber did not contribute to the conductivity. Nevertheless, the effect of fiber fracture on the longitudinal conductivity is significant, so that the longitudinal conductivity can indicate damage in the form of fiber fracture.

The through-thickness volume resistance of a laminate is the sum of the volume resistance of each of the laminae in the through-thickness direction and the contact resistance of each of the interfaces between adjacent laminae (i.e., the interlaminar interface). For example, a laminate with eight laminae has eight volume resistances and seven contact resistance, all in the through-thickness direction. Thus, to study the interlaminar interface, it is better to measure the contact resistance between two laminae rather than the through-thickness volume resistance of the entire laminate.

A laminate of continuous fibers consists of laminae (plies). The interlaminar interface refers to the interface between the adjacent laminae. This region is rich in the matrix (polymer) (Fig. 1.8(a)). The contact resistance of an interlaminar interface is in the direction perpendicular to the plane of the laminae.

Measurement of the contact resistance between laminae can be made by allowing two laminae (strips) to contact at a junction and using the two ends of each strip for making four electrical contacts. An end of the top strip and an end of the bottom strip serve as contacts for passing current. The other end of the top strip and that of the bottom strip serve as contacts for voltage measurement. The fibers in the two strips can be in the same direction or in different directions. This method is a form of the four-probe method of electrical resistance measurement. The configuration is illustrated in Fig. 1.8(b) for a crossply laminate. To make sure that the volume resistance within a lamina in the through-thickness direction does not contribute to the measured resistance, the fibers at each end of a lamina strip should be electrically shorted together by using silver paint or other conductive media. The measured resistance is the contact resistance of the junction. This resistance, multiplied by the area of the junction, gives the contact resistivity, which is independent of the area of the junction and just depends on the nature of the interlaminar interface. The unit of contact resistivity is $\Omega.m^2$, whereas that of volume resistivity is $\Omega.m$.

The structure of the interlaminar interface tends to be more prone to change than the structure within a lamina. For example, damage in the form of delamination is much more common than damage in the form of fiber fracture. Moreover, the structure of the interlaminar interface is affected by the interlaminar stress (whether thermal stress or curing stress), which is particularly significant when the laminae are not unidirectional (as the anisotropy within each lamina enhances the interlaminar stress). The structure of the interlaminar interface also depends on the extent of consolidation of the laminae during composite fabrication. The contact resistance provides a sensitive probe of the structure of the interlaminar interface. The self-sensing of strain (reversible) has been achieved in continuous carbon fiber epoxy-matrix composites without the use of embedded or attached sensors [8,9]. Upon flexure, the tension surface resistance increases reversibly (Fig. 1.9, due to decrease in the penetration of the surface current) while the compression surface resistance decreases reversibly (Fig. 1.10, due to the increase in the current penetration) [8]. The effect of flexure on the surface current penetration is consistent with the observation that the through-thickness resistance increases upon uniaxial tension (Fig. 1.11) and decreases upon uniaxial compression (Fig. 1.12) [8].

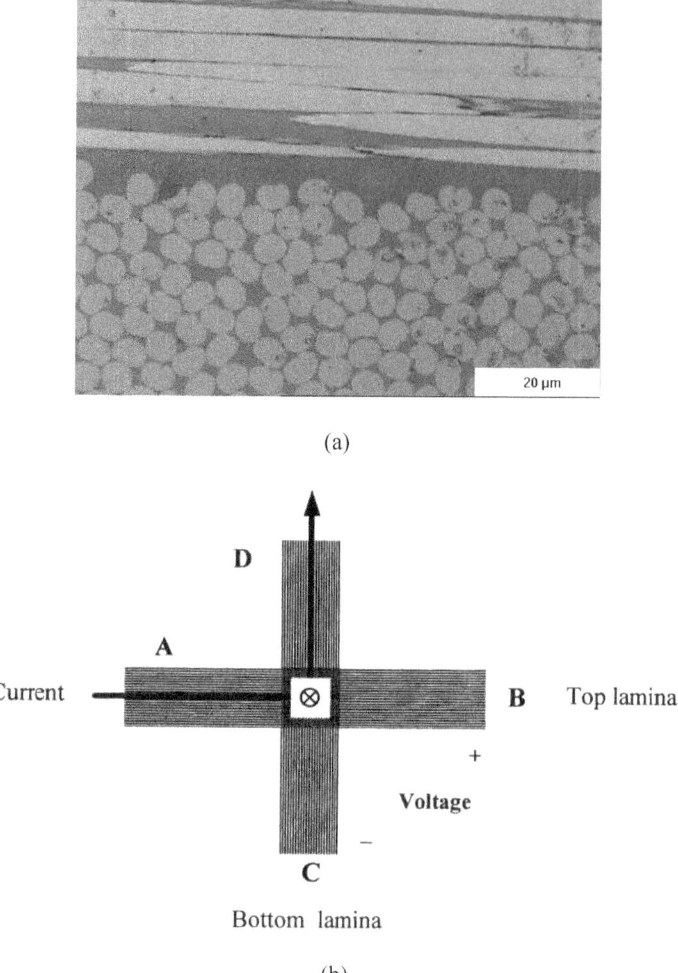

Fig. 1.8 (a) An optical microscope image of the cross-sectional view of a part of a continuous carbon fiber epoxy-matrix composite. Only an interlaminar interface and its two adjacent laminae are shown. The fibers in the top lamina are in the plane of the image; those in the bottom lamina are perpendicular to the plane of the image. (b) Specimen configuration for measurement of contact electrical resistivity between the laminae. (c) A two-dimensional array of sensors, with each sensor in the form of an interlaminar interface between fiber groups that are in different directions. In this illustration, the laminae are 90° apart in the fiber orientation.

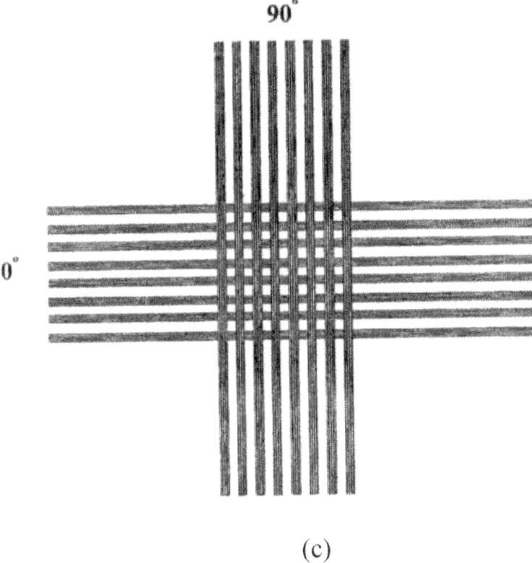

(c)

Fig. 1.8 (*Continued*).

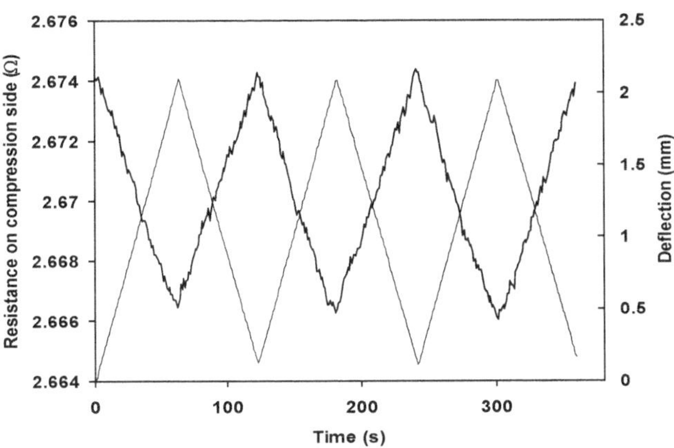

Fig. 1.9 Compression surface resistance (thick curve) during deflection (thin curve) cycling at an maximum deflection of 2.098 mm (stress amplitude of 392.3 MPa) for a 24-lamina quasi-isotropic continuous carbon fiber epoxy-matrix composite [8].

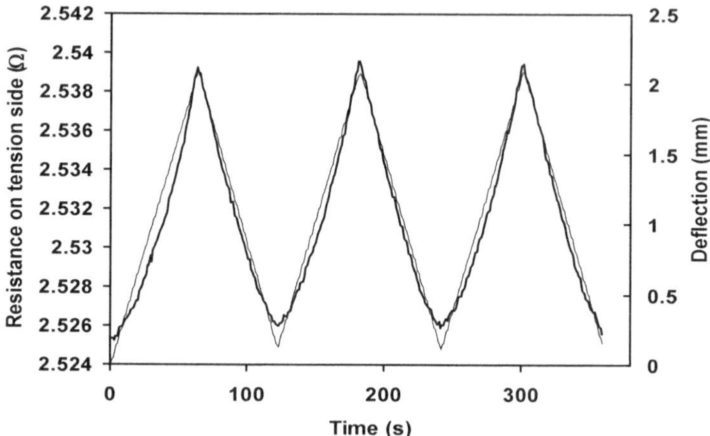

Fig. 1.10 Tension surface resistance (thick curve) during deflection (thin curve) cycling at an maximum deflection of 2.098 mm (stress amplitude of 392.3 MPa) for a 24-lamina quasi-isotropic continuous carbon fiber epoxy-matrix composite [8].

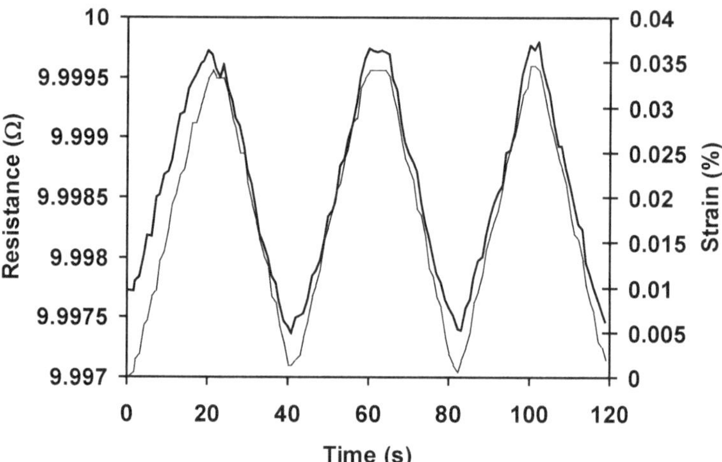

Fig. 1.11 Through-thickness resistance (thick curve) during longitudinal uniaxial tension (strain shown by the thin curve) at a stress amplitude of +17.5 MPa for a 24-lamina quasi-isotropic continuous carbon fiber epoxy-matrix composite [8].

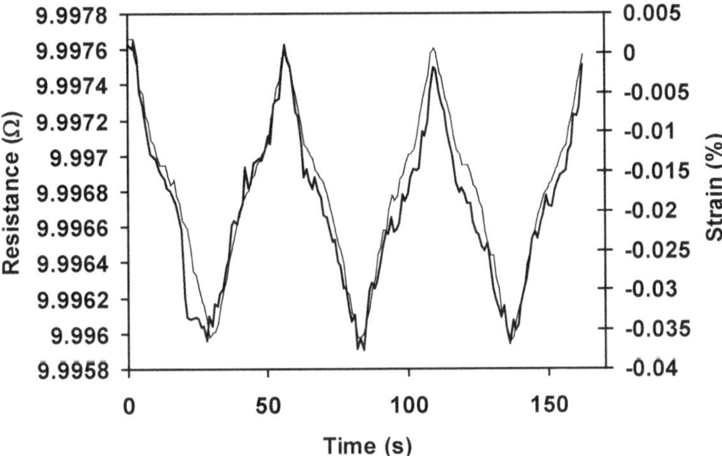

Fig. 1.12 Through-thickness resistance (thick curve) during uniaxial longitudinal compression (strain shown by the thin curve) at a stress amplitude of -17.4 MPa for a 24-lamina quasi-isotropic continuous carbon fiber epoxy-matrix composite [8].

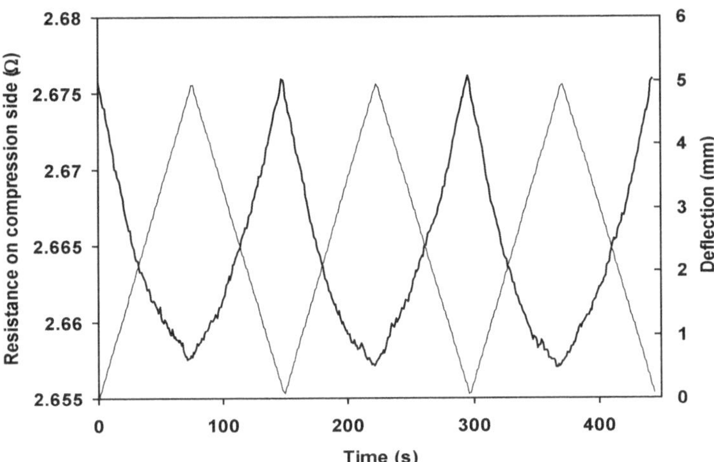

Fig. 1.13 Compression surface resistance (thick curve) during deflection (thin curve) cycling at an maximum deflection of 4.945 mm (stress amplitude of 996.2 MPa) for a 24-lamina quasi-isotropic continuous carbon fiber epoxy-matrix composite [8]. The resistance decreases upon flexure. The nonlinearity in the resistance change is in contrast ot the linearity at a lower stress amplitude in Fig. 1.9.

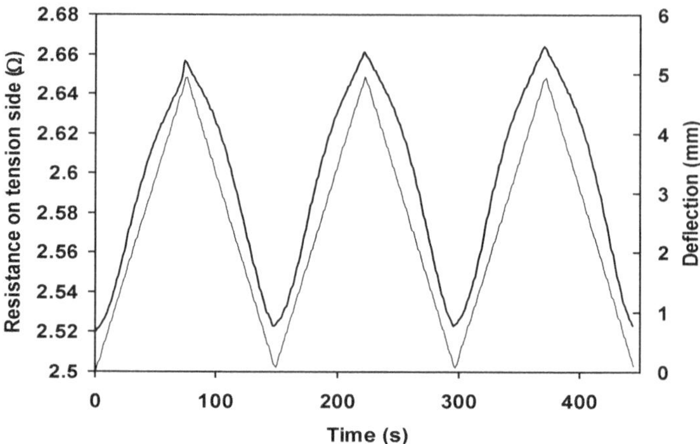

Fig. 1.14 Tension surface resistance (thick curve) during deflection (thin curve) cycling at a maximum deflection of 4.945 mm (stress amplitude of 996.2 MPa) for a 24-lamina quasi-isotropic continuous carbon fiber epoxy-matrix composite [8]. The resistance increases upon flexure. The nonlinearity in the resistance change is in contrast ot the linearity at a lower stress amplitude in Fig. 1.10.

The self-sensing of damage (whether due to stress or temperature, under static or dynamic conditions) has been achieved in continuous carbon fiber polymer-matrix composites, as the electrical resistance of the composite changes with damage. Minor damage under flexure is indicated by the curve of the resistance of the tension or compression surface vs. deflection becoming nonlinear, as shown in Fig. 1.13 and 1.14 for the compression and tension surface resistances respectively. Upon major damage, all resistances abruptly and irreversibly increase.

Damage in the form of delamination or interlaminar interface degradation is indicated by the through-thickness volume resistance (or more exactly the contact resistivity of the interlaminar interface) increasing due to decrease in the number of contacts between fibers of different laminae. Major damage in the form of fiber breakage is indicated by the longitudinal volume resistance increasing irreversibly.

1.5.2 *Self-actuation*

Actuation refers to the generation of strain and/or stress due to a certain stimulus. Self-actuating materials refer to structural materials that provide strain or stress in response to an input. Sensing allows a structure to know its situation, whereas actuation is a way of allowing a structure to respond to what has been sensed. Thus, the presence of both self-sensing and self-actuating abilities enables a structural material to be really smart.

The stimulus for actuation can be the temperature (as in the shape-memory effect and as resulting from the difference in the coefficient of thermal expansion between two materials that are bonded together to form a bi-strip), an electric field (as in the converse piezoelectric effect, the electrostrictive effect, and the diffusion of ions in an electrolyte in the presence of an electric field), and a magnetic field (as in the magnetostrictive effect). Actuation of a stiff material such as a structural composite is difficult. Artificial muscles are soft materials, which provide activation through any of the abovementioned mechanisms.

Electric-field-activated actuation based on the converse piezoelectric effect has been reported in the through-thickness direction (which is the soft direction) of a continuous carbon fiber thermoplastic-matrix composite. The matrix is nylon-6 (a polyamide). The composite is unmodified, with no attachment or embedment of any functional material. The effect is attributed to the polarization associated with the orientation of the electric dipoles in the matrix.

The converse piezoelectric effect converts electrical energy to mechanical energy, so it is valuable for actuation. The attachment of a piezoelectric material such as lead zirconotitanate (PZT, a Perovskite ceramic, Sec. 3.8.4) to a carbon fiber polymer-matrix composite for the purpose of actuation (such as bending) has been reported, but the carbon fiber composite needs to be relatively soft in the direction of the deformation. Furthermore, this method of actuation does not make use of the intrinsic actuation ability of the structural composite without the PZT.

Temperature-activated actuation based on the difference in the coefficient of thermal expansion between carbon fiber polymer-matrix composite and an attached aluminum plate in the fiber direction of the composite has been reported. This effect does not involve the actuation ability of the carbon fiber composite itself.

Electric-field-activated ion-movement-based actuation has been reported using a soft ion-containing polymer material as the electrolyte and carbon materials (e.g., carbon black, carbon nanofibers, carbon nanotubes, graphene, etc.) as the electrodes sandwiching the electrolyte layer. However, such material systems are too soft to function as structural materials.

1.5.3 *Self-healing*

Self-healing is a function by which a structural material heals itself automatically (without any human intervention) when damage is inflicted to it. This is known as autonomic healing. The function is valuable for strategic structures and for structures (such as those in outer space) that are hard to repair. In addition, this function helps reduce hazards. Self-healing requires the ability to sense the occurrence of damage, as well as the ability to heal. There are four techniques of self-healing, namely self-healing using encapsulated healing agents, self-healing using remendable polymers, self-healing using thermoplastic polymers and self-healing using solvents.

1.5.3.1 *Self-healing using encapsulated healing agents*

Self-healing using encapsulated healing agents is the most common method of self-healing. It involves the introduction of the monomer (pre-polymer) and catalyst to a composite so as to fill the damage region (such as the delaminated region) with the polymer resulting from the polymerization of the monomer in the presence of the catalyst. The encapsulation involves microcapsules (with each microcapsule having a thin shell made of urea-formaldehyde, for example) to contain the monomer (e.g., dicyclopentadiene, or DCPD) and embedding the catalyst (e.g., Grubb's catalyst for ring-opening metathesis polymerization (ROMP)) outside the microcapsules, such that both the microcapsules and catalyst are in the composite. Upon damage that causes the breaking of the shell of the microcapsules, the monomer oozes out of the microcapsules, thereby meeting the catalyst and becomes polymerized. The resulting polymer fills the cracks, thus resulting in healing. The breaking of the shell serves to sense the occurrence of damage.

The healing has been shown both in polymers (e.g., epoxy) and continuous fiber epoxy-matrix composites. The microcapsules are of size ranging from 10 to 1000 μm in diameter, with a mean diameter of 170 μm, for example. A small size of the microcapsules allows a lower volume fraction of the microcapsules to be used, thereby reducing the negative effect of the microcapsules on the mechanical properties. Of concern are the stress concentrations around the microcapsules and the delamination cracking tendency in case of a continuous fiber composite laminate. In contrast to microcapsules, nanocapsules of diameter as small as 220 nm and wall thickness 77 nm have been reported.

This method of self-healing has not been shown in cement-based materials, due to the pores in cement-based materials acting as sinks for the polymer. Furthermore, the microcapsule method suffers from the high cost of the catalyst, which needs to be able to promote polymerization at room temperature (without heating). It also suffers from the toxicity of the monomer.

1.5.3.2 *Self-healing using remendable polymers*

Self-healing using remendable polymers involves the polymer matrix in the absence of the embedment of any constituent, such that the polymer matrix itself is able to mend itself. Such a polymer is known as a remendable polymer, also known as a mendomer.

In one of the mechanisms, the remendable polymer concept is based on the inherently thermally reversible bonds of the polymer, with the healing activated by heating. In other words, the polymer is thermally remendable. The heating may be provided by using the reinforcing carbon fibers in the composite material as a resistive heating element.

The polymers are supramolecular polymers (SP), which refer to polymers having monomeric units that are held together by directional and reversible non-covalent interactions. The SP is positioned in the interlaminar interface. This healing mechanism is attractive for its ability to withstand multiple healing cycles without substantial loss of performance. It is also attractive for the absence of embedded constituents.

1.5.3.3 *Self-healing using thermoplastic polymers*

Self-healing using thermoplastic polymers is relatively simple. It involves the incorporation of a thermoplastic polymer (the thermally activated healing agent) at the interlaminar interface. The thermoplastic polymer melts upon heating, thereby enabling healing. The thermoplastic polymer may be blended with the matrix polymer, such as epoxy. Through this blending, it can serve as a toughening additive in addition to a healing agent. Examples of the thermoplastic polymers include poly(ethylene-co-methacrylic acid) (EMAA), poly(bisphenol A-co-epichlorohydrin) (PBE) and poly(bisphenol A-co-epichlorohydrin) (PBAE).

1.5.3.4 *Self-healing using solvents*

Microencapsulated solvents may be introduced to provide healing. For the self-healing of epoxy, epoxy-solvent microcapsules are more effective than microcapsules that contain the solvent alone. Solvents include aromatic esters, which are less toxic than chlorobenzene. The use of a solvent for healing has also been shown for the healing of polymethyl methacrylate (PMMA).

1.5.4 *Self-heating*

A conductive structural composite can serve as a resistance heating element, which uses the Joule effect to convert electrical energy to thermal energy (Sec. 2.7). Carbon materials are even more suitable for serving as a resistance heating element than metals, because the electrical resistivity of carbon materials is higher than that of metals by several orders of magnitude. Due to the low resistivity of metals, a heating element in the form of a metal is commonly in the form of a coiled wire, which provides a long length and a small cross-sectional area, as needed to provide a high enough resistance. Due to the relatively high resistivity of carbon materials, a heating element in the form of a carbon material can be in a bulk form rather than a coil form. Therefore, carbon fiber polymer-matrix composites with continuous carbon fibers can be utilized as a heating element, which is useful for the deicing and anti-icing of aircraft and other structures. Chemical deicing that is currently used for aircraft in airports causes much pollution. Deicing by resistance heating does not have this pollution problem.

Fast heating requires a heating element with a low thermal mass, because a heating element needs to get hot by resistance heating before it can emit heat to the environment. The interlaminar interface of a continuous carbon fiber composite is a resistance heating element that exhibits exceptionally low thermal mass and hence fast heating response. By having a two-dimensional array of these interfaces, as provided by two laminae that are orthogonal (or an angle other than 90°)

to one another in the plane of the array, a two-dimensional array of heating elements is provide for the purpose of spatially resolved heating (Fig. 1.8(c)).

The carbon fibers themselves are effective as heating elements. The coating of the fibers with a metal, such as a Ni/Cu/Ni trilayer (with Cu being very conductive and Ni being more oxidation resistant than Cu), decreases the resistivity of the fibers, thus causing the temperature increase due to the applied current to be smaller, but resulting in faster response.

A commonly neglected problem in resistance heating is the quality of the two electrical contacts used to pass current to the heating element. The resistance associated with the two electrical contacts should be small compared to that of the heating element itself. Otherwise, the electrical contacts rather than the heating element are the main source of heat, which then spreads to the heating element by thermal conduction.

1.5.5 *Electromagnetic interference (EMI) shielding by structural materials*

EMI shielding involves the absorption and/or reflection of electromagnetic radiation, particular in the radio wave regime (Sec. 4.4). The reflection is enhanced by the mobile charges, such as free electrons, in the material. The absorption is enabled by the interaction of the electric field in the radiation with the electric dipoles and with the mobile charges in the material, and by the interaction of the magnetic field in the radiation with the magnetic dipoles in the material.

Continuous carbon fiber epoxy-matrix composites (even without modification) are highly effective for EMI shielding. The shielding effectiveness averaged over the frequency range from 0.3 MHz to 1.5 GHz is 115 dB and 98 dB for the unmodified composite and the composite modified with 2.9 vol% carbon nanofibers (originally known as carbon filaments) in the interlaminar interface. In general, nanofillers are attractive for shielding because of (i) the large surface area, (ii) the Skin Effect (which refers to the phenomenon in which high-frequency radiation interacts with only the near surface region of a conductive material) and (iii) the multiple reflections (which refer to the bouncing back and forth of the radiation at various conductive surfaces in the material). The unexpected decrease in shielding effectiveness due to the nanofiber addition is attributed to the decrease in the continuous fiber volume fraction and is consistent with the slight increase in both through-thickness and in-plane electrical resistivities upon nanofiber incorporation. Although electrical conduction continuity is not required for shielding, the continuity helps. The continuity of the carbon fibers in the composites mentioned above contributes to resulting in the very high shielding effectiveness. The shielding effectiveness values mentioned above are very high compared to values (e.g., 25 dB) typically obtained with polymer-matrix composites rendered conductive by the presence of discontinuous conductive fillers, although values as high as 90 dB have been reported for polymer-matrix composites containing nickel nanofibers. The use of lightly activated continuous carbon fibers helps increase the shielding effectiveness, due to the enhanced multiple reflections.

1.5.6 *Energy storage in structural materials*

Energy storage can be provided by a capacitor or a battery. A capacitor stores the energy as electrical energy, whereas a battery stores the energy as chemical energy. Capacitors are attractive for the high-power capability, whereas batteries are attractive for the high energy capability. Capacitors can be dielectric capacitors or supercapacitors. Dielectric capacitors are purely electrical devices, whereas supercapacitors are electrochemical devices. Batteries are also electrochemical devices. In contrast to supercapacitors, a battery requires the anode and cathode to be different materials, so that oxidation occurs at the anode and reduction occurs at the cathode. Both supercapacitors and batteries require an electrolyte (an ionic conductor) and a separator that avoids short circuiting of the two electrodes. The most effective electrolytes are ionic liquids, but for structural composites, the electrolyte needs to be a solid, which poses issues in relation to the relatively low ionic conductivity of the solid electrolyte, the inadequate quality of the interface between the electrolyte and electrode, and the inadequate mechanical properties of the solid electrolyte. Due to the abovementioned requirements of an electrochemical device, the realization of supercapacitors and batteries that are in the form of structural composites with adequate mechanical performance is challenging. A compromise between the structural properties and electrochemical performance is necessary.

1.5.6.1 *Structural battery*

A structural battery refers to a battery in the form of a structural material. For lithium-ion batteries, carbon materials are commonly used as the anode, due to the electronic conductivity of carbon materials and the capability of carbon materials to be intercalated with lithium in the electrochemical process that occurs in the lithium-ion battery. Thus, carbon fibers may be used as an anode. In spite of the strain in the carbon fibers due to the lithium intercalation, the battery performance is adequate. Much research has been conducted to coat or modify the carbon fibers in order to enhance the effectiveness of the anode for Li-ion batteries and Na-ion batteries. Moreover, carbon fibers have been used as a conductive rigid support for sulfur used in lithium-sulfur batteries, and as a fibrous conductive network for electrically connecting particulate electrode materials.

The electrolyte polymer tends to be inadequate in the mechanical properties compared to the typical polymers used as the matrix of the composites. The higher is the ionic conductivity, the lower is the modulus of elasticity. Thus, the presence of the electrolyte polymer as at least a part of the matrix of the composite tends to reduce the mechanical performance of the composite.

Ideally, the battery is a part of the laminate. However, the incorporation of the battery in the core of a sandwich composite has been reported. This core is normally occupied by a honeycomb or a foam material and sandwiched by fiber composite face sheets for the purpose of greatly lowering the density of the composite. Due to the relatively large space available in the core, it is possible to incorporate a relatively large battery in the core. However, the density of the sandwich composite is increased due to the battery incorporation.

1.5.6.2 *Structural capacitor*

A structural capacitor refers to a capacitor in the form of a structural material. It is commonly a polymer-matrix structural composite with a dielectric film between the electrodes, which are an electronic conductor, commonly the continuous carbon fiber laminae that serve to reinforce the composite. The electrode preferably has a high specific surface area. The dielectric film is preferably small in thickness and should be well bonded to the polymer matrix. It serves to avoid short circuiting of the two electrodes and is commonly a polymeric film.

For a structural dielectric capacitor, the dielectric film should be an electrical insulator that exhibits high electrical resistivity in relation to electronic conduction. For a structural supercapacitor, the dielectric film (known as a separator) should be an ionic conductor and an electronic insulator.

In order to maximize the capacitance by having the capacitor constitute capacitors in parallel, the dielectric film is preferably positioned at every interlaminar interface of the composite, such that alternating electrodes in the stack are connected to opposite polarities of the AC electric field source.

A structural supercapacitor requires the matrix to be a solid electrolyte, which tends to reduce the strength and modulus of the composite relative to the values of the composite with a conventional polymer matrix. In contrast, a dielectric capacitor does not have stringent requirement on the polymer matrix.

Although supercapacitors can have high values of the capacitance per unit area, they cannot operate at high frequencies, due to the sluggishness of the ion movement. In contrast, dielectric capacitors can operate at high frequencies, which are encountered by high-speed computers and other electronics. From the viewpoints of structural performance, safety, service life and high frequency capability, structural dielectric capacitors are closer to commercialization readiness than structural supercapacitors.

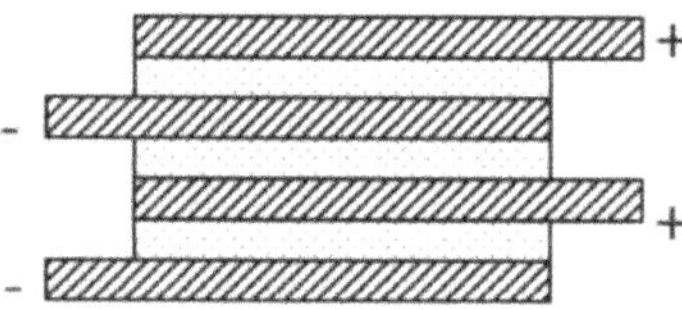

Fig. 1.15 A configuration of a structural capacitor that is in the form of a composite consisting of electrode and dielectric layers. An electrode can be a conductive structural material, such as a carbon fiber polymer-matrix composite lamina. A dielectric layer can be a dielectric structural material, such as a glass fiber polymer-matrix composite lamina. The configuration has a dielectric film positioned between every two adjacent electrode layers, such that alternate electrodes in the stack are connected to opposite polarities of the AC electric field source. [1]

1.5.7 *Energy generation by structural materials*

Energy generation refers to the conversion of a certain form of energy to electrical energy. This is also known as energy harvesting, whether the energy harvested is mechanical, thermal, photovoltaic, magnetic, etc. The use of a structural material for energy generation is attractive, since the structure does not need to rely totally on the electricity that is transmitted from a power station. For structures that are very far from power stations, as for a satellite, this is valuable. Furthermore, with the structural material providing the electricity, the need for batteries and capacitors is reduced, thus saving weight and volume.

Two main mechanisms of energy generation for structural materials are piezoelectricity and thermoelectricity. In spite of the feasibility of energy generation using these mechanisms, the amount of energy generated is small, whether the material is a structural material or not. For example, the amount of energy generated may be adequate for powering some sensors or other devices that do not use a lot of energy.

1.5.7.1 *Piezoelectricity-based energy generation*

The direct piezoelectric effect converts mechanical energy to electrical energy and is responsible for piezoelectricity-based energy generation (Sec. 3.8.1). It is useful for mechanical energy harvesting (e.g., converting the energy associated with vibrations or movements to electricity) and for sensing strain or stress. This effect has been reported in the through-thickness direction of continuous carbon fiber fabric epoxy-matrix composite that has ZnO nanowires grown on the fiber fabric prior to composite fabrication. Zinc oxide is a semiconductor. The adjacent carbon fiber laminae are electrically insulated from one another by an intervening glass fiber lamina. The piezoelectric effect is in the direction of the ZnO nanorods, which are perpendicular to the carbon fiber substrate. A similar effect has also been reported for a composite with the carbon fiber coated with barium titanate, which is piezoelectric.

1.5.7.2 *Thermoelectricity-based energy generation*

The thermoelectric effect converts thermal energy to electrical energy, thereby allowing thermal energy harvesting (e.g., converting waste heat to electricity) (Sec. 2.11). A high value of the dimensionless thermoelectric figure of merit ZT (with Z being the thermoelectric figure of merit and T being the average temperature in K in the range of temperature) requires high thermoelectric power (Seebeck coefficient), low electrical resistivity and low thermal conductivity. In spite of their electrical conductivity, carbon fibers are not good thermoelectric materials, due to their high thermal conductivity and low thermoelectric power. In a continuous carbon fiber polymer-matrix composite, the behavior in the fiber direction (which is in the plane of the composite laminate) is dominated by the fibers, whereas the behavior in the through-thickness direction is dominated by the polymer matrix. Therefore, an approach for enhancing the thermoelectric behavior of a carbon fiber polymer-matrix composite involves tailoring the through-thickness behavior through the use of interlaminar fillers. For example, a combination of fillers are used, with (i) tellurium particles (a well-known thermoelectric material) used in the amount of 13 vol% for increasing the thermoelectric power, (ii) bismuth telluride particles (a well-known thermoelectric material that is more expensive than tellurium) used in the amount of 2 vol% for increasing the electrical conductivity and decreasing the thermal conductivity, and (iii) carbon black (which is conductive and conformable, due to its being in the form of porous aggregates of nanoparticles) used in the amount of 2 vol% for increasing the electrical conductivity. Thus, the through-thickness thermoelectric power is increased from 8 to 163 μV/K, the through-thickness electrical resistivity is decreased from 0.17 to 0.02 Ω cm, the through-thickness thermal conductivity is decreased from 1.31 to 0.51 W/m.K, and ZT at 70 °C is increased from 9×10^{-6} to 9×10^{-2} (i.e., increased by 4 orders of magnitude). In spite of the large increase in ZT, the ZT value remains too low to be practical for thermoelectric power generation. The low value is partly due to the temperature being not high (as limited by the temperature ability of the polymer matrix) and the consequent low value of T in ZT.

Electrically conductive polymers are attractive as matrices for thermoelectric composite materials. The highest ZT among composite materials with inorganic particles is 0.39, as exhibited by a PEDOT matrix and tellurium nanowires as the filler. This ZT value remains too low for practical implementation.

A thermoelectric energy generator needs both n-type and p-type thermoelectric materials (i.e., materials with opposite signs of the thermoelectric power). The series connection of the alternating n-type and *p*-type elements results in a relatively large electrical output, in spite of the small output of an individual element. Carbon fiber polymer-matrix composites with opposite signs of the thermoelectric power have been obtained by using thermoelectric particles of opposite signs (namely tellurium and bismuth telluride) positioned in the interlaminar interface of the two composites.

1.6 Extrinsic smartness

Extrinsically smart structures are structures containing embedded or attached devices; the devices rather than the structural materials render the smartness. The smart materials involved are non-structural materials that provide certain non-structural functions. Due to the large variety of devices, extrinsic smartness is more common than intrinsic smartness. However, the devices are associated with high cost, low durability, poor repairability, limited functional volume and frequently structural performance degradation.

1.6.1 *Sensing using optical fibers*

Optical fibers are well-known for communication use, but they are also used as sensors (Sec. 5.3). An optical fiber is a fiber that is highly transparent to light and that allows the light to travel inside the fiber even when the fiber is bent. For transparency, glass is commonly used. The ability for an optical fiber to guide a light beam is because of total internal reflection, which results from the fact that the fiber has a core and a concentric cladding, such that the refractive index is higher for the core than the cladding.

When an optical fiber is embedded in a structure, the fiber can sense the strain and damage of the structure. This is because the light throughput and other characteristics of the light beam traveling through the fiber are affected by the strain and damage of the fiber. The strain and damage of the structure relate to those of the optical fiber, provided that the fiber is well bonded to the structure. The use of optical fibers requires a light source (a laser or a light-emitting diode) and a light detector.

1.6.2 *Sensing and actuation using piezoelectric devices*

A piezoelectric device is one which converts between mechanical energy and electrical energy (Sec. 3.8.1 and 3.8.2). It is a type of transducer. The conversion from mechanical energy to electrical energy (i.e., from strain/stress to electrical current/voltage) is based on the direct piezoelectric effect and allows application in strain/stress sensing and energy harvesting (i.e., converting mechanical vibration energy to electricity). The conversion from electrical energy to mechanical energy is based on the reverse piezoelectric effect and allows application in actuation.

Piezoelectric materials are dielectric materials, including ceramics and polymers. The most common piezoelectric material is lead zirconotitanate (abbreviated PZT, a $PbZrO_3$-$PbTiO_3$ solid solution).

The severity of the piezoelectric effect is described by the piezoelectric coupling coefficient d, which is defined as the electric polarization divided by the applied stress. In general, d is a tensor quantity, because the polarization and stress can be in different directions. A low value of the relative dielectric constant enhances the direct piezoelectric effect, due to the relationship between the polarization and the relative dielectric constant.

1.6.3 *Actuation using electrostrictive and magnetostrictive devices*

Electrostriction refers to the strain resulting from applying an electric field (Sec. 3.13). Magnetostriction refers to the strain resulting from applying a magnetic field (Sec. 6.11). These effects commonly stem from the interaction of the electric/magnetic field with the electrons in the solid and the consequent effect on the shape of the atoms in the solid. They can also be due to change in the orientation of electric/magnetic dipoles in the solid. In addition, electrostriction can be due to bonds between ions changing in length. These strain effects mean that electro-striction and magnetostriction can be used for actuation.

An electrostrictive material is a dielectric material, akin to a piezoelectric material. Compared to the reverse piezoelectric effect, electriction is advantageous in its smaller dependence on the history of electric field application. A magnetostrictive material is a magnetic material, most commonly a ferromagnetic material.

1.6.4 *Actuation using the shape-memory effect*

The shape-memory effect refers to the ability of a material to transform to a phase having a twinned microstructure (the phase called the martensite) that, after a subsequent plastic deformation, can return the material to its initial shape when heated, as illustrated in Fig. 1.16. If the shape-memory alloy (SMA) is constrained from recovering (say, within a composite material), a recovery stress is generated. The recovery stress builds up upon heating and it increases with increasing prior strain of the martensite. The return in shape upon heating can be used for actuation that is activated by heat.

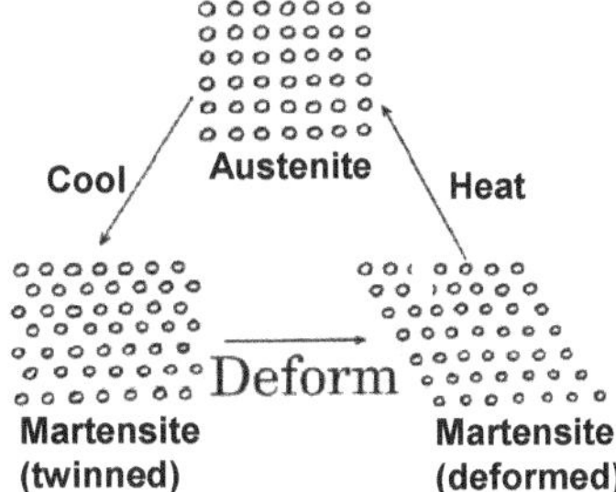

Fig. 1.16 The shape-memory effect illustrated in the crystal lattice level.

Another manifestation of the shape-memory effect involves the use of stress rather cooling to transform the material to martensite. Accompanying this change in phase is elastic deformation that exceeds the elasticity of ordinary alloys by a factor of 10 or more. Upon removal of the stress, the martensite changes back to the original phase and the strain (shape) returns to the value prior to the martensitic transformation [10]. This phenomenon is known as superelasticity or pseudoelasticity, and is illustrated in Fig. 1.17. It can be used for providing a nearly constant stress when strained (typically between 1.5% and 7%). The near constancy of the stress during unloading is exploited in orthodontal braces, where the shape-memory alloy (SMA) is used for the archwire, which applies forces according to the unloading plateau in order to restore the teeth to their proper location. The large hysteresis between loading and unloading in Fig. 1.17 means that a significant part of the strain energy put into the SMA is dissipated as heat. This energy dissipation provides a mechanism for vibration damping.

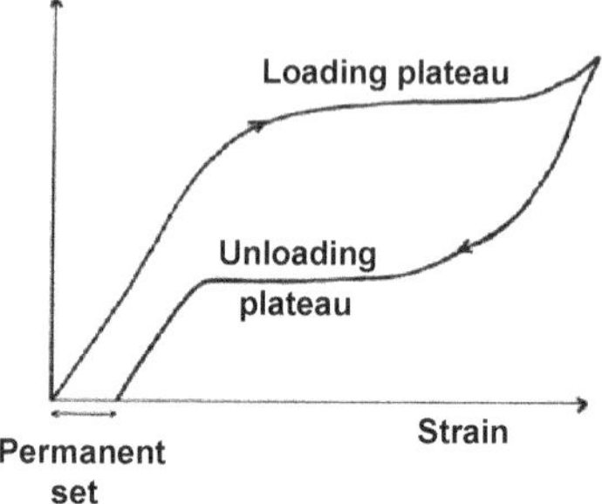

Fig. 1.17 Stress-strain curve of a shape-memory alloy, showing a small permanent set after unloading.

1.6.5 *Actuation using magnetorheological fluids*

Magnetorheology refers to the phenomenon in which the rheological behavior changes upon application of a magnetic field (Sec. 6.14). A magnetorheological fluid (abbreviated MR fluid) is a non-colloidal dispersion of fine magnetic (ferromagnetic or ferrimagnetic) particles in

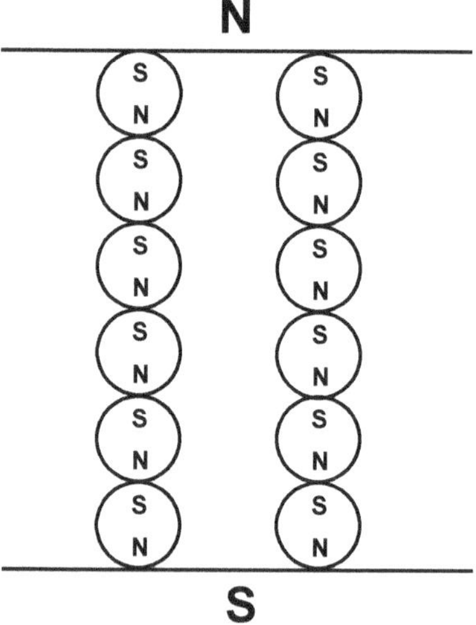

Fig. 1.18 Columns of magnetic particles in a magnetorheological fluid in the presence of a magnetic field in the direction of the columns.

a liquid medium. The most common particles are iron, which is ferromagnetic. It exhibits shear thinning (i.e., a decrease in the viscosity upon increase in the shear strain rate). Upon application of a magnetic field, the magnetic dipole moments of different particles align, resulting in columns of particles in the direction of the magnetic field (Fig. 1.18). Thus, the viscosity of the fluid increases with increasing magnetic field at any shear strain rate.

1.7 Functional applications

1.7.1 *Electronic applications*

Electronic applications include electrical, optical and magnetic applications, as the electrical, optical and magnetic properties of materials are largely governed by electrons. There is overlap among these three areas of application.

Electrical applications pertain to computers, electronics, electrical circuitry (e.g., resistors, capacitors and inductors), electronic devices (e.g., diodes and transistors), optoelectronic devices (e.g., solar cells, light sensors and light-emitting diodes for conversion between electrical energy and optical energy), thermoelectric devices (heaters, coolers and thermocouples for conversion between electrical energy and thermal energy), piezoelectric devices (strain sensors and actuators for conversion between electrical energy and mechanical energy), robotics, micromachines (or microelectromechanical systems, MEMS), ferro-electric computer memories, electrical interconnections (e.g., solder joints, thick-film conductors and thin-film conductors), dielectrics (i.e., electrical insulators in bulk, thick-film and thin-film forms), substrates for thin films and thick films, heat sinks, electromagnetic interference (EMI) shielding, cables, connectors, power supplies, electrical energy storage, motors, electrical contacts and brushes (sliding contacts), electrical power transmission, eddy current inspection (the use of a magnetically induced electrical current to indicate flaws in a material), etc.

Optical applications pertain to lasers, light sources, optical fibers (materials of low optical absorptivity for communication and sensing), absorbers, reflectors and transmittors of electromagnetic radiation of various wavelengths (for optical filters, low-observable or Stealth aircraft, radomes, transparencies, optical lenses, etc.), photography, photocopying, optical data storage, holography, color control, etc.

Magnetic applications pertain to transformers, magnetic recording, magnetic computer memories, magnetic field sensors, magnetic shielding, magnetically levitated trains, robotics, micromachines, magnetic particle inspection (the use of magnetic particles to indicate the location of flaws in a magnetic material), magnetic energy storage, magnetostriction (strain in a material due to the application of a magnetic field), magnetorheological fluids (for vibration damping that is controlled by a magnetic field), magnetic resonance imaging (MRI, for patient diagnosis in hospitals), mass spectrometry (for chemical analysis), etc.

The large range of applications mentioned above means that all classes of materials are used for electronic applications. Semiconductors are at the heart of electronic and optoelectronic devices. Metals are used for electrical interconnections, EMI shielding, cables, connectors, electrical contacts, electrical power transmission, etc. Polymers are used for dielectrics and cable jackets. Ceramics are used for capacitors, thermoelectric devices, piezoelectric devices, dielectrics and optical fibers.

Microelectronics typically refers to electronics that involve integrated circuits. Due to the availability of high-quality single-crystalline semi-conductors, the most critical problems that the microelectronic industry faces do not pertain to semiconductors, but they are related to electronic packaging, which refers to the chip carriers, electrical interconnections, dielectrics, heat sinks, etc.

Because of the miniaturization and increasing power of micro-electronics, heat dissipation is critical to the performance and reliability of microelectronics. Thus, materials for heat transfer from electronic packages are needed. Ceramics and polymers are both dielectrics, but ceramics are advantageous in their higher thermal conductivity compared to polymers. Materials that are electrically insulating but thermally conducting are needed. Diamond is the best material that exhibits such properties, but it is expensive.

Because of the increasing speed of computers, signal propagation delay needs to be minimized by the use of dielectrics with a low value of the relative permittivity (relative dielectric constant). (A dielectric with a high value of the relative permittivity and used to separate two conductor lines acts like a capacitor, thereby causing signal propagation delay.)

Polymers tend to be advantageous over ceramics in their low value of the relative permittivity.

Electronic materials are in bulk (single-crystalline, polycrystalline or, less commonly, amorphous), thick-film (typically over 10 μm thick, obtained by applying a paste on a substrate by, say, screen printing, such that the paste contains the relevant material in particle form, together with a binder and a vehicle) or thin-film (typically less than 1,500 Å thick, obtained by vacuum evaporation, sputtering, chemical vapor deposition, molecular beam epitaxy or other techniques) forms. Semiconductors are typically in bulk single-crystalline form (cut into slices called wafers, each of which may be subdivided into "chips"), although bulk polycrystalline and amorphous forms are emerging for solar cells, due to the importance of low cost for solar cells. Conductor lines in microelectronics are mostly in thick-film and thin-film forms.

The dominant material for electrical connections is solder (e.g., Sn-Pb alloy). However, the difference in CTE between the two members that are joined by the solder causes the solder to experience thermal fatigue upon thermal cycling, which is encountered during operation. The thermal fatigue can lead to failure of the solder joint. Polymer-matrix composites in paste form and containing electrically conducting fillers are being developed to replace solder. Another problem lies in the lead (poisonous) used in solder to improve the rheology of the liquid solder. Lead-free solders are being developed.

Heat sinks are materials with a high thermal conductivity. They are used to dissipate heat from electronics. Because they are joined to materials of a low CTE (e.g., a printed circuit board in the form of a continuous fiber polymer-matrix composite), they need to have a low CTE also. Hence, materials exhibiting both a high thermal conductivity and a low CTE are needed for heat sinks. Copper is a metal with a high thermal conductivity, but its CTE is too high. Therefore, copper is reinforced with continuous carbon fibers, molybdenum particles, or other fillers of low CTE.

Thermoelectric materials are a category of electrically conductive materials. They allow conversion of thermal energy to electrical energy and vice versa. The thermoelectric effect behind the conversion of thermal energy to electrical energy is known as the Seebeck effect. That behind the conversion of electrical energy to thermal energy is known as the Peltier effect. Applications include temperature sensing (based on the Seebeck effect) and heating and cooling (based on the Peltier effect).

1.7.2 *Thermal applications*

Thermal applications refer to applications that involve heat transfer, whether the heat transfer is by conduction, convection or radiation. Heat transfer is needed in heating (heating of buildings, heating involved in industrial processes such as casting and annealing, cooking, de-icing, etc.) and in cooling (cooling of buildings, refrigeration of food and industrial materials, cooling of electronics, removal of heat generated by chemical reactions such as the hydration of cement, removal of heat generated by friction or abrasion as in a brake and in machining (cutting), removal of heat generated by the impingement of electromagnetic radiation, removal of heat from industrial processes such as welding, etc.).

Conduction refers to the heat flow from points of higher temperature to points of lower temperature in a material. It typically involves metals, due to their high thermal conductivity.

Convection is attained by the movement of a hot fluid. If the fluid is forced to move by a pump or a blower, the convection is known as forced convection. If the fluid moves due to differences in density, the convection is known as natural or free convection. The fluid can be a liquid (e.g., oil) or a gas (e.g., air). It must be able to withstand the heat involved. Fluids are outside the scope of this book.

Radiation, i.e., blackbody radiation, is involved in space heaters. It refers to the continual emission of radiant energy from the body. The energy is in the form of electromagnetic radiation, typically infrared radiation. The dominant wavelength of the emitted radiation decreases with increasing temperature of the body. The higher the temperature, the greater is the rate of emission of radiant energy per unit area of the surface of the body. This rate is proportional to T^4, where T is the absolute temperature. It is also proportional to the emissivity of the body. The emissivity depends on the material of the body. In particular, it increases with increasing roughness of the surface of the body.

Heat transfer can be attained by the use of more than one mechanism. For example, both conduction and forced convection are involved when a fluid is forced to flow through the interconnected pores of a solid, which is a thermal conductor. Conduction is more tied to material development than convection or radiation. Radiative heat transfer may be deterred by using a material of high reflectivity, e.g., aluminized

Mylar (biaxially-oriented polyethylene terephthalate [abbreviated PET] film with aluminum coating).

Thermal conduction can involve electrons, ions and/or phonons. Electrons and ions move from a point of higher temperature to a point of lower temperature, thereby transporting heat. Due to the high mass of ions compared to electrons, electrons move much more easily than ions. Phonons are lattice vibrational waves, the propagation of which also leads to the transport of heat. Metals conduct by electrons, because they have free electrons. Diamond conduct by phonons, because free electrons are not available and the low atomic weight of carbon intensifies the lattice vibrations. Diamond is the material with the highest thermal conductivity. In contrast, polymers are poor conductors, because free electrons are not available and the weak secondary bonding (van der Waals forces) between the molecules makes it difficult from the phonons to move from one molecule to another. Ceramics tend to be more conductive than polymers, due to the ionic and covalent bonding making it possible for the phonons to propagate. Moreover, ceramics tend to have more mobile electrons or ions than polymer, and the movement of electrons and/or ions contributes to thermal conduction. On the other hand, ceramics tend to be poorer than metals in the thermal conductivity, due to the low concentration of free electrons (if any) in ceramics compared to metals.

Heating is needed for space heating, water heating, industrial processes and deicing. For example, a water heater commonly involves a flame (i.e., combustion of gases) for heat generation. The decrease of thermal resistance of the heat path from the flame to the water pipe via the heater body requires consideration of the thermal resistance of the heater body as well as the thermal contact between the body and the pipe.

Anti-icing (prevention of icing) and deicing (ice removal) are needed for aircraft, bridges, roads, driveways, etc., since (i) ice and snow frequently cause traffic accidents, (ii) deicing salts cause environmental pollution and (iii) snow removal is costly and is often not timely enough. The embedment of a resistance heating element in a structure is a method of deicing. However, this causes mechanical property loss and the thermal contact between a metal coil heating element and the structural material tends to be inadequate. Structural materials that are also heating elements for resistance heating are desired. The development of self-heating structural materials requires the tailoring of the electrical resistivity of the material.

Materials for thermal applications include thermal conductors, thermal insulators, heat retention materials (i.e., materials with a high heat capacity), thermal interface materials (i.e., materials for improving thermal contacts) and materials for heating and cooling.

1.7.3 *Energy harvesting*

Energy harvesting (also known as energy scavenging) refers to the generation of electricity from various sources of energy that are available in the environment [11]. Due to the high cost of oil, energy harvesting is a topic that is receiving growing research attention. Methods of energy harvesting include the thermoelectric method (Seebeck effect, which converts thermal energy to electricity), the pyroelectric method (which converts thermal energy to electricity), the piezoelectric/piezoelectret method and the effect of strain/stress on the relative dielectric constant (which convert mechanical energy to electricity), the photovoltaic method (i.e., solar cells, which convert light energy to electricity) and the mechanical method (e.g., the use of electromagnetic effect known as the Faraday effect to convert mechanical energy to electrical energy).

An example of the mechanical method pertains to a windmill, which comprises wind turbine blades, the turning of which results in electricity through the Faraday effect. The blades are large in order to be efficient in harvesting the wind energy. They are made of lightweight composites, particularly carbon fiber (continuous) polymer-matrix composites.

The ability of a structure to harvest energy allows self-powered structures, i.e., structures that generate power for their use. Self-powering is particularly valuable for satellites and other aerospace structures. This may be achieved by incorporating materials that are capable of energy harvesting in a structure. A more attractive but scientifically more challenging avenue involves modifying the structural material, so that it exhibits self-powering ability without the incorporation of materials that are themselves capable of energy harvesting, i.e., intrinsic smartness. The latter avenue is more attractive, because of the low material cost (due to the absence of conventional energy harvesting materials), high durability (due to the structural material providing a durable framework), large functional volume (due to the entire structure having the functional capability, if desired) and absence of mechanical property loss (which tends to occur in case of the embedment of a device in a structure). Composites science is at the heart of the latter avenue.

1.7.4 *Energy storage*

Due to the intermittent nature of the generation of electrical energy by photovoltaics, wind and other renewable energy sources, the storage of the generated electrical energy is needed. By storing the energy, energy can be available at times when no energy is generated by the renewable energy source. For example, energy harvested by solar cells in the daytime needs to be stored for use in the nighttime. In addition, energy storage is needed to help manage the energy utilization of the electrical grid. The demand for electricity varies by hour, day and season. The energy management involves monitoring the electricity demand, supply, reserve margins and the mix of electricity generating technologies. For such energy management, large-scale energy storage is needed. Large-scale energy storage is to be distinguished from small-scale energy storage that is used to power devices such as sensors.

Current technologies for energy storage include the following.

(i) Batteries – electrochemical devices in which energy is stored in the form of chemical energy, e.g., lithium-ion, sodium-sulfur and lead-acid batteries.

(ii) Flow batteries – rechargeable batteries that involve two chemical components dissolved in liquids that are contained within the system and most commonly separated by a membrane, such that each liquid circulates in its own respective space, e.g., vanadium redox and zinc-bromine flow batteries.

(iii) Supercapacitors – electrochemical devices in which energy is stored in the form of electrical energy.

(iv) Flywheels – devices involving accelerating a rotor or flywheel to a very high speed and maintaining the energy in the system as rotational energy.

(v) Compressed air energy storage – device based on the principle that the compression of air creates heat and the expansion of air removes heat.

(vi) Thermal energy storage – using materials of high specific heat and phase change materials, which store the energy as thermal energy, with the phase change material such as wax melting upon heating and storing energy by absorbing the latent heat of melting, so that, upon solidication during subsequent cooling, the latent of solidification is released, thereby releasing heat.

(vii) Pumped hydropower – energy derived from falling water or fast running water.

For large-scale energy storage, scale-up is necessary. Compressed air energy storage and pumped hydropower are relatively suitable for scale-up. However, both underground compressed air energy storage and pumped hydropower energy storage are limited by geological and environmental constraints. Intense research is ongoing concerning energy storage using batteries, supercapacitors, flywheels and thermal means.

In order for batteries and capacitors to be able to provide large-scale energy storage, the size of these devices must be increased greatly. Such an increase can be enabled by the use of large structures as the form of these devices. This means that the structure serves not only the structural function, but also the energy storage function. Such multifunctional structures are known as structural batteries and structural capacitors. A further advantage of such multifunctional structures is that weight and volume are not issues, as the devices effectively vanish in the structures. Furthermore, the multifunctional structures enable energy storage to occur in a spatially distributed fashion (as in community energy storage that is grid-connected and utility owned and operated) rather than being limited to centralized locations, thereby increasing the total capacity of energy storage.

The operation of a capacitor does not involve energy conversion, as only electrical energy is involved, but the operation of a battery involves the conversion between chemical energy and electrical energy. Due to this energy conversion, a disadvantage of batteries is that the electro-chemically active species can be depleted after a period of use and hence the battery ceases operation. This disadvantage is particularly serious for structural batteries, as a structure may be required to be in service for decades.

Capacitors excel in their high power capability, whereas batteries excel in their high energy capability. High power is needed for electric vehicles, hybrid vehicles, electric drive-trains, aircraft powertrains and shipboard power systems that need high power for acceleration. It is also needed for electric and hybrid propulsion systems, for regenerative braking (which helps improve the fuel efficiency of a vehicle that is operated under a stop-and-go urban driving condition), and for equip-ment that needs high pulse power, such as rail guns, electromagnetic armor and airborne lasers. Energy storage is also needed for self-powered structures and for structures that need emergency power.

There are two main types of capacitors, namely dielectric capacitors (also known as electrostatic or electrolytic capacitors) and super-capacitors (also known as ultracapacitors). The former is purely electrical in nature, whereas the latter is electrochemical in nature. Due to its electrochemical nature, a supercapacitor requires an electrolyte, which complicates multifunctional structural material design, limits the service life and poses safety concerns. In contrast, a dielectric capacitor does not involve any electrolyte. Since service life and safety are essential for structural capacitors, dielectric structural capacitors are more promising than structural supercapacitors, in spite of the fact that the capacity for small-scale energy storage tends to be greater for a supercapacitor than a dielectric capacitor.

Among supercapacitors, there are two types, which are the electric double layer capacitor (symmetrical, with the two electrodes being the same in composition) and the pseudocapacitor (asymmetrical, with the two electrodes being different in composition, so that a redox reaction as in a battery contributes to the capacitance). In relation to structural supercapacitors, both types suffer from the problems related to service life and safety.

1.7.5 *Sensing and healing*

Sensing (monitoring) is needed for structural vibration control (i.e., the sensing of stress or strain), structural health monitoring (i.e., the sensing of damage, as needed to avoid hazards associated with aging structures), temperature monitoring (as needed for detecting people and avoiding temperature related hazards) and homeland and building security (i.e., chemical sensing and people/vehicle detection). Real-time monitoring, which refers to continuous automated monitoring, is particularly attractive. Stress/strain/damage sensing can be achieved by using elec-trical (piezoresistivity), dielectric (piezoelectric/piezoelectret effect and effect of strain/stress/damage on the relative dielectric constant) and optical (optical fiber) methods and is important for smart structures. Temperature monitoring can be achieved by using electrical (Seebeck effect), dielectric (pyroelectric effect) and optical (optical fiber) methods. Chemical sensing can be achieved by using electrochemical and optical (optical fiber) methods.

The ability of a structure to sense allows self-sensing structures, i.e., structures that can sense certain attributes. Self-sensing is particularly valuable for aircraft, satellites and other aerospace structures, as well as bridges and other civil infrastructures. This may be achieved by incorporating materials that are capable of sensing in a structure. A more attractive but scientifically more challenging avenue involves modifying the structural material, so that it exhibits self-sensing ability without the incorporation of materials that are themselves capable of sensing, i.e., intrinsic smartness.

Healing or repair should follow damage in a timely fashion. It is desirable for a structure to be able to heal itself automatically upon damage. This is known as self-healing.

1.7.6 *Actuation*

Actuation refers to the providing of stress, strain and/or modulus change, as needed for actuators, manipulators, motors, movements and structural control. Such functions are important for smart structures to respond to what they have sensed. Sensors, actuators and a control system constitute the main components of a system that makes a structure smart in the conventional sense (like a robot). Actuation can be achieved by dielectric (piezoelectric/electrostrictive), rheological (electrorheological/magnetorheological) and magnetic (magnetostrictive) methods.

The ability of a structure to actuate allows self-actuating structures, i.e., structures that can actuate on their own. Self-actuation is particularly valuable for small unmanned aircraft. This function may be obtained by incorporating materials that are capable of actuation in a structure. A more attractive but scientifically more challenging avenue involves modifying the structural material, so that it exhibits self-actuating ability without the incorporation of materials that are themselves capable of actuation, i.e., intrinsic smartness. The strain that can be provided by a structural material is small, due to the stiffness of the structural material. However, the stress that can be provided can be substantial.

1.7.7 *Thermal management*

Heat dissipation is the most critical problem in the electronic industry. It is currently limiting the further miniaturization, power, performance and reliability. Thermally conductive materials are required for heat sinks, encapsulations, printed wiring boards, substrates and housings. In addition, a material called a thermal interface material is needed to improve the heat transfer across a thermal contact, such as the contact between the microprocessor (CPU) and the heat sink. The thermal interface material needs to be conformable, so that it conforms to the topography of the proximate surfaces. Heat dissipation is also needed for aircraft, the thermal load of which is rapidly increasing. The thermoelectric effect known as the Peltier effect converts electrical energy to thermal energy, thereby causing either heating (heat evolution) or cooling (heat absorption). Peltier cooling is valuable for microelectronic cooling. On the other hand, thermal insulation materials are needed for energy conservation. For example, construction materials that are thermally insulating are needed to decrease the cost of heating or cooling buildings.

1.7.8 *Electromagnetic applications*

Electromagnetic applications include EMI shielding (as needed to protect electronics from electromagnetic radiation in the radio frequency regime, such as the radiation emitted from cellular phones and microwave ovens) and low observability (Stealth, as needed for aircraft and ships to avoid detection by radar). Composites science is at the heart of the development of shielding and low observability materials. This development is largely focused on electrically conductive fillers, particularly nanofillers.

Current Stealth aircraft attains low observability mainly by having jagged shapes of its body, so that the radio wave from the radar does not reflect back to the radar. The development of structural materials with low observability is greatly needed to improve the Stealth design versatility and the Stealth performance.

The level of shielding required in practice depends on the application. For buildings, a shielding effectiveness of 20 dB is usually adequate, though a much higher value is desired for military applications, such as the deterring of electromagnetic spying. The shielding effectiveness tends to be limited by the windows. Rooms and structures involving sensitive electronics need to be shielded, so shielding materials are not only used for computer housings and electric cables, but are also needed for large structures. For example, underground concrete vaults containing large transformers need to be shielded. Telephones, fax machines and computers emit electromagnetic waves that may be analyzed for information, so EMI shielding is needed for buildings and other structures for the purpose of protection from electromagnetic methods of spying. Cell-phone proof capability (i.e., the capability to cause cell phones to be not able to operate) is desired for concert halls, as the ringing of cell phones during a concert is annoying.

1.7.9 *Charge dissipation*

The dissipation of electrical charges is necessary for lightning protection (e.g., the protection of aircraft and tall structures from lightning strike damage) and electrostatic discharge protection (e.g., the protection of electronic equipment from disturbance due to the static charges). Charge dissipation requires electrically conductive materials.

The increasing use of composite materials instead of aluminum for aircraft makes lightning protection of aircraft increasing challenging. This is because composites (even those with carbon fibers) are less conductive than metals. Thus, the modification or coating of composites to make them more conductive is necessary. The coating method is most common; it typically involves the bonding of a metal film in the form of a mesh on the surface of the composite. Metal coated carbon fibers are also attractive for increasing the electrical conductivity of the composites.

1.7.10 *Microelectronic packaging*

With the increasing degree of miniaturization of microelectronics, the packaging of electronic components associated with integrated circuits is becoming critical and challenging. The packaging involves electrical interconnections (conduction lines and electrical contacts), electrical insulation (encapsulations and interlayer dielectrics). In case of avionics (i.e., aerospace electronics), low density is also desired for the purpose of weight (fuel) saving. Copper is a common heat sink material, but it is heavy. Thus, metal-matrix composites, such as copper-matrix composites containing continuous carbon fibers (particularly carbon fibers that are even more thermally conductive than copper), are attractive.

1.7.11 *Data storage*

Data storage refers to the storage of data using various types of memory. With the explosion of information that needs to be stored, data storage is in high demand. Methods of data storage include dielectric (ferroelectric memory), magnetic (magnetic recording tape and memory), optical (compact disks or CDs) and thermal (phase-change memory) methods. The incorporation of memories in a structure is attractive for space saving.

1.7.12 *Environmental protection*

Environmental applications refer to applications that pertain to the protection of the environment from pollution. The protection can involve the removal of a pollutant or the reduction in the amount of pollutant generated. Pollutant removal can be attained by extraction of the pollutant through adsorption of the pollutant on the surface of a solid (e.g., activated carbon) with surface porosity. It can also be attained by planting trees, which take in CO_2 gas. Pollutant generation can be reduced by changing the materials and/or processes used in industry, by using biodegradable materials (i.e., materials that can be degraded by nature, so that their disposal is not necessary), by using materials that can be recycled, or by changing the energy source from fossil fuels to batteries, fuel cells, solar cells and/or hydrogen. For this reason, composites with biodegradable polymer matrices are attractive. An example of a biodegradable polymer is starch.

Materials have been mainly developed for structural, electronic, thermal or other applications without much consideration of the disposal or recycling problems. It is now recognized that such considerations must be included during the design and development of materials, rather than considering about disposal or recycling after the materials have been developed.

Materials for removal of certain molecules or ions in a fluid by adsorption are central to the development of materials for environmental applications. They include carbons, zeolites, aerogels and other porous materials in particulate and fibrous forms. The particles or fibers can be loose, be held together by a small amount of binder, or, in the case of fibers, be held together by stapling, in which mechanical interlocking between fibers occurs due to a process involving poking of the fiber agglomerate by a needle. The loose form is least desirable because the loose particles or fibers may be pulled into the fluid stream. Continuous fibers that are woven together do not need a binder or stapling, but they are expensive compared to discontinuous fibers.

Desirable qualities of an adsorption material include large adsorption capacity, pore size being large enough for relatively large molecules and ions to lodge in, ability to be regenerated or cleaned after use, fluid dynamics for fast movement of the fluid from which the pollutant is to be removed, and, in some cases, selective adsorption of certain species.

Activated carbon fibers are superior to activated carbon particles in the fluid dynamics, due to the channels between the fibers. However, they are much more expensive.

Pores on the surface of a material must be accessible from the outside in order to serve as adsorption sites. In general, the pores can be macropores (> 500 Å), mesopores (20–500 Å), micropores (8–20 Å) or micro-micropores (< 8 Å). Activated carbons typically have pores that are micropores and micro-micropores.

Materials for removal of particles in a fluid by filtration are typically macroporous materials. They are usually in the form of fibers or particles that are held together by using a small amount of a binder. Fibers are preferred in that they allow the pores to be larger, so that the fluid flow is faster. However, fibers tend to be more expensive than particles. Particularly challenging is the situation in which the fluid is hot. The challenge involves mainly the choice of the binder, which should be effective for binding even at a small concentration and should be stable at high temperatures. These porous materials are composites, although they are not designed for mechanical performance. Without the binder, the material can still be used as a filter, but, as loose particles or fibers, it cannot be conveniently handled, and the incorporation of the loose particles or fibers in the flowing fluid can occur and is not desirable. A filter is also known as a membrane.

Electronic pollution is an environmental problem that has started to be important. It refers to the electromagnetic waves (particularly radio waves) that are present in the environment, due to radiation sources such as cellular telephones. Such radiation can interfere digital electronics such as computers, thereby causing hazards and affecting society operation. To alleviate this problem, radiation sources and electronics are shielded by materials that reflect and/or absorb radiation. For reflection, electrically conducting polymer-matrix composites are important.

Review questions

1. What is meant by a composite material?

Answer: A composite material is a material having multiple components that are combined artificially.

2. What is meant by a semiconductor?
   Answer: A semiconductor is a material that is intermediate in conductivity between a conductor and an insulator. More specifically, it is a material with an energy band gap below 4 eV.

3. What are the differences among cement paste, mortar and concrete?
   Answer: Concrete has both sand and stones. Mortar has sand but no stones. Cement paste has neither sand nor stones.

4. What is difference in chemical bonding between graphite and diamond?
   Answer: Graphite has in-plane covalent and metallic bonding and out-of-plane van der Waals forces. Diamond is a covalent network solid.

5. From the chemical bonding point of view, explain why graphite is a good lubricant.
   Answer: Graphite is highly anisotropic, with much stronger bonding in-plane than out-of-plane. This enables easy sliding between the carbon layers in graphite.

6. What are the advantages of intrinsic smartness compared to extrinsic smartness?
   Answer: Intrinsic smartness is advantageous in the low cost, high durability, large functional volume and the mechanical properties being not compromised.

7. What is meant by the interlaminar interface of a composite material?
   Answer: The interlaminar interface is the interface between the laminae (plies) in a composite material in the form of a laminate.

8. How is a carbon fiber polymer-matrix composite able to sense its damage in the form of fiber breakage?
   Answer: The fiber breakage increases the volume electrical resistivity in the fiber direction of the composite.

9. What is a main problem with the method of self-healing involving microcapsules?
   Answer: The problems relate to the high cost of the catalyst and the stress concentrations around a microcapsule causing reduction in the mechanical properties of the material.

10. What is meant by the shape-memory effect?
    Answer: The shape-memory effect refers to the ability of a material to transform to a phase having a twinned microstructure (the phase called the martensite) that, after a subsequent plastic deformation, can return the material to its initial shape when heated.

11. What is a main application of the magnetostriction phenomenon?
    Answer: Magnetostriction provides actuation.

12. What is meant by a magnetorheological fluid?
    Answer: A magnetorheological fluid is a suspension of magnetic particles in a liquid, such that the fluid responds to an applied magnetic field by changing its fluidity.

13. What is the main application of optical fibers in smart structures?
    Answer: Optical fibers are for sensing.

14. What is the function of a thermal interface material?
    Answer: A thermal interface material is for enhancing thermal contacts.

15. What is the main principle in the design of the ability of a Stealth aircraft to escape radar detection?
    Answer: The main principle involves having a jagged shape for the aircraft so that the radar signal is not reflected back to the radar.

16. What is the main advantage of activated carbon that is mesoporous?
    Answer: The mesopores are larger than the micropores, thus allowing larger molecules or ions to reside in the pores for the sake of purification.

17. Why is a biodegradable polymer attractive for environmental protection?
    Answer: A biodegradable polymer can be degraded naturally, leaving no waste material.

Example problems

The example problems in this chapter are intended to help the readers review the basic concepts of introductory materials science. These basic concepts are not explicitly covered in this chapter, but they are useful for the remaining chapters of this book. For example problems that pertain to the specific topics of the electrical conduction, dielectric, electro-magnetic, optical and magnetic behavior, please refer to Ch. 2, 3, 4, 5 and 6, respectively.

Problem 1 (Diamond)

The density of diamond is 3.52 g/cm³. Calculate the atomic radius in Å. The atomic mass of carbon is 12.01 g/mol. Hint: Consider the crystal structure. For the diamond structure, $\sqrt{3}\, a = 8r$, where a is the edge length of the cubic unit cell and r is the atomic radius.

Solution:

Density = Mass of unit cell/volume of unit cell

$$= \text{Mass of 8 atoms}/a^3$$

$$= [8\,(12.01)/(6.02 \times 10^{23})]/a^3 = 3.52 \text{ g/cm}^3$$

$$a^3 = [8\,(12.01 \text{ g})/(6.02 \times 10^{23})]/(3.52 \text{ g/cm}^3) = 4.534 \times 10^{-23} \text{ cm}^3$$

$$= 45.34 \times 10^{-24} \text{ cm}^3$$

$$a = 3.566 \times 10^{-8} \text{ cm} = 3.566 \text{ Å}$$

$$r = \sqrt{3}\, a/8 = \sqrt{3}\,(3.566 \text{ Å})/8 = 0.77 \text{ Å}$$

Problem 2 (Graphite)

Graphite exhibits the crystal structure shown in Fig. 1.18. It shows half of the full height of the unit cell. The full unit cell spans from plane A to the next plane A (i.e., A-B-A). However, Fig. 1.18 shows only A-B.

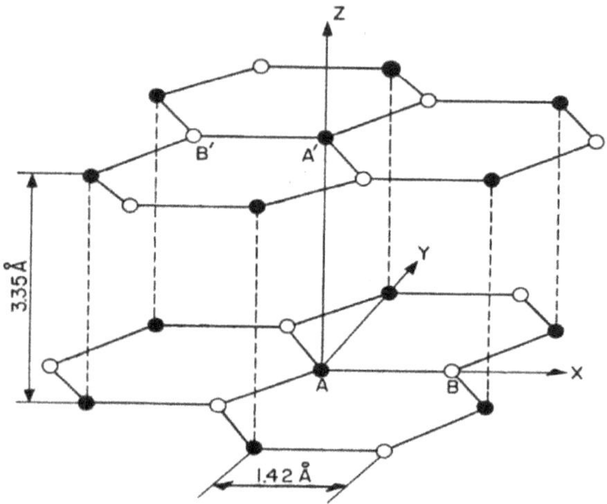

Fig. 1.18 The crystal structure of graphite.

(a) Describe the chemical bonding in graphite.

(b) What is the number of nearest neighboring atoms around a particular atom in graphite?

(c) What is the number of atoms per unit volume in graphite? Give your answer in cm^{-3}.

Solution:

(a) Covalent and metallic in-plane; van der Waals out-of-plane.

(b) 3

(c) No. of atoms per unit cell = 4

Area of hexagon = 3 (1.42 Å)^2 sin 60° = 5.239 Å^2

Volume of unit cell = area of hexagon × height of unit cell

= (5.239 Å^2) (3.35 Å) (2) = 35.1 Å^3 = 35.1 × 10^{-24} cm^3 = 3.51 × 10^{-23} cm^3

No. of atoms per unit volume

    = no. of atoms per unit cell/volume of unit cell

    = $4/(3.51 \times 10^{-23}\ cm^3) = 1.14 \times 10^{23}\ cm^{-3}$

Problem 3 (Graphite)

Calculate the density (in g/cm^3) of ideal graphite, based on the crystal structure (Fig. 1.18). The atomic mass of carbon is 12.01 g/mol.

Solution:

Area of hexagon $= 3\ (1.42\ \text{Å})^2 \sin 60° = 5.239\ \text{Å}^2$

Volume of unit cell = area of hexagon × height of unit cell

$= (5.239\ \text{Å}^2)\ (3.35\ \text{Å})\ (2) = 35.1\ \text{Å}^3 = 35.1 \times 10^{-24}\ cm^3 = 3.51 \times 10^{-23}\ cm^3$

There are 4 atoms per unit cell.

Density = Mass of unit cell/volume of unit cell

    $= 4\ [(12.01\ g)/(6.02 \times 10^{23})]/(3.51 \times 10^{-23}\ cm^3) = 2.27\ g/cm^3$

A more exact calculation gives the value of 2.26 g/cm^3.

Problem 4 (Fullerene)

(a) Describe the chemical bonding in fullerene.

(b) Calculate the molecular mass of fullerene. The atomic mass of carbon is 12.01 g/mol.

Solution:

(a) Covalent bonding.

(b) Molecular mass $= 60\ (12.01\ g/mol) = 721\ g/mol$

Problem 5 (Solid solution)

Consider a substitutional solid solution with Al_2O_3 as the solute and MgO as the solvent (host). It exhibits the Rocksalt Structure. In order to maintain charge neutrality, cation vacancies must be present. The ratio of the number of Al^{3+} ions to the number of Mg^{2+} ions in this solid solution is 0.289.

(a) In order to maintain electrical neutrality, how many cation vacancies must accompany ten Al^{3+} ions?

(b) What is the fraction of the cation sites that are vacant?

Solution:

(a) $10/2 = 5$

(b) Consider 100 cation sites.

No. of cation sites $= n$

No. of Al^{3+} ions $= 2n$

No. of cations $= 100 - n$

No. of Mg^{2+} ions $= (100 - n) - 2n = 100 - 3n$

No. of Mg^{2+} ions $=$ No. of Al^{3+} ions$/0.289 = 2n/0.289$

Hence,

$100 - 3n = 2n/0.289$

$100 = 3n + (2n/0.289) = n\,(3 + 2/0.289) = n\,9.920$

$n = 100/9.920 = 10.08$

Fraction of cation sites that are vacant $= 10.08/100 = 0.101$

Problem 6 (Solid solution)

This problem concerns a solid solution of MgO in NiO. The ratio of the number of Mg^{2+} ions to the number of Ni^{2+} ions in the solid solution is 0.37.

(a) Describe the chemical bonding in this solid solution.

(b) Give the name of the crystal structure exhibited by this solid solution.

(c) What is the fraction of cation sites that are occupied by Ni^{2+} ions?

Solution:

(a) Ionic.

(b) Rocksalt structure.

(d) Consider 100 cation sites. It contains x Mg^{2+} ions. Then it contains $(100-x)$ Ni^{2+} ions.

$x/(100-x) = 0.37$

$x = 37 - 0.37x$

$0.63x = 37$

$x = 37/0.63 = 58.7$

Fraction of cation sites occupied by Ni^{2+} ions

$= (100-x)/100 = (100-58.7)/100 = 41.3/100 = 0.41$

Problem 7 (Non-stoichiometric compound)

A non-stoichiometric compound is a compound with non-integral atom proportions of the elements in the chemical formula. Consider a non-stoichiometric iron oxide with chemical formula $Fe_{0.96}O$. It exhibits the

Rocksalt Structure. In order to maintain charge neutrality, both Fe^{2+} and Fe^{3+} ions need to be present. Due to the non-stroichimetry with the cations being deficient, there are cation vacancies.

(a) What is the ratio of the number of cation vacancies to the number of Fe^{3+} ions?

(b) What is the ratio of the number of Fe^{3+} ions to the number of Fe^{2+} ions?

Solution:

(a) For electrical neutrality, every cation vacancy must be accompanied by two Fe^{3+} ions. Hence,

the ratio of the number of cation vacancies to the number of Fe^{3+} ions $= 1/2 = 0.5$

(b) Consider 100 cation sites.

The chemical formula indicates that there are 96 cations and 4 cation vacancies.

For electrical neutrality, every cation vacancy must be accompanied by two Fe^{3+} ions.

With 4 cation vacancies, there are thus $4 \times 2 = 8$ Fe^{3+} ions.

With 96 cations, there are thus $96 - 8 = 88$ Fe^{2+} ions. Thus,

the ratio of the number of Fe^{3+} ions to the number of Fe^{2+} ions is $8/88 = 0.091$

Problem 8 (Phase transformation)

A material transforms from HCP crystal structure to BCC structure upon heating. The lattice constants of the HCP structure are $a = 2.951$ Å and

$c = 4.684$ Å. Assume that the atomic radius is not affected by the phase transformation. What is the fractional change in volume upon the transformation from HCP to BCC?

Solution:

Let the lattice constant be a_H and a_B for the HCP and BCC structures, respectively.

$a_H = 2r$, where r is the atomic radius. Hence,

$r = a_H / 2$

For HCP, volume of a primitive unit cell (with 2 atoms) = volume of the prism based on a parallelogram $= (a_H^2 \sin 60°)\, c = (2.951^2 \sin 60°)$ (4.684) Å$^3 = 35.33$ Å^3

For BCC, $\sqrt{3}\, a_B = 4r$, and

$$\text{volume of a cubic unit cell (with 2 atoms)} = a_B^3 = (4/\sqrt{3})^3\, r^3$$
$$= (4/\sqrt{3})^3\, (2.951/2)^3\ \text{Å}^3 = 39.56\ \text{Å}^3$$

Hence, there is a volume increase upon the change from HCP to BCC.

Fractional change in volume $= (39.56 - 35.33)/35.33 = 0.12 = 12\%$

Problem 9 (Phase transformation of titanium)

Titanium is BCC in crystal structure (lattice constant $a = 3.32$ Å) at high temperatures and transform from BCC to HCP (lattice constants $a = 2.978$ Å and $c = 4.735$ Å) upon cooling at about 890°C.

(a) What is the fractional change in volume when BCC titanium transforms to HCP titanium?

(b) Is it an expansion or a contraction?

(c) The HCP form is less ductile than the BCC form. Explain from the viewpoint of the scientific origin.

Solution:

(a) BCC: 2 atoms/cell (cubic unit cell)
HCP: 2 atoms/cell (primitive unit cell, which is based on a parallelogram)

Let a_B be the lattice constant of BCC.
Let a_H and c be the lattice constants of HCP.

Vol. of a BCC unit cell $= a_B^3 = (3.32 \text{ Å})^3 = 36.5944 \text{ Å}^3$

Vol. of an HCP unit cell $= a_H^2 c \sin 60°$

$$= 2.978^2 \, 4.735 \sin 60°$$

$$= 36.3664 \text{ Å}^3$$

Fractional change in volume $= (36.3664 - 36.5944)/36.5944 = -0.62\%$

(b) Minus sign means contraction.

(c) There are fewer slip systems for HCP than BCC.

Problem 10 (Angle between crystallographic planes)

The lattice constants of HCP titanium are c and a, with the ratio $c/a = 1.587$. What is the angle between the (101) and (001) crystallographic planes of HCP titanium?

Solution:

$$\tan \theta = c/a = 1.587$$

$$\theta = 57.78°$$

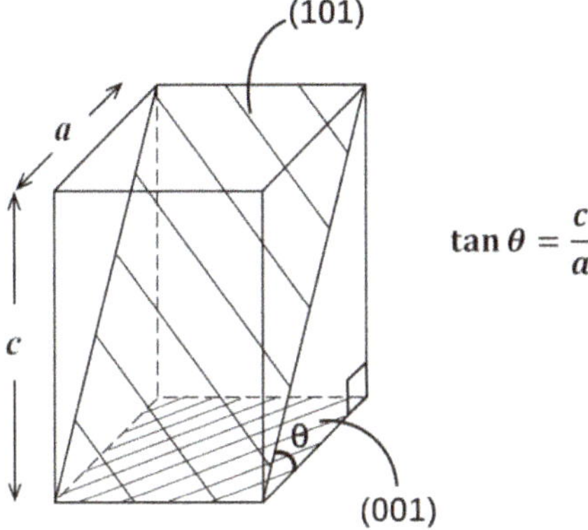

Problem 11 (Metal alloy)

Figure 1.19 shows the aluminum-rich end of the Al-Cu phase diagram.

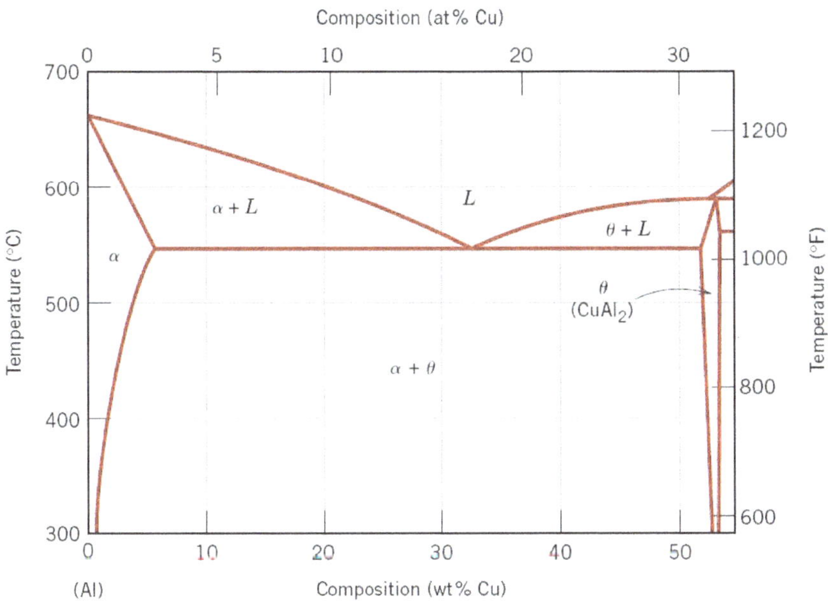

Fig. 1.19 The aluminum-rich end of the Al-Cu phase diagram.

(a) What is the range of composition that is suitable for precipitation hardening?

(b) What is the phase of the precipitate?

(c) For an alloy containing 4 wt.% Cu and equilibrated at 300°C, what is the weight fraction of the precipitate?

Solution:

(a) 2–5 wt.% Cu

(b) θ

(c) Using the Lever Rule,

$$\text{Weight fraction of the precipitate} = (4 - 1)/(53 - 4)$$

$$= 3/49$$

$$= 0.061 = 6 \text{ wt.\%}$$

Problem 12 (Intermetallic compound)

An intermetallic compound is a compound involving two or more metal elements in the chemical formula. In the Al-Ni phase diagram (Fig. 1.20), Al_2Ni is an intermetallic compound.

(a) What is the composition of this compound in at.% Ni?

(b) At what temperature does this compound start to melt upon slow heating?

(c) To what phases does this compound change upon heating past the temperature for the start of melting?

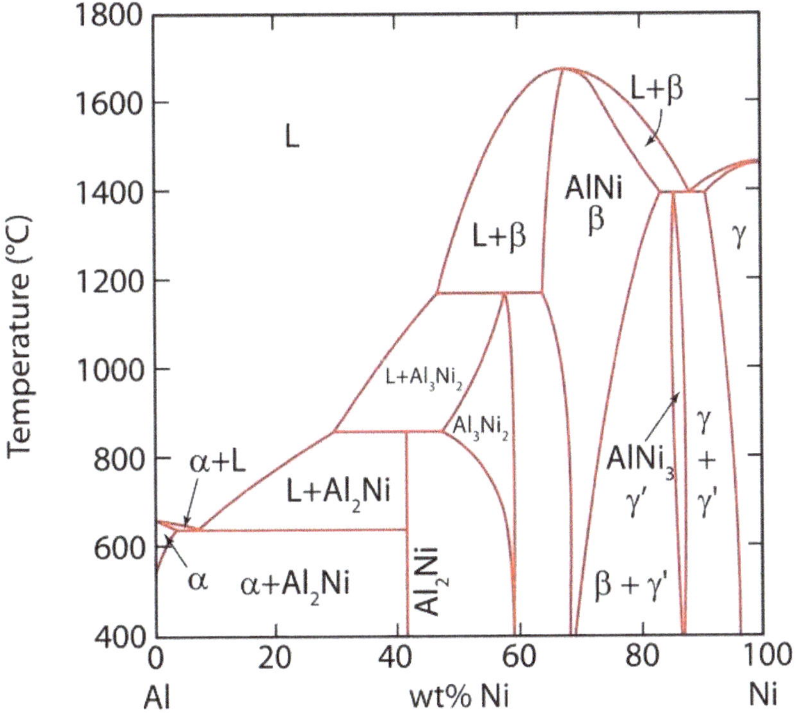

Fig. 1.20 The Al-Ni phase diagram.

Solution:

(a) 33 at.% Ni

(b) 850°C

(c) $L + Al_3Ni_2$

Problem 13 (Intermetallic compound)

In the Al-Ni phase diagram (Fig. 1.20), Ni_3Al is an intermetallic compound.

(a) What is the composition of this compound in wt.% Ni?

(b) At what temperature does this compound start to melt upon slow heating?

(c) At what temperature does this compound become completely molten upon slow heating?

Solution:

(a) 87 wt.% Ni

(b) 1400°C

(c) 1450°C

Problem 14 (Intermetallic compound)

In the Al-Ni phase diagram (Fig. 1.20), AlNi is an intermetallic phase.

(a) Calculate the composition of AlNi in wt.% Ni? The atomic mass of Ni is 58.69 g/mole. The atomic mass of Al is 26.98 g/mole.

(b) What is the maximum nickel content (in wt.% Ni) that can exist in this intermetallic phase?

(c) At what temperature does the maximum nickel content in (b) occur?

Solution:

(a) $58.69/(58.69 + 26.98) = 0.685 = 68.5$ wt.%

(b) 83 wt.% Ni

(c) 1390°C

Problem 15 (Titanium aluminide)

The crystal structure of the TiAl intermetallic compound is shown in Fig. 1.21. The unit cell is tetragonal, with the Al atoms at the 8 corners and two face centers, and the Ti atoms at four face centers.

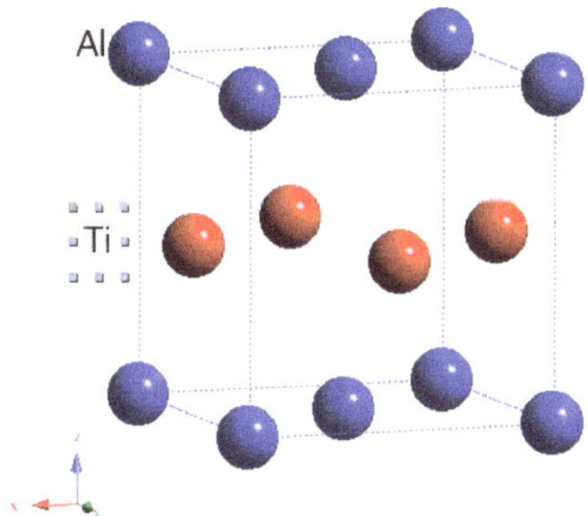

Fig. 1.21 The TiAl crystal structure.
(https://www.phase-trans.msm.cam.ac.uk/2004/titanium/titanium.html)

(a) How many Al atoms are nearest to a Ti atom?

(b) How many Ti atoms are nearest to an Al atom?

(c) How many atoms of each type are there per unit cell?

(d) Where are the lattice points?

(e) How many atoms of each type are associated with each lattice point?

Solution:

(a) 8

(b) 4

(c) 2 of each

(d) The 8 corners

(e) 1 of each

Problem 16 (Nickel aluminide)

The intermetallic compound Ni_3Al has a crystal structure described by a cubic unit cell, with the Ni atoms located at the six face centers and the Al atoms located at the eight corners.

(a) Describe the chemical bonding in this material.

(b) The mechanical strength of this material is high, though the material is brittle. Explain from the viewpoint of the scientific origin.

(c) This material can withstand very high temperatures. Explain from the viewpoint of the scientific origin.

(d) This material has lower entropy than a solid solution, but it is thermodynamically stable. Explain the stability from the viewpoint of the scientific origin.

(e) What is the number of Ni atoms per unit cell?

(f) How many atoms of each type are associated with each lattice point? Hint: A lattice point is a mathematical point in space, such that all lattice points in an infinite structure have the same environment.

Solution:

(a) Metallic bonding

(b) The Ni and Al atoms want to maintain their relative positions, thus causing the material to resist plastic deformation.

(c) The Ni and Al atoms want to maintain their relative positions, thus causing the material to maintain its structural order even at very high temperatures.

(d) The structure is associated with low enthalpy.

(e) 3

(f) 1 Al atom and 3 Ni atoms

Problem 17 (Nickel aluminide)

The intermetallic compound Ni_3Al has a crystal structure described by a cubic unit cell (length of each edge of the cube $= a = 3.572$ Å), with the Ni atoms (atomic radius r) located at the six face centers and the Al atoms (atomic radius R) located at the eight corners. The atomic mass of Ni is 58.69 g/mol. The atomic mass of Al is 26.982 g/mol. Avogadro's number $= 6.02 \times 10^{23}$

(a) What is the composition of this material in at.% Ni? (Hint: at.% means atomic %.)

(b) What is the composition of this material in wt.% Ni? (Hint: weight % is the same as mass %.)

(c) Derive the relationship among a, r and R.

(d) Calculate the density (in g/cm^3) of this material.

Solution:

(a) $3/4 = 75$ at.% Ni

(b) Consider 100 atoms in the material. It consists of 75 Ni atoms and 25 Al atoms.

Mass of 75 Ni atoms $= 75 \, (59.69 \, \text{g}/6.02 \times 10^{23})$

Mass of 25 Al atoms $= 25 \, (26.982/6.02 \times 10^{23})$

Ni mass fraction $= (75 \times 58.69)/[(75 \times 58.69) + (25 \times 26.982)]$

$$= 86.7\%$$

(c) The atoms touch along the face diagonals, so

$$2r + 2R = \sqrt{2}\,a$$

(d) Density = Mass of unit cell / Volume of unit cell

$$= \{[3 \times 58.69 \text{ g}/(6.02 \times 10^{23})]$$
$$+ [26.982 \text{ g}/(6.02 \times 10^{23})]\}/(3.572 \text{ Å})^3$$

$$= 7.40 \text{ g/cm}^3$$

A more exact calculation gives the value of 7.5 g/cm^3.

Problem 18 (Silicon carbide)

Silicon carbide of the cubic allotrope exhibits the Zinc Blende structure and density 3.21 g/cm^3. The atomic masses of C and Si are 12.009 and 28.084 g/mol, respectively.

(a) Describe the chemical bonding in SiC.

(b) What is the lattice constant (i.e., the length of an edge of the cubic unit cell) in Å?

Solution:

(a) Covalent network solid, with slightly polar SiC bonds.

(b) There are 4 Si atoms and 4 C atoms per unit cell.

Density = Mass of unit cell/volume of unit cell

Volume of unit cell = Mass of unit cell/density

$$= [4\,(12.009 + 28.084 \text{ g})/(6.02 \times 10^{23})]/(3.21 \text{ g/cm}^3)$$

$$= 8.30 \times 10^{-23} \text{ cm}^3$$

Let the lattice constant be a.

Volume of unit cell $= a^3 = 8.30 \times 10^{-23}$ cm^3 $= 83 \times 10^{-24}$ cm^3

$$a = 4.36 \times 10^{-8} \text{ cm} = 4.36 \text{ Å}$$

Problem 19 (Silicon carbide)

A silicon carbide (SiC) particle aluminum-matrix composite contains 61 vol.% SiC. The density of SiC is 3.21 g/cm^3. The density of Al is 2.70 g/cm^3.

(a) What are the main attractive properties of this composite?

(b) Describe the method of making this composite by squeeze casting.

(c) Describe the method of power metallurgy (involving sintering) for making metal-matrix composites.

(d) Conventional power metallurgy is not effective for making this particular composite. Why?

(e) Calculate the density (in g/cm^3) of the composite.

Solution:

(a) High modulus, high strength, low thermal expansion.

(b) Infiltrate a SiC particle preform with liquid aluminum.

(c) Sinter together SiC particles and Al particles by using heat and pressure.

(d) The SiC volume fraction is too high for the Al matrix to be sufficient for binding the SiC particles together.

(e) Density $-$ (0.61) (3.21 g/cm^3) + (0.39) (2.70 g/cm^3) $= 3.01$ g/cm^3

Problem 20 (Silicon dioxide)

Consider the cristobalite crystal structure of SiO_2.

(a) Describe the chemical bonding in SiO_2. Hint: Consider the hybridization of the orbitals.

(b) What is the number of O^{2-} ions that are nearest neighbors to a Si^{4+} ion?

(c) What is the number of Si^{4+} ions that are nearest neighbors to an O^{2-} ion?

Solution:

(a) The Si atoms are sp^3 hybridized. It is a covalent network solid, with the Si-O bond being partly ionic.

(b) 4

(c) 2

Problem 21 (Zirconium dioxide)

The cubic unit cell of the Fluorite Structure, as exhibited by cubic zirconia (ZrO_2) is shown in Fig. 1.22. In cubic ZrO_2, the Zr^{4+} ions

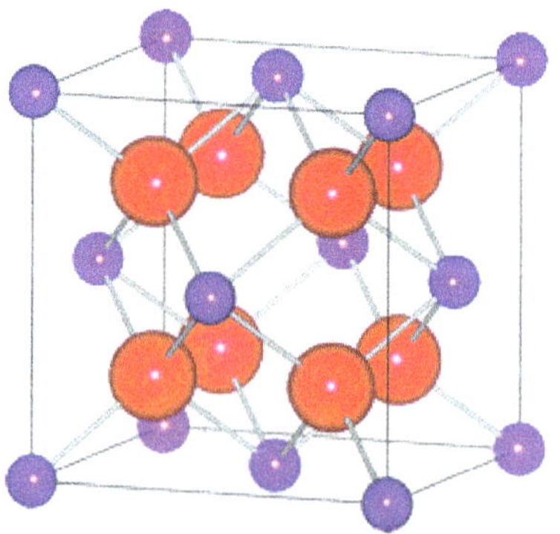

Fig. 1.22 The Fluorite Structure.

occupy the corners and face center positions of the cubic unit cell, while the O^{2-} ions occupy the tetrahedral interstitial sites of the cation structure.

(a) How many Zr^{4+} ions are nearest to an O^{2-} ion?

(b) How many O^{2-} ions are nearest to a Zr^{4+} ion?

(c) How many ions of each type are there per cubic unit cell of ZrO_2?

Solution:

(a) 4

(c) 8

(d) 4 Zr^{4+} ions and 8 O^{2-} ions

Problem 22 (Zirconium dioxide)

The cubic unit cell of the Fluorite Structure, as exhibited by cubic zirconia (ZrO_2) is shown in Fig. 1.22. In cubic ZrO_2, the Zr^{4+} ions occupy the corners and face center positions of the cubic unit cell, while the O^{2-} ions occupy the tetrahedral interstitial sites of the cation structure. The ionic radii of Zr^{4+} and O^{2-} are 0.86 Å and 1.26 Å, respectively.

(a) Calculate the lattice parameter a (in Å) of ZrO_2. The lattice parameter refers to the length of each edge of the cubic unit cell in Fig. 1.22. Hint: The ions touch along the body diagonal of the cubic unit cell.

(b) Yttria (Y_2O_3) is dissolved in zirconia to form a solid solution in order to stabilize the zirconia. What is the scientific meaning (purpose) of the stabilization? Hint: Consider the phase diagram.

Solution:

(a) Let r be the radius of a Zr^{4+} ion. Let R be the radius of an O^{2-} ion.

$$2r + 4R = (\sqrt{3})\, a$$

$$a = (2/\sqrt{3})\,(r + 2R) = (2/\sqrt{3})\,[0.86 + 2\,(1.26)]\ \text{Å} = 3.90\ \text{Å}$$

(b) The solid solution remains cubic over a wide temperature range, so that cubic zirconia does not transform to another crystal structure upon cooling. The phase transformation involves a change in volume and can cause cracking.

Problem 23 (Zirconium dioxide)

Cubic zirconia (ZrO_2) consists Zr^{4+} and O^{2-} ions. It exhibits the Fluorite structure and density 5.68 g/cm^3. The atomic masses of Zr and O are 91.224 and 15.999 g/mol, respectively.

(a) What is the mass (in g) of a conventional cubic unit cell of zirconia?

(b) In yittra-stabilized zirconia, Y^{3+} ions substitute some of the Zr^{4+} ions to form a substitutional solid solution. To maintain electrical neutrality, this substitution must be accompanied by the generation of anion vacancies. What is the ratio of the number of Y^{3+} ions to the number of anion vacancies?

Solution:

(a) A unit cell contains 4 Zr^{4+} and 8 O^{2-} ions. Hence,

$$\text{Mass of a unit cell} = [4\,(91.224) + 8\,(15.999)]\ \text{g}/(6.02 \times 10^{23})$$

$$= 81.9 \times 10^{-23}\ \text{g} = 8.19 \times 10^{-22}\ \text{g}$$

(b) Consider 4 cation sites and 8 anion sites. In case of no substitution, these are occupied by 4 Zr^{4+} ions and 8 O^{2-} ions. With two Y^{3+} ions substituting two of these four Zr^{4+} ions, there would be 2 Zr^{4+} and 2 Y^{3+} ions, which total $8 + 6 = 14$ units of positive charge. Thus, there will need to be 7 O^{2-} ions, which total $7 \times 2 = 14$ units of negative charge.

With 7 O^{2-} ions occupying the 8 anion sites, there is one anion vacancy.

Therefore,
the ratio of the number of Y^{3+} ions to the number of anion vacancies is $2/1 = 2$.

Problem 24 (Polystyrene)

A polystyrene molecule (with the mer shown below) has the degree of polymerization equal to 750. Calculate the molecular mass (in the unit g/mole). The atomic masses of carbon and hydrogen and 12.0 g/mole and 1.0 g/mole, respectively. Hint: The degree of polymerization refers to the number of mers in a molecule.

Solution:

Note that the benzene ring has a carbon atom at each of its six corners, with five of corner carbon atoms bonded to hydrogen atoms. The above molecular structure is according to conventional abbreviation. Thus, there are $2 + 6 = 8$ carbon atoms and $3 + 5 = 8$ hydrogen atoms per mer.

Mer mass = [8(12) + 8(1)] g/mole = 104 g/mole

Molecular mass = Mer mass × degree of polymerization

$$= (104 \text{ g/mole}) (750) = 78000 \text{ g/mole}$$

Problem 25 (Molecular weight of a polymer)

A polymer solid has molecules of various sizes coexisting. A polystyrene solid (mer shown in Problem 24) has
625,000 molecules with degree of polymerization 620,
930,000 molecules with degree of polymerization 810, and
780,000 molecules with degree of polymerization 990.
Calculate the number-average molecular mass (in the unit g/mole).

Hint: Number-average molecular mass = $\Sigma_i(X_i M_i)$

Hint: The degree of polymerization refers to the number of mers in a molecule.

Solution:

Total no. of molecules = 625000 + 930000 + 780000 = 2335000

$$X_1 = 625000/2335000 = 0.268$$

$$X_2 = 930000/2335000 = 0.398$$

$$X_3 = 780000/2335000 = 0.334$$

Number-average molecular mass

$$= [(0.268) \, 620 + (0.398) \, 810 + (0.334) \, 990] \, 104 = 85195$$

Problem 26 (Polymers and fabrication methods)

Explain each of the following in terms of the science.

(a) The elastic modulus of a bulk polymer solid is lower than that of the corresponding polymer in the form of a fiber.

(b) Polyvinylchloride (PVC) is mechanically stronger than polyethylene (PE).

(c) Powder metallurgy (involving sintering) is a process that is applicable to the fabrication of small objects only.

(d) Stir casting is not an effective method for fabricating particulate metal-matrix composites.

(e) Directional solidification is an attractive process for fabricating gas turbine blades.

Solution:

(a) The molecules are relatively aligned along the fiber axis, whereas they are randomly oriented in a bulk polymer.

(b) PVC has stronger inter-molecular bonding (van der Waals forces) than PE.

(c) The need to apply pressure limits the size.

(d) The particles either float or sink in the liquid metal (due to the difference in density between the particles and liquid metal), causing non-uniform particle distribution in the composite.

(e) Directional solidification results in aligned grains after solidification.

Problem 27 (Carbon fiber, precipitate, polymer, etc.)

Explain each of the following in terms of the science.

(a) High-modulus carbon fiber exhibits low strength.

(b) The precipitate particle size is preferably small for effective precipitation hardening.

(c) Having longer molecules are preferred for polymers.

(d) The fabrication process of carbon-carbon composites is expensive.

(e) Ceramics are typically superior to carbons in the ability to withstand high temperatures.

Solution:

(a) The carbon layers are well aligned, making the modulus high while causing the shear between the layers to be easy.

(b) The small size promotes the precipitate particles to be coherent.

(c) The bonding is strong within a molecule than between the molecules, so the mechanical and thermal properties are promoted by having longer molecules.

(d) The process requires densification through multiple cycles of impregnation and carbonization, or chemical vapor infiltration.

(e) Ceramics are compounds, whereas carbons are an element.

Problem 28 (Composite material – longitudinal modulus)

A polymer-matrix composite contains 56 vol.% unidirectional continuous carbon fibers and 44 vol.% matrix. The axial tensile modulus of the fiber is 325 GPa. The modulus of the polymer is 2.3 GPa. Calculate the modulus of the composite in the fiber direction of the composite.

Hint:

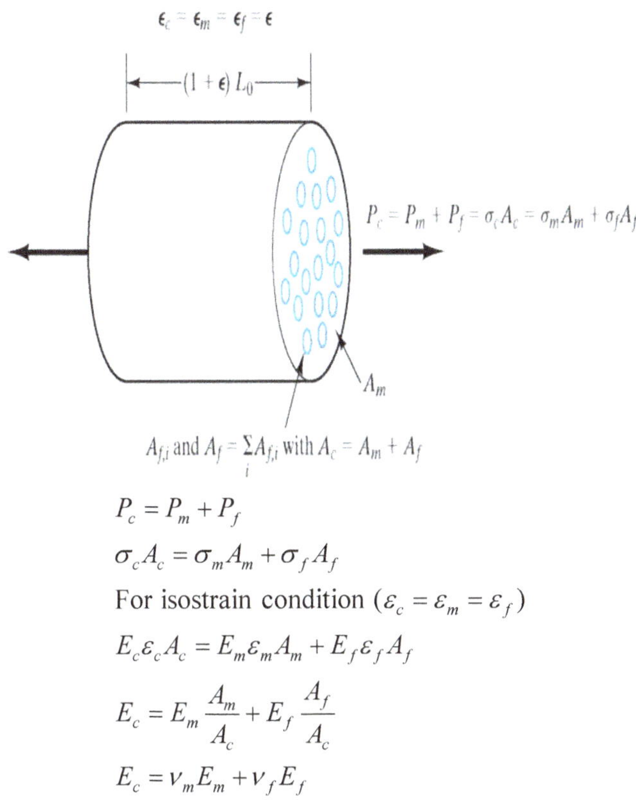

$$P_c = P_m + P_f$$

$$\sigma_c A_c = \sigma_m A_m + \sigma_f A_f$$

For isostrain condition $(\varepsilon_c = \varepsilon_m = \varepsilon_f)$

$$E_c \varepsilon_c A_c = E_m \varepsilon_m A_m + E_f \varepsilon_f A_f$$

$$E_c = E_m \frac{A_m}{A_c} + E_f \frac{A_f}{A_c}$$

$$E_c = v_m E_m + v_f E_f$$

Solution:

Modulus of composite $= [(325\ \text{GPa})(0.56) + (2.3)(0.44)]\ \text{GPa}$

$$= (182 + 1.01)\ \text{GPa}$$

$$= 183\ \text{GPa}$$

Problem 29 (Composite material – transverse modulus)

A polymer-matrix composite contains 56 vol.% unidirectional continuous carbon fibers and 44 vol.% matrix. The transverse tensile modulus of the fiber is 13.4 GPa. The modulus of the polymer is 2.3 GPa. Calculate the modulus of the composite in the transverse direction of the composite.

Hint:

$$\sigma_c = \sigma_m = \sigma_f$$

$$\Delta L_c = \Delta L_m + \Delta L_f$$

$$\frac{\Delta L_c}{L_c} = \frac{\Delta L_m}{L_c} + \frac{\Delta L_f}{L_c}$$

$$L_m = A_m L_c$$

$$L_f = A_f L_c$$

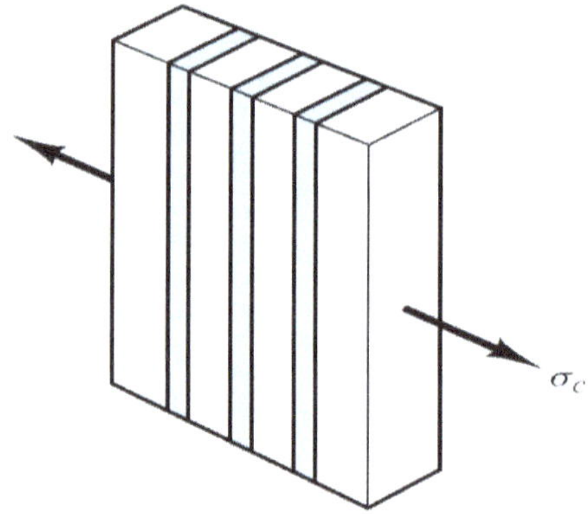

$$\frac{\Delta L_c}{L_c} = \frac{A_m \Delta L_m}{L_m} + \frac{A_f \Delta L_f}{L_f}$$

$$\varepsilon_c = V_m \varepsilon_m + V_f \varepsilon_f$$

For isostress condition $(\sigma = E_c \varepsilon_c = E_m \varepsilon_m = E_f \varepsilon_f)$,

$$\frac{\sigma}{E_c} = V_m \frac{\sigma}{E_m} + V_f \frac{\sigma}{E_f}$$

$$\frac{1}{E_c} = \frac{V_m}{E_m} + \frac{V_f}{E_f}$$

$$E_c = \frac{E_m E_f}{V_m E_f + V_f E_m}$$

Solution:

Transverse modulus of the composite $= E_c$

$1/E_c = [(0.44)/(2.3\ \text{GPa})] + [(0.56)/(13.4\ \text{GPa})]$

$\qquad = (0.191 + 0.042)\ \text{GPa}^{-1} = 0.233\ \text{GPa}^{-1}$

$E_c = 4.3\ \text{GPa}$ (which is relatively close to the modulus of the polymer)

Problem 30 (Composite material – tensile modulus)

(a) Carbon fiber has a high tensile modulus. Explain this from a scientific viewpoint.

(b) Calculate the tensile modulus of a unidirectional carbon fiber composite that contains 58 vol.% carbon fiber and 42 vol.% epoxy. The modulus of the carbon fiber is 347 GPa. The modulus of the epoxy is negligible compared to that of the carbon fiber.

Hint: Rule of Mixtures, i.e.,

Modulus of composite
     = (volume fraction of fiber) (modulus of fiber)
        + (volume fraction of epoxy) (modulus of epoxy)

Solution:

(a) The carbon layers are preferentially aligned along the fiber axis and each layer exhibits strong covalent bonding.

(b) Modulus of composite = (0.58) (347 GPa) + (0.42) (0) = 201 GPa

Problem 31 (Composite material – fraction of load carried by the fibers)

A unidirectional continuous fiber polymer-matrix composite consists of fibers with elastic modulus 560 GPa and a matrix of elastic modulus 16 GPa, such that the fiber volume fraction is 57%. What is the fraction of longitudinal load carried by the fibers? Assume perfect bonding between the fibers and the matrix.
Hint:

$$\frac{P_f}{P_c} = \frac{\sigma_f A_f}{\sigma_c A_c} = \frac{E_f \varepsilon_f A_f}{E_c \varepsilon_c A_c} = \frac{E_f}{E_c} V_f$$

Solution:

Composite modulus = [(560) (0.57) + (16) (0.43)] GPa = 326 GPa

Fraction of load carried by the fibers = (560/326) (0.57) = 0.98

Problem 32 (Polymer)

Polyethylene (Fig. 1.23) has a C-C bond length of 1.54 Å. The C-C-C bond angle is 109.5°. There are 5000 mers in a molecule. The atomic mass of C is 12.0107 g/mol. The atomic mass of H is 1.00794 g/mol.

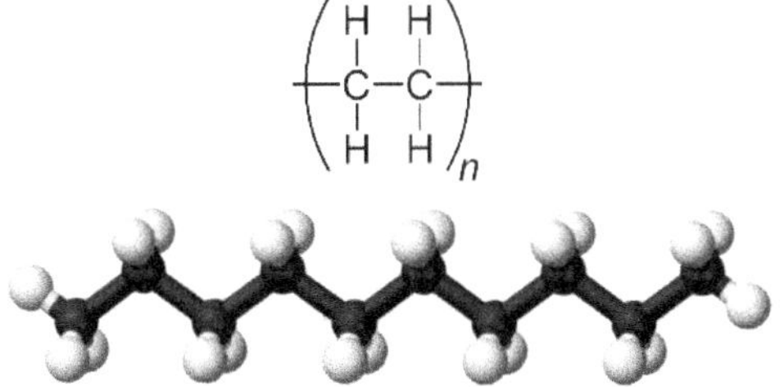

Fig. 1.23 The molecular structure of polyethylene.
(https://upload.wikimedia.org/wikipedia/commons/thumb/b/b0/Decane-3D-balls-
B.png/800px-Decane-3D-balls-B.png)

(a) Describe the chemical bonding in a molecule of polyethylene, including a description of the hybridization of the electron orbitals.

(b) Polyethylene is a relatively soft polymer. Explain from the viewpoint of the scientific origin.

(c) Calculate the molecule mass.

(d) Calculate the length of the molecule. Assume that there is no bending of the molecule.

Solution:

(a) Covalent bonding, with the C atoms sp^3 hybridized.

(b) The bonding between the molecules is weak (Van der Waals forces).

(c) Molecular mass = [2 (12.0107) + 4 (1.00794)] g/mol × 5000

$$= 140300 \text{ g/mol}$$

(d) Molecular length = 5000 × 2(1.54 Å) cos [(180 − 109.5)/2] = 12600 Å

Problem 33 (Diamond)

Diamond is an allotrope of carbon. Its crystal structure is shown in Fig. 1.24.

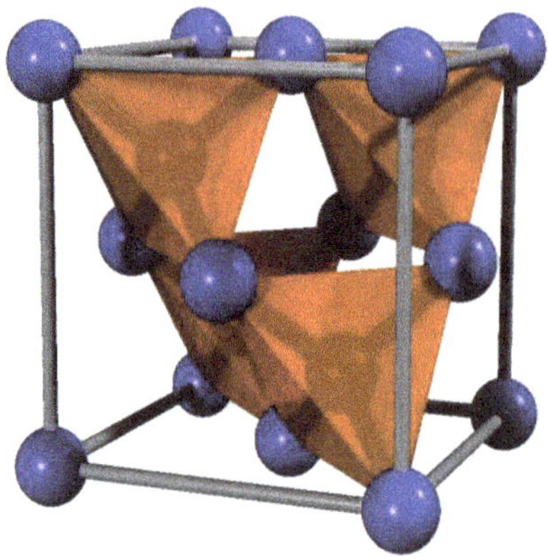

Fig. 1.24 The diamond crystal structure.
(https://en.wikipedia.org/wiki/Diamond, public domain)

(a) Describe the chemical bonding in diamond, including a description of the hybridization of electron orbitals.

(b) Diamond is a thermal conductor. Explain from the viewpoint of the scientific origin.

(c) Diamond is mechanically very hard. Explain from the viewpoint of the scientific origin, with consideration of the chemical bonding.

(d) Diamond has a high dielectric strength. Explain from the viewpoint of the scientific origin, with consideration of the chemical bonding. Hint: The dielectric strength is the electric field above which a material ceases to be an electrical insulator.

(e) Upon heating to a sufficiently high temperature, diamond changes to graphite. Explain from the viewpoint of the scientific origin.

(f) What is the main difference in material structure between diamond and diamond-like carbon?

Solution:

(a) Covalent bonding, with each C atom being sp^3 hybridized.

(b) The phonons (lattice vibrational waves) transport the heat.

(c) The covalent network with directional bonding (tetrahedral coordination) resists plastic deformation.

(d) All the electrons are trapped in the covalent bonds.

(e) Diamond is not thermodynamically stable, whereas graphite is.

(f) Diamond-like carbon is amorphous, whereas diamond is crystalline.

References

1. Chung DDL. A review of multifunctional polymer-matrix structural composites. Composites, Part B, 2019;160:644-660.

2. Chung DDL. A critical review of piezoresistivity and its application in electrical-resistance-based strain sensing. J Mater Sci 2020;55(32): 15367-15396.

3. Chung DDL. Self-sensing concrete: from resistance-based sensing to capacitance-based sensing (Invited Paper). Int J Smart Nano Mater (2020). DOI:10.1080/19475411.2020.1843560.

4. Wen S, Chung DDL. Piezoresistivity-based strain sensing in carbon fiber reinforced cement. ACI Mater J 2007;104(2):171-179.

5. Wen S, Chung DDL. Uniaxial tension in carbon fiber reinforced cement, sensed by electrical resistivity measurement in longitudinal and transverse directions. Cem Concr Res 2000;30(8):1289-1294.

6. Wen S, Chung DDL. Effect of moisture on the piezoresistivity of carbon fiber reinforced cement. ACI Mater J 2008;105(3):274-280.

7. Wen S, Chung DDL. Electrical-resistance-based damage self-sensing in carbon fiber reinforced cement. Carbon 2007;45(4):710-716.

8. Wang S, Chung DDL. Self-sensing of flexural strain and damage in carbon fiber polymer-matrix composite by electrical resistance measurement. Carbon 2006;44(13):2739-2751.

9. Wang S, Chung DDL. Negative piezoresistivity in continuous carbon fiber epoxy-matrix composite. J Mater Sci 2007;42(13):4987-4995.

10. Duerig TW, Melton KN, Stockel D, Wayman CM. Engineering aspects of shape memory alloys. Butterworth Heinemann, London, 1990.

11. Chung DDL. Development, design and applications of structural capacitors. Applied Energy 2018;231:89-101.

Chapter 2

Electrical Conduction Behavior

This chapter covers the science and applications that are based on electrical conduction behavior, which relates to the movement of charged species, such as electrons. Applications include electronics, diodes, resistance heating (e.g., deicing), thermistors, thermocouples and thermoelectric energy generation.

2.1 Origin of electrical conduction

Unless noted otherwise, an electrically conductive material discussed in this book involves electrons and/or holes as the charge carrier (mobile charged particle) for electrical conduction. A hole is an electronic vacant site. A hole is positively charged, since the removal of an electron, which is negatively charged, leaves something that is positively charged. Although a hole is a vacant site, it can move due to an electron moving from an electron site to fill this vacancy, thereby causing the hole to move to the original position of this electron. Conventional metals (e.g., copper) have electrons, but no holes. On the other hand, semiconductors (e.g., silicon) and semimetals (e.g., graphite) have both electrons and holes. Both electrons and holes contribute to electrical conduction.

The abundance of the carrier (i.e., a high carrier concentration) and the high mobility of the carrier contribute to causing high conducitivity in a material. With high carrier concentration and high carrier mobility, both resulting from the delocalized (free) valence electrons, as characteristic of metallic bonding, metals tend to have high conductivity. Semiconductors and semimetals tend to have lower conductivity than metals, due to their relatively low carrier concentration and relatively low carrier mobility. Ceramics (e.g., silica, i.e., silicon dioxide or SiO_2) and polymers (e.g., polyethylene) tend to be insulators, with very low conductivity, due to the very low values of the carrier concentration and carrier mobility.

Ions include cations (positively charged) and anions (negatively charged). They can also serve as charge carriers, but their high mass compared to electrons causes their movement to be sluggish. As a consequence, ionic conductors tend to be less conductive than electronic conductors. In general, an electrical conductor may be an electronic conductor (with electrons/holes as the main carrier), an ionic conductor (with ions as the main carriers), or a mixed conductor (with both ions and electrons/holes as the main carriers).

Examples of electronic conductors are metals (e.g., copper) and carbons (e.g., graphite). An example of an ionic conductor is a cement-based material in the absence of admixtures (additives added to the cement mix) in the form of electronic conductors.

In the presence of an electric field (i.e., a voltage gradient), the holes move toward the negative end of the voltage gradient while the electrons move toward the positive end of the voltage gradient, due to the electrostatic attraction. This results in an electric current, which is defined as charge per unit time (not the magnitude of charge per unit time) going through a particular area, with the unit being Ampere = Coulomb per second (i.e., A = C/s). The current density is defined as the current per unit area, i.e., the charge per unit area per unit time, with unit A/m^2. From the viewpoint of the basic science, the current density is more meaningful than the current.

Although the holes and electrons move in opposite directions in response to the same electric field, they contribute to the current in the same direction. In other words, the movement of electrons in a certain direction results in a current in the opposite direction, due to the negative charge on the electrons. On the other hand, the movement of holes in a certain direction results in a current in the same direction. This means that, in a material with both holes and electrons as carriers, the total current is the sum of the current due to the holes and that due to the electrons. These two currents are additive, not subtractive.

Drift refers to the net movement of charge in response to an applied electric field. The net movement is considered, because electrons/holes are quantum particles that are constantly in motion, even in the absence of an applied electric field. The constant motion of an electron is in line with the concept of an electron cloud.

The drift velocity (abbreviated v) is the velocity of the drift. It is proportional to the applied electric field E, with the constant of proportionality being the mobility (abbreviated μ), which describes the ease of movement of the carrier in the medium under consideration. Hence,

$$v = \mu E. \tag{2.1}$$

The mobility depends on the medium as well as the carrier. In the same medium, the mobility differs between the electrons and holes, with the hole mobility typically lower than the electron mobility. For a given combination of carrier and medium, v depends on the temperature. The v typically decreases with increasing temperature, due to the thermal vibrations of the atoms causing the scattering of the carriers and the increasing amplitude of the thermal vibrations as the temperature increases. Since the unit of E is V/m and the unit of v is m/s, the unit of μ is m^2/(V.s) (Eq. (2.1)).

Percolation refers to the state in which the adjacent conductive filler units (usually short fibers or particles) touch one another, so that a continuous conductive path is formed in the composite material consisting of the conductive filler and a nonconductive matrix. The minimum conductive component volume fraction for percolation to occur is known as the percolation threshold. The electrical conductivity of the composite material increases greatly (typically by orders of magnitude) upon exceeding the percolation threshold.

In order for percolation to take place at a low conductive filler volume fraction, the filler should have a high aspect ratio (i.e., the ratio of the length to the width, which corresponds to the diameter in case of a fiber). A low value of the percolation threshold enables a composite to be conductive at a low filler volume fraction, which facilitates composite fabrication.

Fibers with a small diameter tend to have a high aspect ratio. Thus, microfibers tend to be attractive. Glass fibers and polymer fibers are not conductive, but carbon fibers and metal fibers are. Carbon fibers (also known as carbon microfibers, with typically diameter 7 μm) are dominant as conductive fillers. Although carbon fibers are more expensive than glass fibers, they are less expensive than steel microfibers.

The dispersion of the conductive filler in a composite is important. If the filler units are clumped together, the attaining of percolation is difficult. In case of a cement-matrix composite, the addition of silica fume (submicron silica or SiO_2 particles) to the cement mix helps the dispersion, due to the movement of the fine particles during the mixing promoting the filler dispersion. In general, regardless of the matrix material, the surface treatment of the filler is frequently used to promote the filler dispersion, with the treatment lowering the energy of the filler-matrix interface.

Nanofibers tend to be more expensive than mmicrofibers, but they tend to have higher values of the aspect ratio. As a consequence, nanofibers tend to give lower values of the percolation threshold than the microfibers. However, the dispersion of nanofibers is more difficult than that of microfibers, due to the common intertwined morphology of the nanofibers. Sonication that is commonly used for promoting the dispersion of nanofibers (e.g., carbon nanotubes) is not amenable to scale-up, which is particularly necessary in case of cement-matrix composites for large-scale infrastructure applications.

Example Problem 6 at the end of the chapter is relevant.

2.2 Volume electrical resistivity

The electric current (or current, abbreviated I, with unit Ampere, or A) is defined as the charge (in Coulomb, or C) flowing through the cross section per unit time. The direction of the current is the same as the direction of positive charge flow and is opposite to the direction of negative charge flow.

The presence of a current in a material requires the presence of charges that can move in response of the applied voltage. These mobile charges are called carriers (i.e., carriers of electricity). The carrier concentration (with unit m^{-3}) is defined as the number of carriers per unit volume.

The current due to mobile charges of charge q per carrier is given by

$$I = nqvA, \tag{2.2}$$

where n is the carrier concentration. In case of the carrier being electrons, $q = -1.6 \times 10^{-19}$ C. Equation (2.2) comes from the simple argument that each electron drifts by a distance v in one second and that nvA electrons (the number of electrons in a volume of vA) move through a particular cross section in one second, as illustrated in Fig. 2.1. Dividing Eq. (2.2) by A gives

$$I/A = nqv. \tag{2.3}$$

The current density (abbreviated J, with unit A/m^2) is defined as the current (I) per unit cross-sectional area (A), i.e.,

$$J = I/A. \tag{2.4}$$

Hence, Eq. (2.3) can be written as

$$J = nqv. \tag{2.5}$$

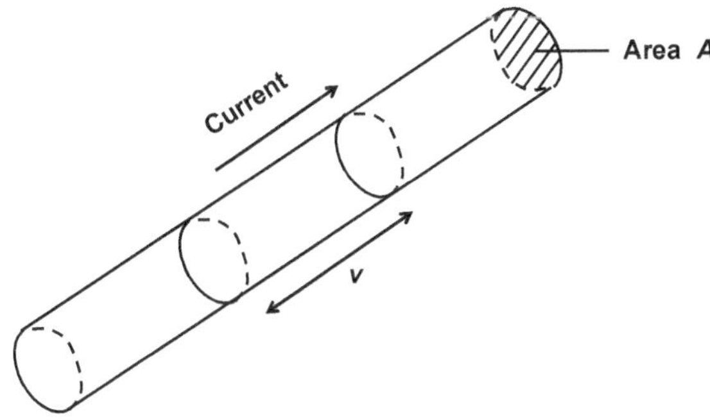

Fig. 2.1 An electric current in a wire of cross-sectional area A. The current results from the drift of a type of charge carrier with charge q per carrier at a drift velocity of v.

Dividing Eq. (2.5) by the electric field E and using Eq. (2.1), we get

$$J/E = nq\mu. \tag{2.6}$$

The electrical conductivity (abbreviated σ) is defined as J/E. In other words, it is defined as the current density per unit electric field. Hence, Eq. (2.6) becomes

$$\sigma = nq\mu. \tag{2.7}$$

From Eq. (2.7), the unit of σ is $1/(\Omega.\text{m})$, i.e., $\Omega^{-1}.\text{m}^{-1}$, since the unit of n is m^{-3}, the unit of q is C (Coulomb) and the unit of μ is m^2/(V.s). An alternate unit for σ is S.m^{-1}, where S (short for Siemens) equals Ω^{-1}. Note that, by definition,

$$\text{Ampere} = \text{Coulomb/second} \tag{2.8}$$

and, from Ohm's Law,

$$\Omega = \text{Volt/Ampere} \tag{2.9}$$

The electrical resistivity (abbreviated ρ) is defined as $1/\sigma$. Hence, the unit of ρ is Ω.m. The electrical resistivity is also known as the volume electrical resistivity, in order to emphasize that it relates to the property of a volume of the material. It is also known as the specific resistance.

Table 2.1 shows the values of the electrical resistivity for various materials at 20°C. Metals have low resistivity values that typically range from 10^{-6} to 10^{-5} Ω.cm. Nichrome is a nickel-chromium alloy (20 wt.% Cr) that has a low resistivity of 1×10^{-4} Ω.cm. Due to the high resistivity and the resistance to oxidation at high temperature, nichrome is widely used as heating elements for resistance heating (i.e., heating by passing an electric current) (Sec. 2.7). Applications include hair dryers, ovens and toasters. In order to attain a required high value of the resistance (not resistivity), as enabled by having a long wire, nichrome for a heating element is in the form of a wire coil, which is a configuration that can accommodate a long wire.

Table 2.1 shows that the resistivity values of carbon, germanium and silicon are much higher than those of metals. These values cover a wide range (7 orders of magnitude being covered by these three materials). Table 2.1 also includes electrical insulators, namely glass, paraffin (e.g., paraffin wax), quartz and Teflon (polytetrafluoroethylene, abbreviated PTFE), with resistivity exceeding 10^{10} Ω.cm (much higher than the values for carbon, germanium and silicon).

The electrical resistivity ρ is related to the electrical resistance (abbreviated R) by the equation

$$R = \rho l/A, \tag{2.10}$$

where l is the length of the material in the direction of the current. This length is perpendicular to the cross-sectional area A. Equation (2.10) means that the resistance R depends on the geometry, such that it is proportional to l and is inversely proportional to A. On the other hand, the resistivity ρ is independent of the geometry and is thus a material property. Similarly, σ is a material property. Equation (2.7) allows the calculation of σ from n, q and μ.

Table 2.1 Volume electrical resistivity at 20°C (ρ_o) and the temperature coefficient of electrical resistivity (α) for various materials.

Material	Resistivity at 20°C (Ω.cm)	Temperature coefficient (/°C)
Silver	1.59×10^{-6}	0.0038
Copper	1.72×10^{-6}	0.0039
Gold	2.44×10^{-6}	0.0034
Aluminum	2.82×10^{-6}	0.0039
Tungsten	5.60×10^{-6}	0.0045
Nickel	6.99×10^{-6}	0.0059
Iron	1.0×10^{-5}	0.005
Tin	1.09×10^{-5}	0.0045
Platinum	1.1×10^{-5}	0.00392
Lead	2.2×10^{-5}	0.0039
Nichrome (Ni-Cr20)	1.1×10^{-4}	0.0004
Carbon	3.5×10^{-3}	−0.0005
Germanium	4.6×10^{1}	−0.048
Silicon	6.40×10^{4}	−0.075
Glass	10^{12}–10^{16}	/
Paraffin	10^{19}	/
Quartz (fused)	7.5×10^{19}	/
Teflon	10^{24}–10^{26}	/

Rearrangement of Eq. (2.10) gives

$$\rho = RA/l. \tag{2.11}$$

Hence,

$$\sigma = l/RA. \tag{2.12}$$

Since, by definition, $\sigma = J/E$,

$$J/E = l/RA. \tag{2.13}$$

Based on Eq. (2.4), Eq. (2.13) becomes

$$I/E = l/R. \qquad (2.14)$$

Since, by definition, $E = V/l$, Eq. (2.14) becomes

$$V = IR, \qquad (2.15)$$

which is known as Ohm's Law. Thus, the equation $\sigma = J/E$ is the same as Ohm's Law.

Example Problems 5, 7–11 and 20–23 at the end of the chapter are relevant.

2.3 Resistivity-density product

Table 2.2 Resistivity-density product of selected electrical conductors

Material	Resistivity ($\mu\Omega$.cm)	Density (g/cm^3)	Resistivity-density product ($\mu\Omega$.g/cm^3)
Aluminum	2.650	2.70	7.2
Copper	1.678	8.96	15.0
Silver	1.587	10.49	16.6

Some applications (such as wires for long distance overhead powerline transmission) require low density (i.e., lightweight) in addition to low electrical resistivity. For such applications, the product of the resistivity and the density should be low. In other words, the ratio of the conductivity to the density should be high. Table 2.2 shows the values of the resistivity-density product for three conductors, which happen to be all face-centered cubic (FCC) in crystal structure. Among the three conductors listed, aluminum has the highest resistivity, but the lowest density. As a result, aluminum has the lowest value of the resistivity-density product. Thus, aluminum is used for lightweight conductor applications.

2.4 Sheet resistance

Consider a square conductive film of thickness t, length l along the current direction and width l in the transverse direction. Using Eq. (5.10), the resistance R of the square film along the length is given by

$$R = \rho l/(lt) = \rho/t, \tag{2.16}$$

since the cross-sectional area of the film is lt. The quantity R in Eq. (2.16) is called the sheet resistance (or sheet resistivity), which refers to the resistance of a film of square film. The unit of the sheet resistance is Ω, or equivalently Ω/square. If the film consists of N squares, so that its length is Nl, while the width remains l, the resistance of the film along its length is N times the sheet resistance. Figure 2.2 illustrates the case where $N = 2$. That is why the unit Ω/square is commonly used. The sheet resistance is commonly used to describe the resistance of a film, though the corresponding value of ρ (which is a quantity that is scientifically more meaningful) is unclear unless the film thickness t is known.
   Example Problem 4 at the end of the chapter is relevant.

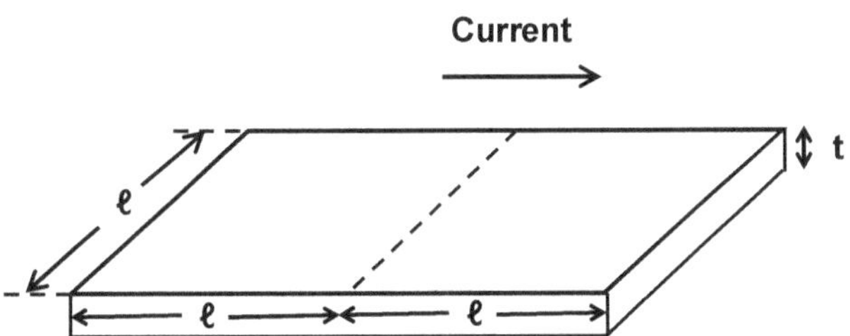

Fig. 2.2 Sheet resistance defined in terms of the resistance per square in the current direction. Two squares are shown, so the resistance is that of two squares.

2.5 Surface resistance

The surface resistance refers to the resistance measured using a surface current, which refers to an electric current that emanates from the surface at one side of the specimen, as illustrated in Fig. 2.3(a), so that the current is not able to penetrate the entire thickness of the specimen. The current path is not straight, as it starts and finishes on the same surface. In contrast, in measuring the volume resistivity, the current path ideally

                     Functional Materials

starts and ends at the two opposite surfaces perpendicular to the current path, so that the current density is uniform throughout the cross-sectional area of the specimen, as illustrated in Fig. 2.3(b).

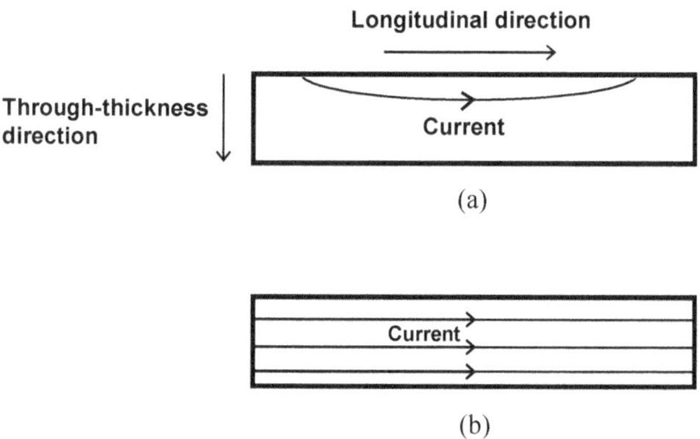

Fig. 2.3 (a) Incomplete penetration of the current in surface resistance measurement. (b) Complete penetration of the current in volume resistance measurement.

The extent of penetration of the current depends on the volume electrical resistivity. The greater is the penetration, the more is the cross-sectional area of the current path and the lower is the measured surface resistance (Eq. (2.10)). Due to the fact that the extent of current penetration is not known, it is difficult to obtain the volume resistivity from the surface resistance. Nevertheless, the surface resistance relates to the volume resistivity and may be used as a rough indicator of the volume resistivity on a relative scale. In spite of its limited scientific value, the surface resistance is attractive in that it can be measured more conveniently than the volume resistivity. In surface resistance measurement, only one surface of the specimen needs to be probed electrically. In case of testing a bridge, for example, it is much simpler to probe just the top surface of the bridge instead of probing multiple surfaces.

In case of a material that is electrically anisotropic, so that the electrical resistivity is different in the longitudinal and through-thickness directions (Fig. 2.3(a)), the extent of current penetration depends on the relative values of the two resistivities. If the through-thickness resistivity is much higher than the longitudinal resistivity, as in the case of a continuous carbon fiber polymer-matrix composite with the fibers in the longitudinal direction, the extent of current penetration is much smaller than that for the case of an isotropic material.

2.6 Contact electrical resistivity

An interface is associated with an electrical resistance in the direction perpendicular to the interface. The contact electrical resistance refers the resistance of an interface between two objects (or between two components in a composite material) when a current is passed across the interface in the direction perpendicular to the interface. When the interface has air voids, for example, the contact resistance will be relatively high. This is because air is an electrical insulator and the objects themselves are more conductive than air. When the interface has an impurity that is less conductive than the objects themselves, the contact resistance will also be relatively high. Thus, the contact resistance is highly sensitive to the condition of the interface.

The contact resistance also depends on the area of the interface. The larger is the area, the lower is the contact resistance. Hence, the contact resistance R_c is inversely proportional to the interface area A and the relationship between these two quantities can be written as

$$R_c = \rho_c \,/A, \tag{2.17}$$

where the proportionality constant ρ_c is known as the contact electrical resistivity, which is a quantity that does not depend on the interface area and only reflects the condition of the interface. The unit of ρ_c is $\Omega.cm^2$. This unit is different from that of the volume resistivity ρ, which has the unit $\Omega.cm$.

Example Problems 24–26 at the end of the chapter are relevant.

2.7 Electric power and resistance heating

The electric power P associated with the flow of a current I under a voltage difference V is given by

$$P = VI. \tag{2.18}$$

This power is dissipated as heat, thus allowing a form of heating known as resistance heating (or Joule heating). Applications include the deicing of aircraft and bridges. This effect, in which electrical energy is converted to thermal energy, is known as the Joule effect. The unit of P is Watts (abbreviated W), with Watts = Volts × Amperes. Based on Eq. (2.15), Eq. (2.18) can be expressed as

$$P = I^2 R \tag{2.19}$$

and as

$$P = V^2/R \tag{2.20}$$

In order to obtain a high power P, both V and I should not be too small (Eq. (2.18)). The values of V and I depend on the resistance R. When R is small, V is small, since $V = IR$. When R is large, I is small, since $I = V/R$. Thus, an intermediate value of R is optimum for obtaining a high power P. It is R rather ρ that governs P. Therefore, for a given material (i.e., a given value of ρ), the dimensions may be chosen to obtain a particular value of R (Eq. (2.9)). Since the current direction does not affect the resistance heating, the heating can be carried out using DC or AC electricity.

Consider an example in which $R = 1$ MΩ (i.e., 1×10^6 Ω) and $V = 120$ V (which is a common voltage from an electrical outlet). Hence,

$$I = V/R = (120 \text{ V})/ (1 \times 10^6 \ \Omega) = 1.2 \times 10^{-4} \text{ A} = 0.12 \text{ mA}, \tag{2.21}$$

and

$$P = VI = (120 \text{ V}) (1.2 \times 10^{-4} \text{ A}) = 1.4 \times 10^{-2} \text{ W} = 14 \text{ mW}. \tag{2.22}$$

Consider another example in which $R = 1$ kΩ (i.e., 1×10^3 Ω) and $V = 120$ V. Hence,

$$I = V/R = (120 \text{ V})/(1 \times 10^3 \ \Omega) = 0.12 \text{ A}, \tag{2.23}$$

and

$$P = VI = (120 \text{ V}) (0.12 \text{ A}) = 14 \text{ W}. \tag{2.24}$$

Consider yet another example in which $R = 1$ Ω and $V = 120$ V. Hence,

$$I = V/R = (120 \text{ V})/(1 \text{ }\Omega) = 120 \text{ A}, \tag{2.25}$$

which is too high a current to be provided by a conventional power source, which may limit the current to, say, 1 A. Thus, in spite of the high V that the power source can provide, the actual V across the heating element is just

$$V = IR = (1 \text{ A})(1 \text{ }\Omega) = 1 \text{ V}, \tag{2.26}$$

and the power is thus merely

$$P = VI = (1 \text{ V})(1 \text{ A}) = 1 \text{ W}. \tag{2.27}$$

Therefore, among the three examples mentioned above, the intermediate R of 1 kΩ gives the highest P.

Nichrome (Ni-20Cr alloy), with $\rho = 1 \times 10^{-4}$ Ω.cm, is a common material for resistance heatin, because the resistivity is high among metals. For a nichrome wire of diameter 1 mm to provide a resistance of 1 kΩ, the length of the wire needs to be the following, where Eq. (2.9) is used.

$$l = RA/\rho = (1 \text{ k}\Omega)\,[\pi\,(0.5 \text{ mm})^2]/(1 \times 10^{-4} \text{ }\Omega.\text{cm})$$

$$= 7.9 \times 10^4 \text{ cm} = 790 \text{ m}. \tag{2.28}$$

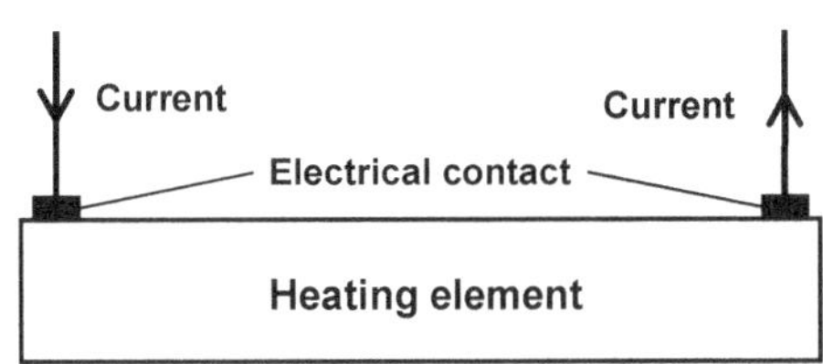

Fig. 2.4 Electrical contacts used for passing current to a heating element contributing to the electrical resistance.

The length of 790 m is quite long, so the wire cannot be packaged straight and needs to be coiled. This is why heating elements are commonly in the form of coils.

In order to pass current to a heating element, two electrical contacts are required, as illustrated in Fig. 2.4. An electrical contact may, for example, be a soldered joint between the heating element and a wire. Each electrical contact is associated with a resistance, which is the sum of (i) the resistance of the joining medium (e.g., solder), (ii) the resistance of the interface between the joining medium and the heating element, and (iii) the resistance of the interface between the joining medium and the wire. Since the joining medium is typically a highly conductive material, the resistance of the joining medium is usually negligible compared to the two interfacial resistances. The interfacial resistance can be substantial, due to the air voids, impurities, reaction products or other species at the interface. The resistance of each of the two contacts contributes to the resistance R encountered by the current I. Therefore, the design of a heating element must include consideration of the contribution to R by the electrical contacts. In particular, the resistance at the two contacts must not be so large that it overshadows the resistance within the heating element material. If it is too large, the electrical contacts become effectively the main heating element, while the actual heating element contributes little to the heat generation. A consequence will be non-uniform heating, as the heating is concentrated at the electrical contacts.

2.8 Effect of temperature on the electrical resistivity

For a given material, the volume electrical resistivity depends on the temperature. The dependence allows the resistivity to be an indicator of the temperature. In other words, by measuring the resistance, one can obtain the temperature. This type of temperature sensor is known as a thermistor.

The dependence of the volume electrical resistivity on temperature is expressed by the temperature coefficient of electrical resistivity (α), which is defined as

$$(\Delta\rho)/\rho_o = \alpha\,\Delta T, \tag{2.29}$$

where ρ_o is the resistivity at 20°C and $\Delta\rho$ is the change in resistivity relative to ρ_o when the temperature is changed from 20°C by ΔT. The unit of α is °C^{-1}, which is the same as K^{-1}, since a unit of temperature in K is the same as a unit of temperature in °C. Values of α are shown in Table 2.1 for various materials.

Since

$$\Delta\rho = \rho - \rho_o \qquad (2.30)$$

Equation (2.29) can be rewritten as

$$\rho = \rho_o(1 + \alpha\,\Delta T). \qquad (2.31)$$

Thus, the plot of ρ vs. ΔT gives a line of slope $\rho_o\alpha$. This line tends to be straight only for metals. However, temperature sensing based on this phenomenon does not require the curve to be a straight line. The greater is the magnitude of α, the more sensitive is the thermistor.

The coefficient of thermal expansion γ is defined as the fractional change in length l per unit change in temperature δT. In other words,

$$\delta\ell/\ell = \gamma\,\delta T. \qquad (2.32)$$

The unit of γ is K^{-1}, which is the same as °C^{-1}.

A change in temperature affects the resistance R because it affects the dimensions due to the phenomen of thermal expansion and it possibly affects the resistivity as well. The cross-sectional area A in Eq. (2.10) is given by

$$A = w_1 w_2, \qquad (2.33)$$

where w_1 and w_2 are the two dimensions that are transverse to the length l. Based on Eq. (2.33),

$$\delta A/A = \delta w_1/w_1 + \delta w_2/w_2, \qquad (2.34)$$

In case that the material is isotropic,

$$\delta w_1/w_1 = \delta w_2/w_2 = \delta w/w. \qquad (2.35)$$

Hence, Eq. (2.34) becomes

$$\delta A/A = 2\,\delta w/w. \tag{2.36}$$

Let v be the Poisson's ratio of the istropic material. By definition of the Poisson's ratio,

$$\delta w/w = -v\,\delta\ell/\ell. \tag{2.37}$$

The combination of Eq. (2.36) and (2.37) gives

$$\delta A/A = 2\,(-v\,\delta\ell/\ell) = -2v\,\delta\ell/\ell. \tag{2.38}$$

Based on Eq. (2.10),

$$\delta R/R = \delta\rho/\rho + \delta\ell/\ell - \delta A/A. \tag{2.39}$$

The combination of Eq. (2.38) and (2.39) gives

$$\delta R/R = \delta\rho/\rho + \delta\ell/\ell - (-2v\,\delta\ell/\ell)$$

$$= \delta\rho/\rho + \delta\ell/\ell + 2v\,\delta\ell/\ell$$

$$= \delta\rho/\rho + \delta\ell/\ell\,(1 + 2v). \tag{2.40}$$

The combination of Eq. (2.32) and (2.40) gives

$$\delta R/R = \delta\rho/\rho + \gamma\,\delta T\,(1 + 2v). \tag{2.41}$$

Based on Eq. (2.29), Eq. (2.41) becomes

$$\delta R/R = \alpha\,\delta T + \gamma\,\delta T\,(1 + 2v)$$

$$= [\alpha + \gamma\,(1 + 2v)]\,\delta T. \tag{2.42}$$

Equation (2.42) can be used to calculate $\delta R/R$ based on the values of α, γ, v and δT.

The temperature coefficient of resistance (β) is to be distinguished from the temperature coefficient of resistivity (α). It refers to the fractional change in resistance per unit change in temperature. In other words,

$$\beta = (\delta R/R)/\delta T. \tag{2.43}$$

The combination of Eq. (2.42) and (2.43) gives

$$\beta = \alpha + \gamma\,(1 + 2v). \tag{2.44}$$

Equation (2.44) is useful if one wants to determine the fractional change in ρ from the measured fractional change in R due to a change in temperature, provided that the values of γ and v are known. From the viewpoint of the basic science, α is more meaningful than β.

For metals, α is positive, i.e., the resistivity increasing with increasing temperature. This is because the amplitude of thermal vibration of the atoms in the metal increases with increasing temperature, thereby decreasing the mobility μ (as defined in Eq. (2.1)). A decrease in mobility in turn results in a decrease in the conductivity (Eq. (2.7)), i.e., an increase in the resistivity. The carrier concentration n of a metal does not change with temperature, due to the availability of the valence electrons as mobile charges without any need for exciting them. When the temperature is 0 K, there is no thermal energy for providing thermal vibration, so the resistivity at this temperature is not due to thermal vibrations, but is due to the scattering of the electrons (carrier) at the defects that may be present. The resistivity at 0 K is known as the residual resistivity (Fig. 2.5). The greater is the defect concentration, the higher is the residual resistivity.

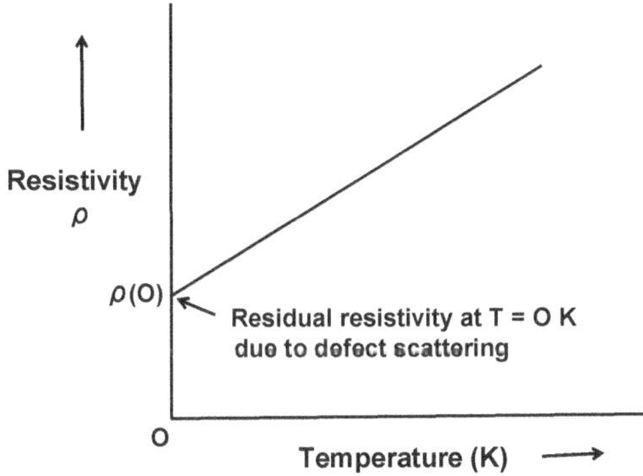

Fig. 2.5 Dependence of the volume electrical resistivity on the temperature near 0 K, showing the residual resistivity at 0 K.

Semiconductors (e.g., silicon and germanium) and semimetals (e.g., graphite) together constitute a class of materials known as metalloids. For metalloids, α is negative. This is because, for these materials, n increases significantly with increasing temperature. This trend of n is due to the thermal excitation of electrons to form mobile carriers. For a semiconductor, the valence electrons need to be excited across the energy band gap in order for them to become mobile. The larger is the energy band gap, the more is the effect of temperature on the resistivity. For a semimetal, there is no energy band gap, but there is a small band overlap, thus resulting in a small effect of temperature on the resistivity. The increase of n with increasing temperature overshadows the decrease in mobility μ with increasing temperature (Eq. (2.7)), thus resulting in an overall effect in which the conductivity increases with increasing temperature (i.e., the resistivity decreasing with increasing temperature).

Table 2.1 shows the values of ρ_o and α for various materials. For metals, the value of α is typically around +0.004/°C. For semiconductors and semimetals, α is negative, such that the magnitude is higher for semiconductors than semimetals, due to the energy band gap present in a semiconductor. Thus, the magnitude of α is smaller for carbon (a semimetal) than for silicon or germanium (semiconductors). The magnitude of α is larger for silicon than germanium, due to the larger energy band gap for silicon (1.12 eV, compared to 0.68 eV for germanium). The energy band gap (also known as the energy gap) is the energy between the top of the valance band and the bottom of the conduction band above the valence band and describes the energy needed to excite an electron from the valence band to the conduction band, so that the electron becomes free to respond to an applied electric field.

For a composite with an electrically nonconductive matrix and a conductive discontinuous filler (particles or fibers), such that the CTE of the matrix is higher than that of the filler, an increase in temperature causes more thermal expansion of the matrix than the filler. As a consequence, the degree of contact between adjacent filler units (i.e., adjacent particles or adjacent fibers) is reduced, thus causing the volume resistivity of the composite to increase. The resistivity typically increases abruptly when the increase in temperature has reached a sufficiently large value. This means that the curve of the resistivity vs. temperature is not linear. When the filler volume fraction is around the percolation threshold, the extent of resistivity increase is particularly large. This is because the resistivity is particularly sensitive to the degree of contact between adjacent filler units when the filler volume fraction is around the percolation threshold. This phenomenon is useful for temperature-activated switching, i.e., the switching off of a circuit (due to the high resistivity of the composite used in series in the circuit) when the temperature is higher than a critical value. The composite serves as a fuse that protects the electronics from excessive temperatures.

Example Problems 18 and 19 at the end of the chapter are relevant.

2.9 Electrical conduction evaluation methods

2.9.1 *Volume electrical resistivity measurement*

The measurement of the electrical resistivity of a material involves passing a current through the material and measuring the voltage drop in the direction of the current. According to Ohms' Law, the resistance is given by the voltage divided by the current. The volume electrical resistivity is then calculated from the resistance, the length of the specimen along which the measured voltage drop occurs and the cross-sectional area of the specimen, using Eq. (2.10). Although the concept is simple, there are numerous pitfalls.

(a) The current must penetrate the entire thickness of the specimen in the region where the voltage drop is measured. Otherwise, the area is not equal to the entire cross-sectional area of the specimen.

(b) The voltage drop should be that in the specimen, with exclusion of the voltage drops at the two electrical contacts used for voltage measurement. Otherwise, the measured resistance can be much higher than the true resistance, since the resistance at the electrical contacts can be substantial compared to the specimen resistance. The lower is the specimen resistance compared to the contact resistance, the more severe is error. Degradation of the electrical contacts may occur during loading and exposure to heat and/or moisture. Thus, the error may increase during real-time resistance measurement that is conducted during loading or heat/moisture exposure, thus possibly overshadowing the true effect of loading or heat/moisture exposure on the specimen.

To alleviate the problems mentioned above, the four-probe method rather than the two-probe method should be used. The four-probe method

 Functional Materials

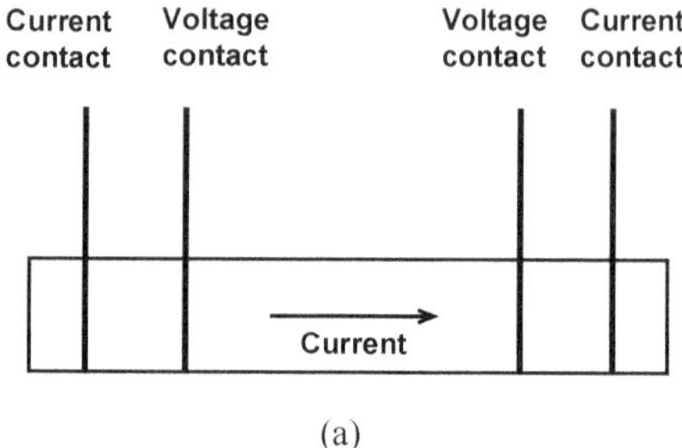

(a)

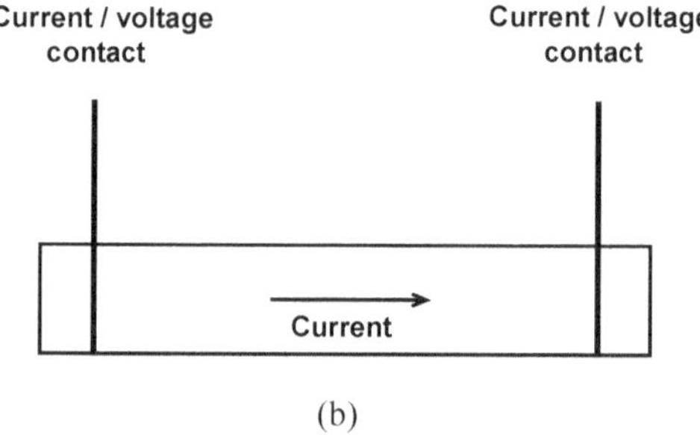

(b)

Fig. 2.6 Methods of volume electrical resistivity measurement. (a) Four-probe method. (b) Two-probe method. All the electrical contacts are embedded in the specimen.

uses four electrical contacts that are positioned at four parallel planes that are perpendicular to the current direction, which is the direction of the resistance measurement, as illustrated in Fig. 2.6(a). The outer two contacts are for passing current and are referred to as current contacts. The inner two contacts are for voltage measurement and are referred to as voltage contacts. The resistance of the part of the specimen between the two voltage contacts is given by the voltage between the two voltage contacts divided by the current directed between the two current contacts. The resistivity is then calculated from this resistance, the specimen length between the two voltage contacts and the cross-sectional area. In contrast, the two-probe method uses two electrical contacts that are positioned at two planes that are perpendicular to the current direction, with each of the two contacts serving as both voltage and current contacts, as illustrated in Fig. 2.6(b).

No current goes through an ideal voltmeter. The current going through a real (not ideal) voltmeter is negligibly small. As a consequence, the voltage drop at each of the two voltage contacts is negligible, i.e., due to Ohm's Law, the voltage drop is zero when the current is zero. Therefore, the measured resistance in the four-probe method essentially excludes the resistances of the two voltage contacts. On the other hand, in the two-probe method, the resistance of each of the two contacts is included in the measured resistance.

In Fig. 2.6(a) and Fig. 2.6(b), each electrical contact is embedded in the specimen, so that it covers the entire cross-sectional area of the specimen. As a result, the current flows from one current contact to the other current contact, with the current density being uniform throughout the cross-sectional area of the specimen. Provided that the current density is uniform throughout the cross section of the specimen, so that the electric potential is uniform in the plane of each voltage contact (i.e., an equipotential situation within the plane), the voltage contacts used in the four-probe method do not need to cover the entire cross-sectional area of the specimen. Under this situation, a voltage contact can be simply on the surface, as illustrated in Fig. 2.7.

The current contacts do not have to be embedded as in Fig. 2.6. A simpler configuration involves the current contacts at the two end surfaces that are perpendicular to the current direction, as shown in Fig. 2.8, where each current contact covers the entire end surface.

In some applications, the end surfaces are not available or not suitable for placing electrical contacts. An example is the use of the end surfaces to impose compressive loading in the direction of the current during resistance measurement. In this case, the configuration of Fig. 2.8 will cause compressive loading to be directed at the current contacts,

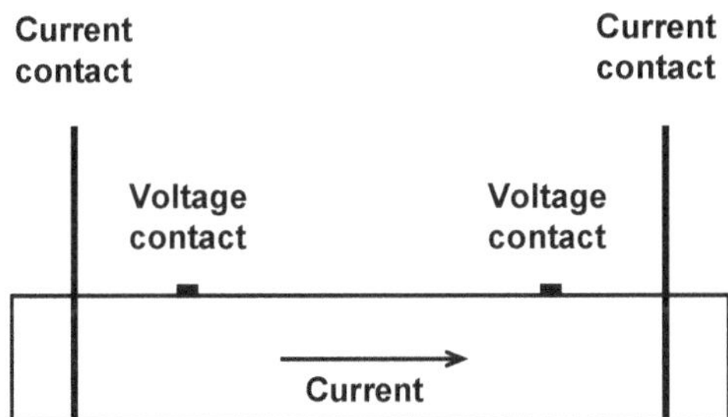

Fig. 2.7 Method of volume electrical resistivity measurement involving the four-probe method, with the voltage contacts on the surface and the current contacts embedded in the specimen.

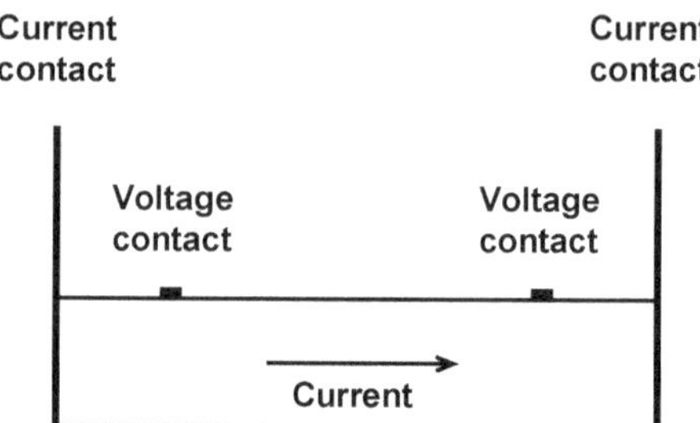

Fig. 2.8 Method of volume electrical resistivity measurement involving the four-probe method, with the voltage contacts on the surface and the current contacts at the ends of the specimen.

which may deform or degrade as the loading occurs, thus affecting the resistance associated with the current contact. The effect of compression on the contact resistance may even overshadow the effect of compression on the resistance of the specimen itself. Therefore, the configuration of Fig. 2.9(a) is commonly used. This configuration involves each of four electrical contacts being all the way around the perimeter of the specimen, such that the four contacts are at four parallel planes that are perpendicular to the current. The main disadvantage of this configuration is that the current density may not be uniform throughout the cross section of the specimen in the part of the specimen between the two voltage contacts. The non-uniformity is because the current penetrates from the surface (the four sides that make up the perimeter) to the interior of the cross section as it flows from one current contact to the other. Unless the spacing between the adjacent current and voltage contacts is sufficient, the current penetration may not be complete in the part of the specimen between the two voltage contacts. Figure 2.9(b) illustrates a case of incomplete current penetration in the part of the specimen between the voltage contacts. Figure 2.9(c) illustrates a case of complete current penetration in the part of the specimen between the voltage contacts. The extent of current penetration can be increased by increasing the distance between the adjacent current and voltage contacts.

In case that the specimen dimension is too small in the direction of resistance measurement for the installation of four electrical contacts at four planes that are perpendicular to the direction of resistance measurement, the four-probe method may be carried out by using a less ideal configuration. This configuration, as illustrated in Fig. 2.10, involves a current contact in the form of a hollow loop (the large hollow rectangle in Fig. 2.10) and a voltage contact in the form of a solid dot (the small solid rectangle in Fig. 2.10) on each of the two opposite surfaces that are perpendicular to the direction of resistance measurement. This configuration is not ideal because (i) the current and voltage contacts are in the same plane perpendicular to the direction of resistance measurement, and (ii) the current density is not uniform throughout the cross section perpendicular to the direction of resistance measurement. By having the current contact cover a larger fraction of the area, the current density becomes more uniform. Nevertheless, the configuration involves separate current and voltage contacts, making it a form of the four-probe method.

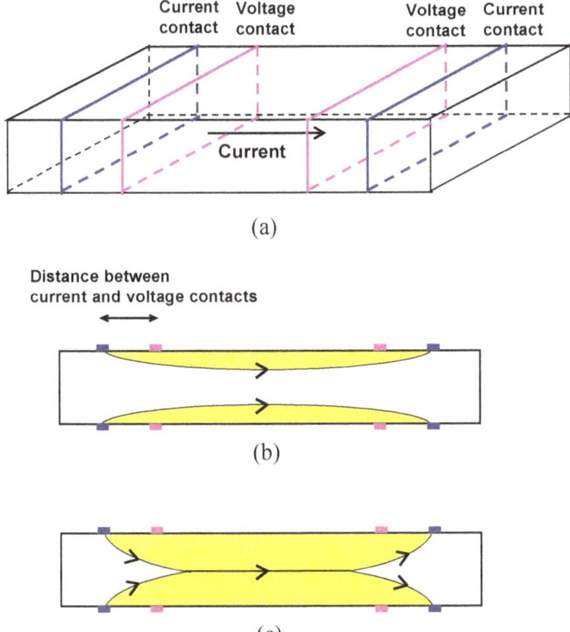

Fig. 2.9 Method of volume electrical resistivity measurement involving the four-probe method, with all the electrical contacts on the surface all the way around the perimeter of the specimen. (a) Three-dimensional view. (b) Cross-sectional view showing the case of incomplete current penetration in the region between the voltage contacts. (c) Cross-sectional view showing the case of complete current penetration in the region between the voltage contacts.

                    Functional Materials

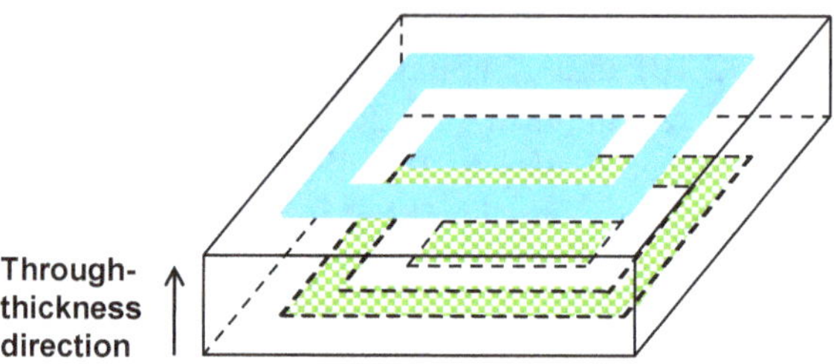

Fig. 2.10 Method of volume electrical resistivity measurement involving the four-probe method, with a current contact and a voltage contact on each of two opposite surfaces. The resistance is measured in the through-thickness direction, which is perpendicular to the two opposite surfaces.

An electrical contact should be a good conductor. The voltage contacts must be electronic conductors, since these contacts are connected to a voltmeter. However, the current contacts may be electronic or ionic conductors. A current contact that is an electronic conductor (i.e., a purely electronic conductor, such as copper, that does not conduct with ions at all) does not allow ions to be injected from the current contact to the specimen and does not allow ions to enter the other current contact. On the other hand, a current contact that is an ionic conductor does not allow electrons to be injected from the current contact to the specimen and does not allow electrons from the specimen to enter the other current contact. An example of a purely ionic conductor is water, which may be held in a sponge on the surface of the specimen, as illustrated in Fig. 2.11, where a metal foil, an electronic conductor, is inserted in each sponge for the purpose of connecting electrically the current contact to a power source. The metal foil is not the electrical contact material, as it does not touch the specimen. It is simply a connector that connects the electrical contact material to the electronics. Water is the electrical contact material. As a consequence of the current contacts that are ionic conductors, the electronic carrier flow in the specimen is enhanced by the use of current contacts that are electronic conductors, such that the ionic carrier flow, if any, is not eliminated, since the ions present in the specimen still respond to the electric field provided by the current contacts. On the other hand, the ionic carrier flow in the specimen is enhanced by the use of current contacts that are ionic conductor, such that the electronic carrier flow, if any, is not eliminated, since the electrons present in the specimen still respond to the electric field provided by the current contacts.

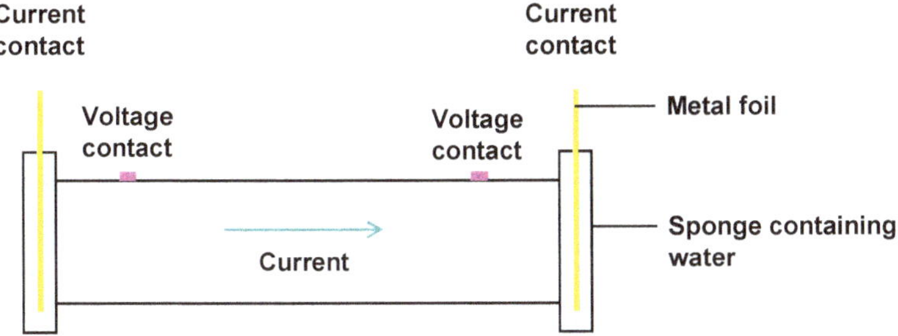

Fig. 2.11 Method of volume electrical resistivity measurement involving the four-probe method, with current contacts in the form of an ionic conductor (water held in the sponge) at the two ends of the specimen and voltage contacts in the form of an electronic conductor on the surface.

A specimen which is a mixed conductor (with both electronic and ionic carriers) has electronic and ionic contributions to the overall electrical conductivity, i.e.,

$$\sigma = \sigma_e + \sigma_i, \tag{2.45}$$

where σ_e is the electronic contribution to the conductivity and σ_i is the ionic contribution to the conductivity. For a mixed conductor, σ_e is enhanced by the use of current contacts that are electronic contacts and σ_i is enhanced by the use of current contacts that are electronic contacts. Comparison of the measured conductivity (which includes both contributions) using the two kinds of current contacts allows indication of the relative importance of the two contributions. If the ionic contribution is small compared to the electronic contribution, the use of current contacts that are ionic conductors will give a measured conductivity that is lower than that obtained by using current contacts that are electronic conductors. If the electronic contribution is small compared to the ionic contribution, the use of current contacts that are electronic conductors will give a measured conductivity that is lower than that obtained by using current contacts that are ionic conductors. The use of current contacts that are electronic conductors does not provide measurement of σ_e; the use of current contacts that are ionic conductors does not provide measurement of σ_i.

An example of a conductor that has a high concentration of ionic carrier (e.g., H^+ ions) is water. An example of a conductor that has a high concentration of electronic carrier is copper. In case of a conductor that has a high concentration of a certain type of carrier, which is essentially the only type of carrier present, the measured conductivity obtained by using the two forms of current contacts may be close. This is because the electric field provided by either form of current contacts is sufficient to cause a significant flow of the carrier, which is plentiful in the specimen, even in the absence of injection of the carrier through a current contact. In other words, the inherent carrier concentration is so high that carrier injection from a current contact essentially does not affect the carrier concentration in the specimen. This means that a significant difference between the measured conductivity values obtained by using the two forms of current contacts occurs only when the conductor is a mixed conductor, with both ions and electrons present at concentrations that are not very high.

Electronic conductors include metals and carbons in the graphite family. An electrical contact that is an electronic conductor may be in the form of a metal mesh that has been embedded in the specimen during specimen fabrication. For example, for a specimen in the form of a cement paste, the electrical contacts may be inserted after pouring the cement paste in a mold and before the paste sets. After the paste has set, the electrical contacts are fixed firmly in the specimen. An advantage of using an embedded mesh rather than an embedded foil for the electrical contact is that the paste can penetrate through the holes in the mesh, thus allowing (i) mechanical interlocking between the mesh and the cement paste after setting, and (ii) movement of ions through the holes in the metal mesh, which is a purely electronic conductor. The ability of ions to move through the mesh is important when the mesh is used as a voltage contact and ions are involved in the conduction. However, when only electrons are involved in the conduction, the use of a metal foil instead of a metal mesh is acceptable, since the electrons can flow through the foil. For the voltage contacts, through which the applied current passes in case of the four-probe method (Fig. 2.6(a)), a metal mesh is preferred to a nonporous metal foil, in case that an ionic carrier is involved. This is because a nonporous metal foil does not allow ions to go through, although it allows electrons to go through. For the current contacts, an embedded metal foil is advantageous over an embedded metal mesh in that, due to the absence of holes, it provides more uniform current injection into the specimen. However, due to the mechanical interlocking, an embedded metal mesh provides a stronger mechanical connection between the contact and the specimen.

For electrical contacts that are electronic conductors on the surface of the specimen, as in the case of Fig. 2.9, the electrical contact material is commonly a solder (e.g., eutectic tin-lead alloy), a conductive adhesive (e.g., silver epoxy) or a conductive paint (e.g., silver paint, nickel paint and graphite colloid). The use of a solder requires that the specimen is (i) solderable (i.e., it allows the solder to adhere to it after soldering) and (ii) resistant to the elevated temperatures encountered during soldering. Solders and conductive adhesives give stronger bond than conductive paints. Silver epoxy and silver paint are expensive, due to the silver. Nickel paint and graphite colloid are less expensive.

2.9.2 *Surface electrical resistance measurement*

Measurement of the surface electrical resistance involves the use of electrical contacts that are on the surface at one side of the specimen, as illustrated in Fig. 2.12. Because of the limited current penetration (Fig. 5.3(a)), the cross-sectional area for the current path is small compared to the cross-sectional area of the specimen. As a result, the surface resistance is high compared to the volume resistance that is obtained under the condition of complete current penetration (Fig. 5.3(b)). The area of an electrical contact in surface resistance measurement (Fig. 2.12(a)) is small compared to that of an electrical contact in volume resistance measurement (Fig. 2.9(a)). As a consequence, the resistance associated with an electrical contact is higher in surface resistance measurement than in volume resistance measurement.

In Fig. 2.12(a), each electrical contact is a strip oriented in the transverse direction (i.e., the direction in the plane of the electrical contacts such that it is perpendicular to the direction of resistance measurement). In Fig. 2.12(b), each electrical contact is a dot. The configuration of Fig. 2.12(a) is advantageous over that of Fig. 2.12(b) in that (i) the area of each electrical contact is larger (hence a lower contact resistance), and (ii) there is no current spreading in the transverse direction (Fig. 2.13(a)). The configuration of Fig. 2.12(b) is associated with current spreading in the transverse direction, as shown in Fig. 2.13(b). Since the resistance in the longitudinal direction is the quantity to be measured, the current should be in the longitudinal direction and a current component in the transverse direction is undesirable.

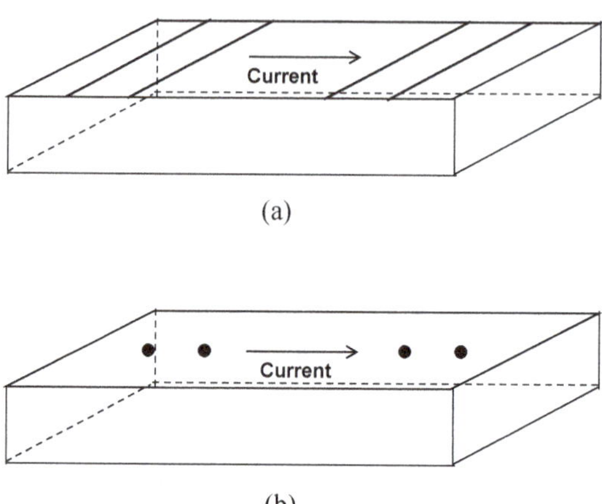

Fig. 2.12 Method of surface resistance measurement involving the four-probe method. (a) Each electrical contact is in the form of a strip oriented perpendicular to the current direction. (b) Each electrical contact is in the form of a dot.

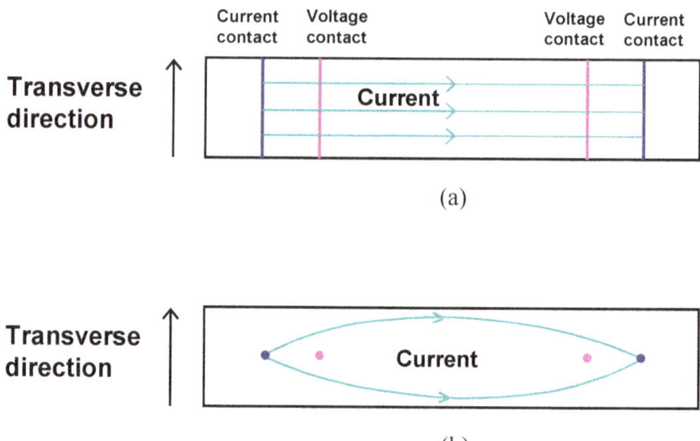

Fig. 2.13 Direction of current flow during electrical resistance measurement. (a) No current in the transverse direction. (b) Current spreading in the transverse direction.

In case that the resistivity of the specimen is anisotropic in the plane of the electrical contacts, the current spreading will be enhanced in the direction of low resistivity within this plane. For example, for a continuous carbon fiber polymer-matrix composite with all the fibers of the surface lamina oriented in the same direction, the resistivity of the surface lamina is much lower in the fiber direction than the direction transverse to it and hence the current spreading is much greater in the fiber direction. Thus, measurement of the surface resistance in the transverse direction using electrical contacts in the form of dots will involve much current spreading in the fiber direction, as illustrated in Fig. 2.14(a). In contrast, measurement of the surface resistance in the fiber direction using electrical contacts in the form of dots will involve little current spreading, as illustrated in Fig. 2.14(b).

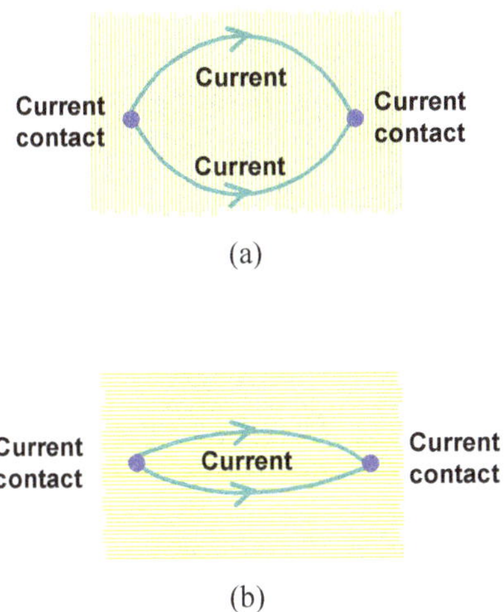

Fig. 2.14 Current spreading in the direction of low resistivity in a carbon fiber polymer-matrix composite with surface fibers in the direction indicated by the parallel lines. (a) Current spreading is significant along the fiber direction, which is the direction of low resistivity. (b) Current spreading is less along the transverse direction, which is the direction of high resistivity.

2.9.3 *Contact electrical resistivity measurement*

The contact electrical resistivity of an interface is given by the product of the contact resistance of the interface and the geometric interface area. The measurement of the contact resistance can involve one of two configurations. In Configuration A, the electrical contacts are on the specimen, such that they are directly above and below the interface area. In Configuration B, the electrical contacts are on the specimen, such that they are not above or below the interface area. Configuration B allows application of compressive stress perpendicular to the interface for squeezing at the interface, without the possibility of degrading the electrical contacts. In contrast, degradation of the electrical contacts during loading is possible in case of Configuration A.

2.9.3.1 *Configuration A*

A method of measuring the contact electrical resistivity of an interface involves the configuration of Fig. 2.15(a). In this configuration, the four electrical contacts are at four different planes that are perpendicular to the current direction. The resistance between the two voltage contacts in Fig. 2.15(a) is measured. This resistance is the sum of (i) the volume resistance of the part of the top object between the voltage contact and the interface, (ii) the contact resistance of the interface, and (iii) the volume resistance of the part of the bottom object between the interface and the voltage contact. Term (ii) can be obtained by subtraction from

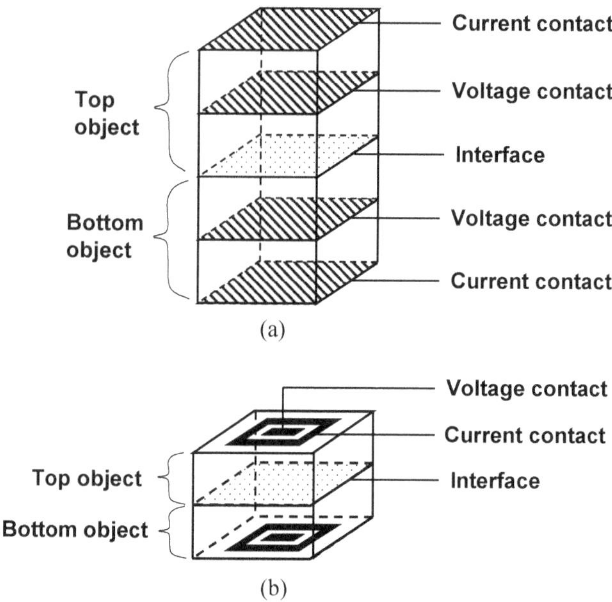

Fig. 2.15 Method of measuring the contact electrical resistivity of the interface between two objects using the four-probe method. (a) The four electrical contacts are on four planes. (b) Two electrical contacts are on each of two planes.

the measured resistance, after calculation of terms (i) and (iii) from the known volume resistivities of the two objects.

The method of Fig. 2.15(a) requires a considerable amount of material for each of the two objects. This is because each object needs to be sufficient long in the direction of the current in order for two contacts to be positioned in each object. A variation of this method that requires less material is shown in Fig. 2.15(b), where the voltage and current contacts on each object are in the same plane, akin to the configuration of Fig. 2.10.

2.9.3.2 Configuration B

The contact electrical resistivity of an interface between two objects may be measured by using the configuration of Fig. 2.16. In this configuration, the two objects are allowed to overlap, so that the overlap region is the interface to be evaluated. The two objects may be bonded or not at the interface.

The parts of each object that extend beyond the overlap region serve to provide room for applying electrical contacts. An electrical contact is applied to cover at least the entire end surface of an object, with the end surface being perpendicular to the interface under investigation. Each of the two objects has a current contact at one end and a voltage contact at the other end. Current flows from the current contact of the top object to the current contact of the bottom object, such that it goes through the interface. The resistance of the current path is the sum of (i) the resistance of the part of the top object from its current contact to the edge of the interface area, (ii) the resistance of the interface area in the direction perpendicular to the interface, and (iii) the resistance of the part of the bottom object from the edge of the interface area to its current contact. Terms (i) and (iii) may be calculated from the known volume resistivity of each object. Hence, term (ii) may be obtained by subtraction from the measured resistance of the entire current path. In order to increase the accuracy of determining term (ii), the values of terms (i) and (iii) should be kept small by minimizing the length of each object in the current direction. The resistance within an object in the through-thickness direction (direction perpendicular to the plane of the object) is not included in this sum, because each current contact covers the entire end surface of an object, thus causing the electric potential to be constant within an object along the through-thickness direction. Since this resistance can be substantial and its inclusion will reduce the accuracy of determining term (ii), it is important to have each current contact cover the entire end surface of an object.

The dominance of term (ii) over terms (i) and (iii) is common when each object is a unidirectional continuous carbon fiber polymer-matrix composite, such that the fiber direction of each object is in the direction of the current path. If the two objects have not been cured at the time of overlapping them and curing is conducted for the assembly under pressure in the direction perpendicular to the plane of the interface, the interface after curing is an interlaminar interface. On the other hand, when each object has been cured prior to the overlapping of the two objects, the interface is the unbonded interface rather than the interlaminar interface. An unbonded interface is relevant to mechanical fastening, as it is the fastened joint interface. The interlaminar interface is important for its significant effect on the properties of a composite.

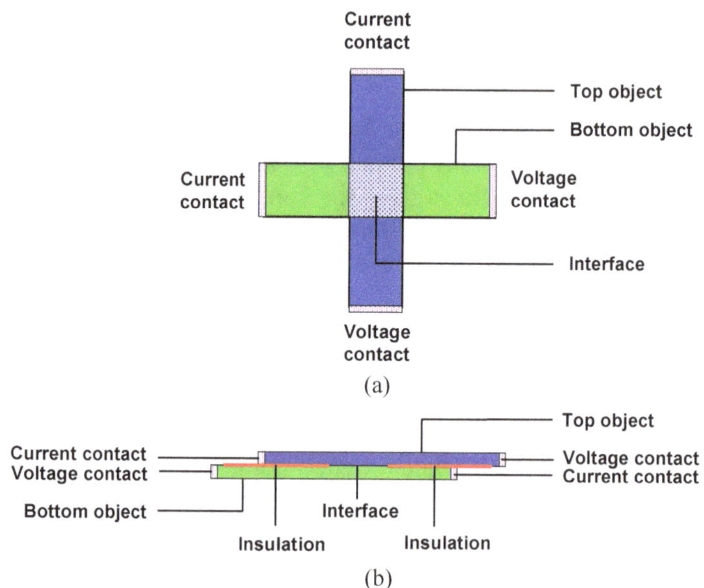

Fig. 2.16 Method of measuring the contact electrical resistivity of the interface between two objects that overlap, with protruded parts to allow room for making electrical contacts associated with the four-probe method. (a) Top view for the case of the two objects being strips that are perpendicular to one another. (b) Side view for the case of the two objects being strips that are parallel to one another, with the use of electrical insulation films to delineate the interface area.

In Fig. 2.16(a), the two objects overlap at 90°. In Fig. 2.16(b), the two objects are in the same orientation, so a thin electrically insulating film needs to be applied between the two objects in areas other than the area of the interface for the purpose of directing the current through a well-defined interface area.

The distinction between Fig. 2.16(a) and 2.16(b) is significant when the objects are anisotropic, as in the case of continuous carbon fiber polymer-matrix composites. For example, in Fig. 2.16(a), one object has fibers that are 90° with respect to the fibers in the other object (i.e., a crossply configuration), whereas, in Fig. 2.16(b), the fibers are in the same direction for the two objects (i.e., a unidirectional configuration). The structure of the interface is different between the crossply and unidirectional configurations.

The method of Fig. 2.16 requires less material than that of Fig. 2.15(a). In addition, it allows loading on the interface area without the possibility of disturbing the electrical contacts. However, Fig. 2.16 does not involve as classical a four-probe method as Fig. 2.15(a).

2.10 Effect of strain on the electrical resistivity (piezoresistivity)

2.10.1 *Origin of piezoresistivity*

Piezoresistivity is an electromechanical phenomenon in which the electrical resistivity of a material changes with strain. It is practically useful for providing strain sensing through electrical resistance measurement. Strain sensing is to be distinguished from damage sensing. Since strain and stress are simply related in the elastic deformation regime, stress sensing can also be accomplished using piezoresistivity. Strain causes reversible effects, whereas damage causes irreversible effects.

Piezoresistivity is distinct from piezoelectricity (Sec. 3.8.1). The former pertains to the electrical conduction behavior (as indicated by the resistivity) changing with strain, whereas the latter pertains to the dielectric behavior (Ch. 3, as indicated by the electric field) changing with strain. Nevertheless, both piezoresistivity and the direct piezo-electric effect are useful for strain/stress sensing.

The resistivity can be the volumetric, interfacial or surface resistivity, though the volumetric resistivity is most meaningful scientifically and is most widely considered in piezoresistivity. Unless it is noted otherwise, the piezoresisitivty pertains to the volume resistivity. An example of a pertinent interfacial resistivity is the contact resistivity of the inter-laminar interface of a continuous carbon fiber polymer-matrix composite (Fig. 1.8). This interface is prone to damage, so zooming in electrically at this interface is advantageous for damage sensing.

Strictly speaking, an irreversible resistivity change due to strain is not piezoresistivity, although it is often referred to in the literature as piezoresistivity. For the sensing to be repeatedly effective (rather than being effective for only one time of loading at a given maximum level), the resistivity change must be reversible upon unloading. Because the irreversible resistivity change due to a microstructural change or damage adds to the reversible change that occurs at lower strains (Fig. 1.6 and 1.7), the inclusion of the irreversible effect as piezoresistivity makes the piezoresistivity appear much stronger than what the inherent piezo-resistivity really is. Furthermore, damage tends to degrade the inherent piezoresistivity by decreasing the extent of resistivity change for the same strain. As a consequence, the resistance-based strain sensing effectiveness tends to be diminished by damage. Of fundamental scientific importance is the inherent piezoresistivity that occurs without irreversible resistivity changes.

The resistance measurement should be the specimen resistance, without inclusion of the contact resistance of the specimen-electrode interface. Thus, the decoupling the specimen resistance from the contact resistance should be conducted. The contact resistance can be large, even when the contact material (e.g., silver) is highly conductive. This is because the resistance of the interface between the contact material and the specimen can be substantial. An effective method to achieve this decoupling is the four-probe method (Fig. 1.1 and 1.2, and Sec. 2.9.1).

Whether the electrodes for measuring the volume resistance are embedded or attached, the interface between each electrode and the specimen tends to change under strain. For example, under compression, in case that the contact is not an intimate one, the contact resistance tends to decrease, due to the increased degree of consolidation of the interface; however, in case that the contact is an intimate one, the contact resistance tends to increase, due to the disturbed degree of intimacy. Thus, under strain, both the contact resistance and specimen resistance can change, whether the change is reversible or not. The chance of the contact resistance changing with strain is particularly high when the electrode is stressed along with the specimen during the strain/stress application. In order to obtain correct information on the piezoresistivity of the specimen, the contact resistance should be removed from the measured resistance. Without the removal, one cannot tell if the observed resistance change is due to the change in the contact resistance and/or the change in the specimen resistance. Under this situation, the interpretation of the observed resistance change as being completely due to the piezoresistivity of the specimen tends to be erroneous. The use of the two-probe method, with the assumption that the contact resistance is negligible, is a common pitfall in piezoresistivity research. Moreover, the piezoresistive behavior is clearer and more pronounced, with the gage factor higher and less dependent on the strain amplitude, if the four-probe method rather than the two-probe method is used.

The piezoresistive effect is described by the gage factor (*GF*), which is defined as the fractional change in resistance per unit strain. The *GF* can be positive or negative. Strong piezoresistivity involves the magnitude of the fractional change in resistivity much exceeding the strain magnitude. The reversible effect of strain on the electrical connectivity is the primary piezoresistivity mechanism.

Giant piezoresistivity means exceptionally strong piezoresistivity. Based on the values of *GF* reported for a large number of different materials, giant piezoresistivity is defined as piezoresistivity with the magnitude of *GF* exceeding 500. This definition is supported by the fact that the vast majority of the reported *GF* values are below 100, and most of the reported *GF* values are below 500. Giant piezoresistivity requires the resistivity to change much with strain. As for piezoresistivity in general, the mechanism for giant piezoresistivity relates to the effect of the microstructure on the strain. In giant piezoresistivity, the resistivity is highly sensitive to the strain, with the microstructural change typically not related to the electrical connectivity, but related to changes in the crystallographic texture and interface structure. Giant piezoresistivity is attractive for providing unusually high effectiveness for strain/stress sensing.

Piezoresistive materials must be electrically conductive, at least to a degree. When the conductivity is so low that the material is almost insulating, the resistance is very high and is thus hard to measure. When the conductivity is so high that the material is almost short-circuited inside, the resistance is very low and the changes in resistance are even smaller, thus causing the piezoresistivity to be hard to quantify. Piezoresistive materials include metals, carbons, and composite materials with essentially nonconductive matrices that are rendered conductive by the incorporation of conductive constituents. Although conductive polymers are available, they are not adequate in their processability or mechanical behavior.

Due to their nonconductivity, glass fiber polymer-matrix composites are not piezoresistive, unless a conductive filler is added to the polymer matrix. However, the filler incorporation causes the composite to be unconventional. Such unconventional materials are not used in existing structures. As a consequence, the application is limited to new structures. This problem does not apply to carbon fiber polymer-matrix composites, which are piezoresistive without the need for filler incorporation.

The formation of a continuous conductive path through the touching of the conductive filler units (such as particles) is known as percolation. The establishing of percolation causes the resistivity to decrease by orders of magnitude. The percolation threshold is the filler volume fraction above which the degree of contact is adequate for the formation of a continuous conductive path. At filler volume fractions in the vicinity of the percolation threshold, the resistivity is particularly sensitive to the strain, thus causing the piezoresistivity to be strong.

For composites with continuous conductive fibers, the mechanism behind the piezoresistivity differs from that for composites with discontinuous conductive fillers. This is because the continuous conductive fibers are typically aligned, thus providing a continuous conductive path in the direction of alignment without the need for percolation. In contrast, percolation is central to the piezoresistivity of a composite with a discontinuous conductive filler.

In case of a composite with a continuous conductive fiber, the piezoresistivity stems for the effect of strain on the microscopic fiber arrangement in the composite. The continuous fibers are not perfectly straight. The fiber waviness (known as marcelling) results in a degree of touching between the fibers in directions perpendicular to the fiber direction. The touching occurs at points. The degree of touching strongly affects the resistivity in the directions perpendicular to the fiber direction. The through-thickness direction is one of these perpendicular directions and it is the direction that is relevant to composite laminates with various fiber directions in various laminae. Unidirectional composites with the fibers in all the laminae oriented in the same direction are not usually used for structures, because they are strong in only one direction.

Because of the fiber-fiber contacts, the resistivity in the through-thickness direction of a continuous carbon fiber epoxy-matrix composite is never infinity, though the epoxy is itself nonconductive. For a unidirectional carbon fiber polymer-matrix composite, the resistivity is similar in the in-plane transverse direction and the through-thickness direction. Regardless of the fiber lay-up configuration, the interlaminar interface region is a polymer-rich region. Thus, this region strongly affects the through-thickness resistivity of the composite. The degree of fiber-fiber contact across the interlaminar interface is enhanced by compressive strain/stress in the through-thickness direction, thereby decreasing the resistivity in this direction.

The degree of fiber-fiber contact is also enhanced by compressive strain/stress in the fiber (longitudinal) direction of the composite, due to the accompanying increase in the degree of fiber waviness, thereby decreasing the through-thickness resistivity. The through-thickness resistivity decreases upon longitudinal compression and increases upon longitudinal tension. In this case, the strain and resistivity are not in the same direction.

The through-thickness and longitudinal resistivities of a continuous carbon fiber polymer-matrix composite correlate. A decrease in the through-thickness resistivity causes a decrease in the longitudinal resistivity, because there are always some defects present in the fibers and the enhanced fiber-fiber contact in the through-thickness direction facilitates the detour of the longitudinal current from a fiber that as a defect to an adjacent fiber, so that the defective fiber still contributes to electrical conduction. In contrast, the degree of fiber-fiber contact is diminished by tensile strain/stress in the longitudinal direction, due to the accompanying decrease in the degree of fiber waviness, thereby causing the through-thickness and longitudinal resistivities to increase.

For a composite panel, flexure is more commonly encountered in practice than uniaxially tension or compression. Under flexure, the tension surface experiences longitudinal tension, which causes the surface resistance on this surface to increase (due to the decreased current penetration), while the compression surface experiences longitudinal compression, which causes the surface resistance on this surface to decrease (due to the increased current penetration) (Fig. 1.9 and 1.10).

A brittle matrix tends to have microcracks. In case that the filler is a discontinuous conductive fiber, some of the fibers tend to bridge the microcracks (Fig. 2.17). Upon tension, the microcrack opens, thus causing a degree of pull-out of the bridging fiber. Along with the pull-out is the slight loosening of the fiber-matrix interface and the consequent increase in the contact resistance at the fiber-matrix interface. The increase in this contact resistance causes the overall volume resistivity of the composite to increase under tension (Fig. 1.4) and decrease under compression (Fig. 1.1 and 1.2). Since the extent of the fiber pull-out is small, the fiber would not break. Under this situation, upon subsequent unloading, fiber push-in occurs and the fiber-matrix interface tightens, thereby causing the resistivity to decrease. This mechanism has been shown for a cement-matrix composite containing discontinuous carbon fiber.

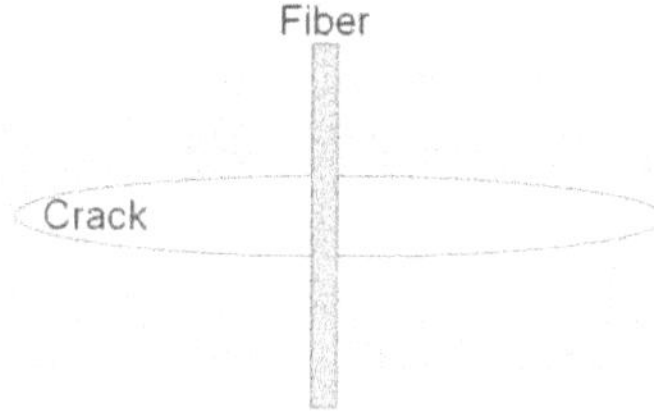

Fig. 2.17 Fiber bridging for explaining the piezoresistivity in short carbon fiber cement-matrix composites.

In case that the material is not a composite but consists of a single phase, the abovementioned mechanisms that pertain to composites do not apply. Nevertheless, a single-phase material may exhibit piezoresistive behavior due to other mechanisms. A relatively common mechanism for a single-phase material relates to the effect of strain/stress on the microstructure, which affects the resistivity. Another mechanism that pertains to single-phase materials in the form of semiconductors relates to the effect of stress on the energy bandgap.

Metals tend to have no damage in the elastic regime. For brittle materials (e.g., carbons, ceramics and cements), damage can occur even in the elastic regime, so that the degree of electrical continuity tends to change with substantial irreversibility even in the high strain part of the elastic regime. As a result, in the study of the piezoresistivity of brittle materials, it is important to limit the strain to the low-strain part of the elastic regime. Damage or plastic deformation results in an irreversible change in the resistivity. For the case of metals, plastic deformation affects the microstructure, such as increasing the dislocation density, thereby increasing the resistivity irreversibly, i.e., the resistivity remaining at the increased value after unloading.

In order to make sure that the resistance change at a given strain is totally reversible, unloading should be conducted after reaching this strain while the resistance is continuously measured. A suitable testing scheme involves measuring the resistance upon repeated loading and unloading at progressively increasing stress/strain amplitudes. Using this scheme, the stress/strain at the onset of partial irreversibility of the resistance change can be determined.

2.10.2 *Piezoresistivity theory*

The resistance R is related to the resistivity ρ, the length ℓ in the direction of resistance measurement and the cross-sectional area A perpendicular to the direction of resistance measurement, i.e.,

$$R = \rho\ell/A \qquad (2.46)$$

The fractional change in resistance is given by the equation

$$\delta R/R = \delta\rho/\rho + (\delta\ell/\ell)(1 + v_{12} + v_{13}), \qquad (2.47)$$

where v_{12} and v_{13} are values of the Poisson's ratio for the transverse and through-thickness strains respectively. Refer to Example Problem 32 at the end of the chapter for the derivation of this equation. If the material is isotropic, so that $v_{12} = v_{13} = v$, Eq. (2.47) becomes

$$\delta R/R = \delta\rho/\rho + (\delta\ell/\ell)(1 + 2v). \qquad (2.48)$$

Refer to Example Problem 32 at the end of the chapter for the derivation of Eq. (2.48).

Positive piezoresistivity refers to the behavior in which the resistivity increases with increasing strain, i.e., $(\delta\rho/\rho)/(\delta\ell/\ell) > 0$. Negative piezoresistivity refers to the behavior in which the resistivity decreases with increasing strain, i.e., $(\delta\rho/\rho)/(\delta\ell/\ell) < 0$. A pitfall relates to not recognizing that compressive strain is negative strain, whereas tensile strain is positive strain.

Positive piezoresistivity is much more common than negative piezoresistivity, because tension tends to increase the resistivity, whereas compression tends to decrease the resistivity. For example, tension decreases the degree of electrical continuity in the material, whereas compression increases this degree. The degree of electrical continuity tends to change reversibly in the elastic regime of deformation, whereas it changes with increasing irreversibility when the elastic limit is increasingly exceeded.

According to Eq. (2.48), both the change in ρ (i.e., $\Delta\rho$) and the change in l (i.e., Δl) contribute to the change in R (i.e., ΔR). Three cases are considered here. In Case 1, $(\Delta\rho)/\rho \gg (\Delta l)/l$. In Case 2, $(\Delta\rho)/\rho \sim (\Delta l)/l$. In Case 3, $(\Delta\rho)/\rho \ll (\Delta l)/l$.

In Case 1, Eq. (2.48) becomes

$$\frac{\Delta R}{R} = \frac{\Delta\rho}{\rho} \tag{2.49}$$

and GF is given by

$$GF = \frac{\dfrac{\Delta R}{R}}{\dfrac{\Delta l}{l}} = \frac{\dfrac{\Delta\rho}{\rho}}{\dfrac{\Delta l}{l}} \tag{2.50}$$

Tensile strain is positive, whereas compressive strain is negative. If $(\Delta\rho)/\rho$ is positive when $(\Delta l)/l$ is positive, or if $(\Delta\rho)/\rho$ is negative when $(\Delta l)/l$ is negative, the piezoresistivity is said to be positive. If $(\Delta\rho)/\rho$ is negative when $(\Delta l)/l$ is positive, or if $(\Delta\rho)/\rho$ is positive when $(\Delta l)/l$ is negative, the piezoresistivity is said to be negative.

In Case 2, $(\Delta\rho)/\rho \sim (\Delta l)/l$. In the specific condition under Case 2 in which $(\Delta\rho)/\rho = (\Delta l)/l$, Eq. (2.48) becomes

$$\frac{\Delta R}{R} = \frac{\Delta l}{l}(2 + 2v) \tag{2.51}$$

so that GF is given by

$$GF = 2 + 2v. \tag{2.52}$$

Since $v \leq 0.5$, Eq. (2.52) indicates that, in this specific situation under Case 2, the maximum magnitude of GF is 3, which is a small value.

In Case 3, $(\Delta\rho)/\rho \ll (\Delta l)/l$, Eq. (2.48) becomes

$$\frac{\Delta R}{R} = \frac{\Delta l}{l}(1 + 2v) \tag{2.53}$$

and *GF* is given by

$$GF = \frac{\dfrac{\Delta R}{R}}{\dfrac{\Delta l}{l}} = 1 + 2v \tag{2.54}$$

In Case 3, with $(\Delta\rho)/\rho$ being negligible, the resistance change is not a manifestation of the piezoresistivity, but is merely due to the dimensional changes. In other words, it is a resistive effect rather than a piezoresistive effect. Since $v \leq 0.5$, Eq. (2.56) indicates that, in Case 3, the maximum magnitude of *GF* is 2, which is a small value.

In Case 1, Eq. (2.50) indicates that *GF* is dictated by $(\Delta\rho)/\rho$, which has a magnitude that depends on the severity of the piezoresistivity of the material and is not limited. Thus, for achieving high sensing effectiveness, Case 1 is much preferred to Case 2 or Case 3. Although Case 2 corresponds to weak piezoresistivity and Case 3 does not even correspond to piezoresistivity, these two cases allow a low-sensitivity form of resistance-based strain sensing. Case 3 pertains to the conventional metal strain gages that are commercially available.

Equation (2.50) suggests that $(\Delta\rho)/\rho$ is proportional to $(\Delta l)/l$, so that *GF* is independent of the strain range. However, in practice, due to the reversible effect of strain on the microstructure and the effect of the strain amplitude on the microstructure, *GF* tends to vary with the strain in the elastic regime. This reversible microstructural change is subtle compared to the irreversible microstructural change that may accompany an irreversible change in the resistance.

The above theory indicates that *GF* decreases with increasing strain (relative to the fractional change in resistivity). Therefore, the strain range should be described in reporting a *GF* value. To attain a high value of *GF*, a small strain helps. However, the measurement of a small strain is more challenging than that of a large strain. For example, the strain based on the crosshead movement of a mechanical testing system is not accurate compared to the strain based on an attached strain gage or extensometer. Under tension, possible slight slippage at the grip that connects the specimen to the testing system causes the crosshead displacement to be not reliable to giving the strain in the specimen. This problem is particularly severe when the strain involved is small. Under compression, possible consolidation of the interface between the compression platen and specimen causes a similar problem.

The resistivity-strain relationship may be non-linear. A pitfall relates to not treating appropriately the nonlinearity of the dependence of the resistivity on the strain in case that such nonlinearity is significant. The *GF* should be obtained from the slope of the initial approximately linear part of the curve of resistivity (or resistance) vs. strain in the elastic regime. For this reason, the strain range for the applicability of a reported value of *GF* should be indicated.

Equation (2.34) shows that *GF* depends both on the fractional change in resistivity per unit strain and the Poisson's ratio. A positive value of the gage factor does not necessarily mean that the piezoresistivity is positive, but a negative value of the gage factor necessarily means that the piezoresistivity is negative.

In order to attain a large fractional change in resistance at a particular strain, positive piezoresistivity is more desirable than negative piezo-resistivity that exhibits the same magnitude of the fractional change in resistivity. When the strain is small, as is the case when the piezoresistive material is a stiff structural material, the fractional change in resistance is essentially equal to the fractional change in resistivity. Under this circumstance, positive and negative piezoresistivities are equally desirable for providing a large magnitude of the fractional change in resistance.

The resistivity is usually DC rather than AC, due to the relative simplicity of DC measurement. However, AC is sometimes used, particularly in case that the material being tested has a tendency to be polarized electrically (Sec. 3.1). The polarization results in an electric field that opposes the applied electric field used for the resistance measurement. As a result, the measured resistance increases as the polarization builds up, so that it is above the true resistance value (Sec. 3.6.4). A method of circumventing this problem involves measuring the resistance in the first few seconds, prior to substantial polarization build-up. High-conductivity materials, such as metals and carbons, have less tendency to be polarized than low-conductivty materials, such as composites. The polarization is much less under AC than DC. However, in case of the electrodes positioned on the surface of the specimen, the degree of current penetration of the applied AC current into the specimen decreases with increasing frequency. Without complete current penetration, the cross-sectional area of the current path is smaller than that of the specimen.

Example Problems 32–36 at the end of the chapter are relevant.

2.11 Seebeck effect (a thermoelectric effect)

The thermoelectric effect entailing the conversion of thermal energy to electrical energy is useful for the generation of renewable energy from waste heat. The effect is also the basis of thermocouples for temperature measurement.

When an electrical conductor is subjected to a temperature gradient, charge carriers move. The movement is typically in the direction from the hot point to the cold point of the temperature gradient, due to the higher kinetic energy of the carriers at the hot point. The carrier flow results in a voltage difference (called the Seebeck voltage) between the hot and cold ends of the conductor. This phenomenon is known as the Seebeck effect, which is a type of thermoelectric effect.

The Seebeck effect allows the conversion of thermal energy (associated with the temperature gradient) to electrical energy and is the basis of thermoelectric energy generators, which are devices for generating electricity from temperature gradients. This type of electricity generation is attractive due to the absence of pollution and the common presence of temperature gradients, such as the temperature difference between a water heater and its surrounding, that between an engine and its surrounding, etc. Many temperature gradients that occur anyway can be used to produce electricity through the Seebeck effect. In other words, waste heat is utilized to generate electricity.

If the dominant carrier is negatively charged (as in the case of electrons), the movement results in a Seebeck voltage which has its negative end at the cold end of the specimen, as illustrated in Fig. 2.18(a). If the dominant carrier is positively charged (as in the case of holes, which refer to electron vacancies), the movement results in a Seebeck voltage which has its negative end at the hot end of the specimen, as illustrated in Fig. 2.18(b). It is possible for a conductor to have both electrons and holes, as in the case of a semiconductor.

                    Functional Materials

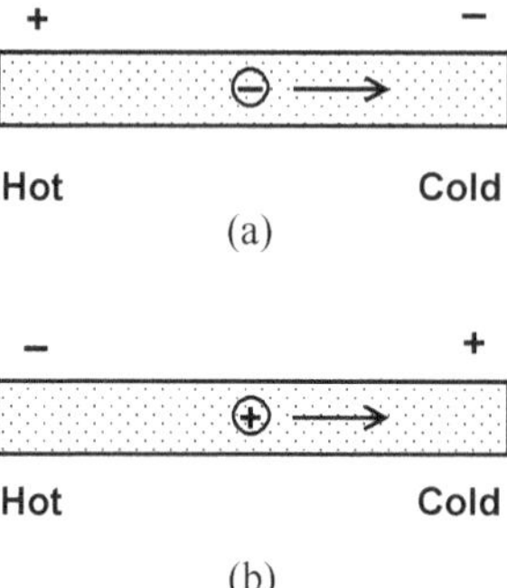

Fig. 2.18 Seebeck effect due to carriers moving from the hot point to the cold point. (a) Negative carrier. (b) Positive carrier.

The Seebeck coefficient (abbreviated S, and also known as the thermoelectric power or the thermopower) is defined as the negative of the Seebeck voltage divided by the temperature difference between the hot and cold ends. In other words,

$$S = -\Delta V/\Delta T, \tag{2.55}$$

where ΔV is the voltage difference and ΔT is the corresponding temperature difference. The unit of S is V/K.

In case of Fig. 2.18(b), i.e., the situation corresponding to the carrier being positive, the Seebeck coefficient is positive, since ΔV and ΔT are opposite in sign. In case of Fig. 2.18(a), i.e., the situation corresponding to the carrier being negative, the Seebeck coefficient is negative, since ΔV and ΔT are the same in sign.

Carriers are prone to be scattered upon collision with the thermally vibrating atoms in the solid. The extent of carrier scattering depends on the relationship of the carrier energy with the carrier momentum (i.e., the energy band structure) near the Fermi energy (the highest occupied energy level in the absence of any thermal agitation, i.e., at the temperature of 0 K). As a result of the scattering, the movement of the carriers does not necessarily follow the direction illustrated in Fig. 2.18. In other words, it is possible for a material have negative carriers to have a positive Seebeck coefficient.

The Seebeck effect also allows the sensing of temperature, since the voltage output relates to the temperature difference. Thus, the Seebeck

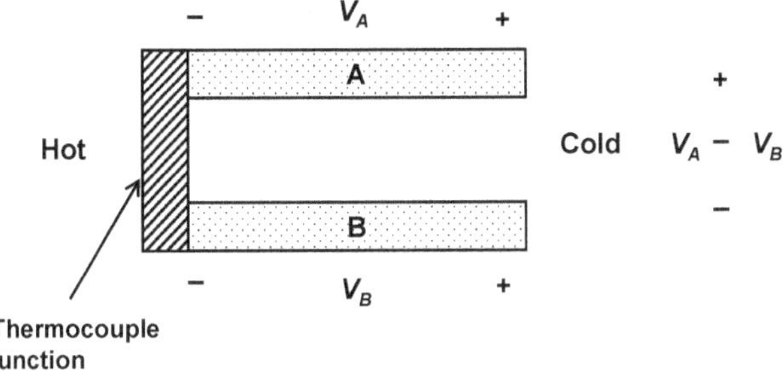

Fig. 2.19 A thermocouple consisting of dissimilar conductors A and B (dotted regions) that are electrically connected at one end. The shaded region at the hot end is the conductor that electrically connects A and B.

voltage can be used to indicate the temperature of one end of the conductor, if the temperature of the other end is known. For measuring the temperature of a hot point, it is more convenient to measure the voltage difference between two cold points rather than that between the hot point and the cold point. Therefore, two conductors that have different values of the Seebeck coeffient are used in the configuration shown in Fig. 2.19. This configuration constitutes a thermocouple, which is a device for temperature measurement. The two conductors, labeled A and B, are electrically connected at one end to form a thermocouple junction, which corresponds to the hot point (in case that the temperature to be measured is above room temperature). The cold ends of A and B are at the same temperature (typically room temperature). The voltage difference between the hot and cold points of A is given by

$$V_A = S_A \, \Delta T, \tag{2.56}$$

where S_A is the Seebeck coefficient of A. The voltage difference between the hot and cold points of B is given by

$$V_B = S_B \, \Delta T, \tag{2.57}$$

where S_B is the Seebeck coefficient of B. The voltage difference between the cold ends of A and B is given by

$$V_A - V_B = (S_A - S_B) \, \Delta T, \tag{2.58}$$

where Eq. (2.56) and (2.57) have been used. Equation (2.58) means that $V_A - V_B$ is proportional to ΔT. The greater is $S_A - S_B$, the more sensitive is the thermocouple. Thus, A and B should be chosen so that they are as dissimilar as possible.

Particularly strong dissimilarity occurs when A and B involve carriers of opposite signs. In other words, S_A and S_B are opposite in sign, so that $S_A - S_B$ is equal to the sum of the magnitudes of S_A and S_B. Due to the consequent large voltage output, the configuration involving A and B with Seebeck coefficients of opposite sign is commonly used for thermoelectric power generation. In order to obtain an even larger voltage output, multiple junctions are used, as illustrated in Fig. 2.20, so that the A and B legs are in series and the output voltage is the sum of the voltages from all the legs.

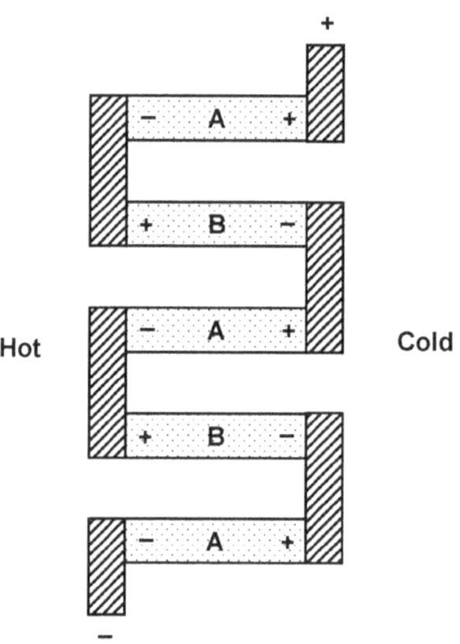

Fig. 2.20 A thermoelectric energy generator consisting of conductors A and B (dotted regions) that are alternating and electrically connected in series. The shaded regions are conductors that electrically connect A and B. The temperature to the left of the device is different from that to the right of the device.

The measurement of the Seebeck coefficient involves using electrical leads (wires), which are made of a certain conductor (e.g., copper). The leads are attached to the hot and cold ends of the conductor under investigation, as illustrated in Fig. 2.21. The lead attached to the hot end has a temperature gradient along its length, while that attached to the cold end does not. Thus, a Seebeck voltage V_w occurs between the hot and cold ends of the wire attached to the hot end of the conductor under investigation, while a Seebeck voltage V_s occurs between the hot and cold ends of the conductor under investigation (i.e., the specimen). Thus, the measured voltage difference between the cold ends of the two leads equals $V_s - V_w$. This means that V_w must be added to this measured voltage difference in order to obtain V_s.

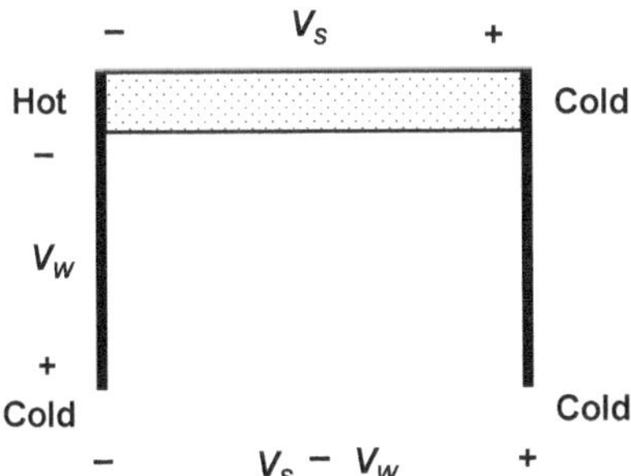

Fig. 2.21 A conductor (dotted region) subjected to a temperature gradient, with electrical leads for measuring the Seebeck voltage.

Based on Eq. (2.20), the electric power provided by the Seebeck effect relates to S^2/ρ, where S is the Seebeck coefficient and ρ is the electrical resistivity. Thus, S^2/ρ is known as the power factor. This means that a high value of S and a low value of ρ are preferred.

In order for a material to sustain a substantial temperature gradient, its thermal conductivity must be sufficiently low. Therefore, both a high power factor and a low thermal conductivity are necessary for a thermo-electric material in a thermoelectric energy generator. As a consequence, the figure of merit (Z) of a thermoelectric material is given by

$$Z = S^2/(\rho k), \tag{2.59}$$

where k is the thermal conductivity. The unit of Z is K^{-1}. The dimensionless figure of merit (ZT) of a thermoelectric material is given by the product of Z and the temperature in K, where the temperature is the average temperature in the temperature gradient. A higher value of T helps to provide a higher value of ZT. Thus, a material (e.g., a ceramic material) that can withstand high temperatures is preferred. The quantity ZT relates to the thermodynamic efficiency of the energy conversion and is commonly used to describe the performance of a thermoelectric material.

The requirement of a low thermal conductivity and a low electrical resistivity is a challenge. Metals do not satisfy this requirement, since they are high in the thermal conductivity, although their electrical resistivity is low. Electrical insulators do not satisfy this requirement, since they are high in the electrical resistivity. Semiconductors, on the other hand, can have reasonably low thermal conductivity and reasonably low electrical resistivity. Thus, thermoelectric materials are commonly semiconductors in the form of alloys or compounds that have been tailored for the abovementioned combination of properties. Examples are bismuth telluride (Bi_2Te_3) and lead telluride (PbTe). Another advantage of semiconductors is that they can be doped (typically through the use of a solute to form a dilute solid solution) to be n-type (i.e., a situation in which electrons are the dominant carrier) or p-type (i.e., a situation in which holes are the dominant carrier) (Sec. 2.12.1), thus providing materials that are needed for the thermoelectric generator configuration of Fig. 2.20. However, disadvantages of the semiconductors include the high material cost and the mechanical brittleness.

The thermoelectric effect is exhibited by thermoelectric materials, which are effective due to the combination of high thermoelectric power, low electrical resistivity and low thermal conductivity. The achievement of this combination of properties is challenging, because the properties are not independent of one another. For the case of electrons dominating the mechanism of electrical and thermal conduction, the electrical conductivity and thermal conductivity are directly proportional to one another, in accordance with the WiedemanneFranz Law. Moreover, the electrical conductivity and thermoelectric power are inversely related, because the electrical conductivity increases with the carrier concentration, whereas the thermoelectric power decreases with the carrier concentration.

The most established thermoelectric materials are inorganic materials, such as PbTe, Bi_2Te_3 and SnSe. These materials are expensive and suffer from low processability, brittleness and toxicity. This situation calls for the investigation of polymers as thermoelectric materials. Polymers tend to be low cost, low in thermal conductivity, high in processability and ductility, and low in density (lightweight). Compared to the traditional inorganic thermoelectric materials, polymer-based thermoelectric materials (including polymers and polymer-matrix composites) are disadvantageously low in *ZT*, electrical conductivity and thermoelectric power, though they are advantageously low in the thermal conductivity. Therefore, the emphasis of research on polymer-based thermoelectric materials is on increasing the electrical conductivity and thermoelectric power.

Due to the requirement of low electrical resistivity, conventional polymers are not effective as thermoelectric materials, unless other constituents are used with them, thus forming a polymer-matrix composite. On the other hand, if the polymer is a conductive polymer, other constituents may not be necessary. Examples of conductive polymers are polyaniline (PANI), polypyrrole (PPY), polyacetylene (PA) polythiophene (PT), poly(3,4-ethylenedioxythiophene) (PEDOT) and poly(phenylene vinylenes) (PPV). However, the *ZT* values of conductive polymers are typically low, such as 0.00008, as exhibited by a composite with conductive polymer poly(3-butylthiophene) (P3BT) and polystyrene. However, a *ZT* value of 0.44 has been reported for a polymer-polymer composite with a PEDOT matrix and PEDOT nanowires.

An attractive constituent used in thermoelectric polymer-matrix composites is carbon nanotubes (CNTs), which are attractive for their electrical conductivity and amenability for being doped. For example, CNT can be doped by nitrogen or treated with polyethyleneimine (PEI) to become n-type. Due to the conductivity of the CNTs, the polymer does not have to be conductive. As the CNT content increases, the electrical conductivity increases greatly, while the thermal conductivity is increased only slightly. Graphene may be used in place of CNT. An issue in the use of CNT or graphene relates to the dispersion in the polymer.

Another class of nonstructural thermoelectric polymer-matrix composites are those that contain discontinuous inorganic fillers, such as $SnSe_{0.8}S_{0.2}$, SnS nanobelt, BiTe-SbTe, TiO_2, silver, tellurium nanowires, etc. Optionally these fillers are used along with CNTs in order to improve the electrical conductivity.

The large variation in *ZT* among the various thermoelectric materials is due to the sensitivity of *ZT* to the composition as well as the method of preparation of the composite. The method of preparation affects the distribution of the constituents, in addition to affecting the interfaces in the resulting composite material.

Nonstructural thermoelectric materials are used for wearable, flexible or stretchable electronics. Structural thermoelectric materials are used for structures, such as airframes and satellites. Nonstructural thermoelectric materials are mainly in the form of polymers and polymer-matrix composites with discontinuous fillers. In contrast, structural thermoelectric materials are structural composites involving continuous fibers as the reinforcement.

A *ZT* value of 1 is considered to be very good. A *ZT* value around 2–3 is needed to provide 15–20% efficiency in the energy conversion. The highest *ZT* values reported are in the range from 2 to 3. However, a *ZT* value above 3 is necessary for a thermoelectric energy generator to compete with mechanical methods of electricity generation. This is why thermoelectric energy generation is currently not playing a substantial role in the energy sector. Nevertheless, due to the severity of the energy crisis, much research is being conducted to develop thermoelectric materials with higher *ZT* values.

Thin-film thermoelectric materials can reach higher *ZT* values than bulk materials. However, for practical application, bulk materials are needed.

Example Problem 31 at the end of the chapter is relevant.

2.12 Semiconductors and their junctions

2.12.1 *Intrinsic and extrinsic semiconductors*

The energy of electrons in an isolated atom is organized in the form of discrete narrow energy levels (Fig. 2.22(a)). The energy range between any two energy levels is forbidden. In other words, an electron may occupy one energy level or another, but not between the two levels. In a solid, the atoms are close together, so they interact with one another. The interaction results in the broadening of an energy level into an energy band (Fig. 2.22(b)). Between adjacent energy bands is an energy range (known as an energy band gap, or simply an energy gap) which is forbidden.

A metal is defined as a solid that has its highest filled energy band partly filled (Fig. 2.23). As a result, all the electrons in the highest filled band are available for electrical conduction, i.e., movement in response to an applied electric field. These electrons just need to move up in

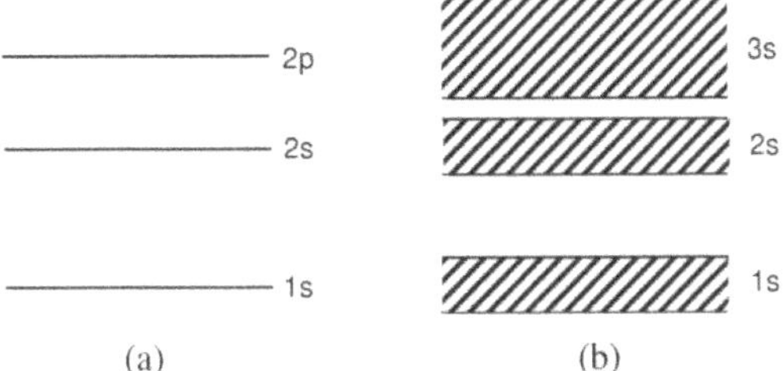

Fig. 2.22 Electron energy. (a) Narrow discrete energy levels for an isolated atom. (b) Energy bands for a solid.

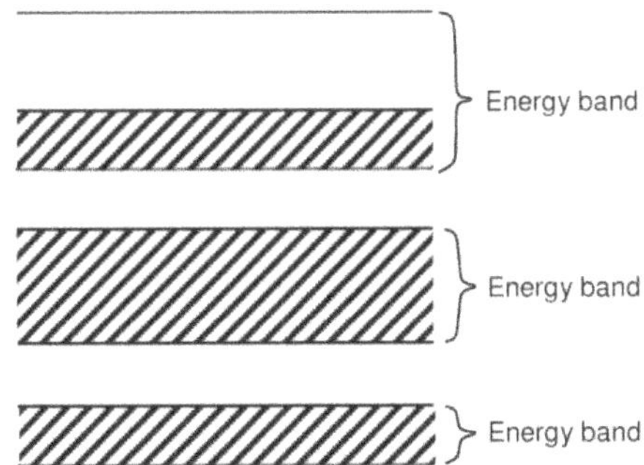

Fig. 2.23 A metal, which is characterized by the highest filled energy band being partially filled.

energy within the same band in order to gain kinetic energy. This is why a metal is a very good electrical conductor. Note that this definition of a metal is consistent with the notion of metallic bonding, which is exhibited by metals.

A semiconductor is defined as a solid that has its highest filled energy band completely filled at temperature 0 K, i.e., in the absence of thermal excitation. Hence, at 0 K, the energy band above this band is completely empty. There is an energy band gap between the highest filled energy band (known as the valence band, since it contains the valence electrons) and the energy band above (known as the conduction band). The gap is bound by the top of the valence band and the bottom of the conduction band, as illustrated in Fig. 2.24(a).

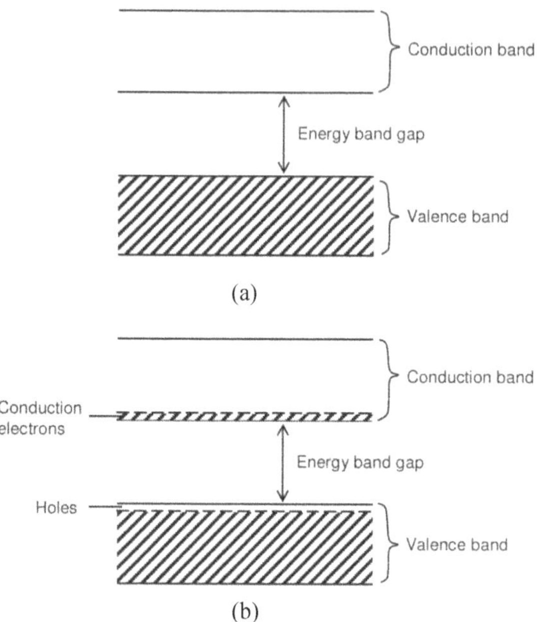

Fig. 2.24 (a) A semiconductor, which is characterized by the highest filled energy band being completely filled at 0 K. (b) Above 0 K, thermal excitation causes some of the valence electrons of a semiconductor to cross the energy band gap and reach the conduction band, where they become conduction electrons.

Above 0 K, the thermal energy present excites a small fraction of the electrons at the top of the valence band to the bottom of the conduction band. The excitation causes the electrons to move up in energy across the energy band gap above the valence band. This excitation results in conduction electrons at the bottom of the conduction band and holes at the top of the valence band, as illustrated in Fig. 2.24(b). The conduction electrons are free to move in response to an applied electric field, in contrast to the immobile valence electrons in the valence band. The holes in the valence band are actually vacant electronic states. The removal of a negative charge from an electrically neutral object results in a positive charge in what remains. Therefore, a hole is positively charged and can respond to an applied electric field, thus contributing to electrical conduction. Under the same electric field (voltage gradient), an electron moves toward the high voltage end of the electric field, whereas a hole moves toward the low voltage end of the electric field. Thus, both electron and hole result in an electric current. These two currents are additive, because the current is defined as the charge per unit time (not the magnitude of charge per unit time). Electrons moving in a particular direction correspond to a current in the opposite direction. Holes moving in a particular direction correspond to a current in the same direction. Hence, based on Eq. (2.7), the electrical conductivity of a semiconductor is given by

$$\sigma = qn\mu_n + qp\mu_p, \tag{2.60}$$

where q is the magnitude of the electronic charge (1.6×10^{-19} C), μ_n and μ_p are the mobilities of the conduction electrons and holes respectively, and n and p are the conduction electron concentration (number of conduction electrons per unit volume) and the hole concentration (number of holes per unit volume) respectively. For the same semiconductor, μ_p tends to be lower than μ_n. Because the excitation of one valence electron across the energy band gap results in a conduction electron in the conduction band and a hole in the valence band.

$$n = p, \tag{2.61}$$

which is true for an undoped (intrinsic) semiconductor. The combination of Eq. (2.60) and (2.61) gives

$$\sigma = qn\,(\mu_n + \mu_p), \tag{2.62}$$

which is true for an undoped (intrinsic) semiconductor.

As the temperature increases, n (equal to p) increases significantly (exponentially), due to the excitation of the electrons from the valence band to the conduction band. This large increase in n overshadows the decrease in mobilities as the temperature increases, so the electrical conductivity of a semiconductor increases with increasing temperature. This trend is opposite to that of a metal, the conductivity of which decreases with increasing temperature, due to the decrease of the mobility with increasing temperature and the availability of the valence electrons to serve as carriers in the absence of thermal excitation. This difference between a semiconductor and a metal is so striking that the temperature dependence of the conductivity may be used to distinguish between a metal and a semiconductor.

A doped (extrinsic) semiconductor is a semiconductor that has been alloyed by using a very small proportion of an alloying element, which is called the dopant. The alloy is typically a substitutional solid solution, with the dopant atoms substituting some of the host atoms. The semiconductor is the host. The electronic configuration of the dopant atoms needs to be such that the number of valence electrons is close to but not the same as that of the host.

If the electronic configuration of the dopant atom is such that the number of valence electrons is greater than that of the host, the dopant is said to be a donor, as it donates the extra electron(s) to the host, thus converting the donor atom to a donor cation and, most importantly, resulting in extra electrons. Since the valence band is already full, the extra electrons prior to their thermal excitation reside at a special energy level (called the donor level) within the energy band gap, but close to the bottom of the conduction band, as illustrated in Fig. 2.25(a). The energy band gap is forbidden only for an intrinsic semiconductor, but a semi-conductor alloy is a different material. Upon thermal excitation, the electrons at the donor level are excited to the bottom of the conduction band, thus resulting in conduction electrons in the conduction band. The excitation from the donor level to the bottom of the conduction takes much less energy that that required for excitation of the valence electrons across the energy band gap, so, at normal temperatures, the conduction electrons are mostly due to excitation of the electrons at the donor level rather than the excitation of the valence electrons in the valence band. As a consequence of the extra conduction electrons, the generation of which does not involve the formation of holes, $n > p$ and the semiconductor doped with a donor is said to be n-type. Furthermore, the high value of n causes the electrical conductivity to be much higher than that of an intrinsic semiconductor. An example of a donor for the semiconductor silicon is phosphorous, which has 5 valence electrons, in contrast to silicon, which has 4 valence electrons. Since phosphorous has one extra electron compared to silicon, each phosphorous atom donates one electron.

If the electronic configuration of the dopant atom is such that the number of valence electrons is smaller than that of the host, the dopant is said to be an acceptor, as it accepts electron(s) from the host, thus converting the acceptor atom to an acceptor anion. The acceptance of an electron by the acceptor atom is associated with the excitation of

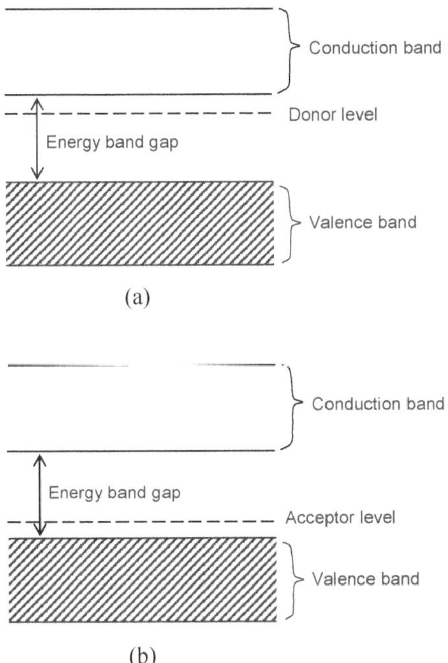

Fig. 2.25 Doped semiconductors. (a) An *n*-type semiconductor. (b) A *p*-type semiconductor.

a valence electron from the top of the valence band to a special energy level known as the acceptor level, which is within the energy band gap and is close to the top of the valence band, as illustrated in Fig. 2.25(b). After the acceptance, extra holes are generated. This acceptance costs much less energy than the excitation of an electron across the energy band gap, so, at normal temperatures, the holes are mostly due to excitation of the electrons to the acceptor level rather than the excitation of the valence electrons from the valence band to the conduction band. As a consequence of the extra holes, the generation of which does not involve the formation of conduction electrons, $p > n$ and the semiconductor doped with an acceptor is said to be p-type. Furthermore, the high value of p causes the electrical conductivity to be much higher than that of an intrinsic semiconductor. An example of an acceptor for the semiconductor silicon is aluminum, which has 3 valence electrons, in contrast to silicon, which has 4 valence electrons. Since aluminum has one less electron compared to silicon, each aluminum atom accepts one electron, hence contributing one hole.

Example Problems 12–17 and 27–30 at the end of the chapter are relevant.

2.12.2 *The pn-junction*

Through thin film deposition, semiconductors of different compositions may be in direct contact, without any impurity (e.g., oxide) at the interface. The resulting semiconductor junctions exhibit various electrical chacteristics, thus making them valuable as electronic devices. The most basic kind of semiconductor junction is the *pn*-junction, which is the junction between *p*-type and *n*-type semiconductors.

Diffusion is an entropy-driven phenomenon that occurs from a point of high concentration to a point of low concentration. Due to the high concentration of holes in the *p*-side compared to the *n*-side of a *pn*-junction, holes diffuse from the *p*-side to the *n*-side. Due to the high concentration of electrons in the *n*-side compared to the *p*-side, electrons diffuse from the *n*-side to the *p*-side. This interdiffusion results in a current known as the diffusion current. When a hole and an electron meet at the interface, the electron drops from the conduction band to the valence band and occupies the hole state at the valence band. As a consequence of this downward transition of the electron, both the conduction electron and the hole cease to exist. This phenomenon is known as recombination. Since recombination involves a release in energy, it occurs spontaneously. Due to recombination occurring at the interface region, this region is depleted of holes and conduction electrons and is thus known as the depletion region. The depletion of the carriers in the depletion region results in the exposure of the donor cations at the *n*-side and of the acceptor anions at the *p*-side and hence an electric field at the interface, as illustrated in Fig. 2.26(a), which is for an open-circuited *pn*-junction. The electric field is a voltage gradient. The difference in potential between the *p*-side and the *n*-side is known as the contact potential, which is labeled V_o in Fig. 2.26(a). This electric field is a barrier to the movement of holes from the *p*-side to the *n*-side, since holes naturally want to move toward the negative end of a voltage gradient. Similarly, the electric field is a barrier to the movement of conduction electrons from the *n*-side to the *p*-side, since electrons naturally want to move toward the positive end of a voltage gradient. On the other hand, the minority electrons on the *p*-side get swept to the *n*-side by this electric field and the minority holes on the *n*-side get swept to the *p*-side by this electric field. This sweeping by the electric field gives rise to a current known as the drift current. In spite of the sweeping, the drift current is limited by the low concentration of minority carrier on each side. The diffusion current and the drift current are in opposite directions. At equilibrium, the diffusion current is exactly balanced by the drift current, so that there is no net current, as expected for an open-circuited situation.

A forward-biased *pn*-junction has the *p*-side connected to the positive end of a voltage supply. The applied voltage is V_b, as shown in Fig. 2.26(b). Due to this biasing, the potential at the *p*-side is raised relative to that of the *n*-side, thus resulting in decrease of the contact potential from V_o (the open-circuited case) to $V_o - V_b$ (the forward-biased case). As a result of the diminished barrier, the diffusion current is enhanced, thereby resulting in the diffusion current exceeding the drift current and hence a net current in the direction from the *p*-side to the *n*-side. The higher is V_b, the higher is the net current.

A reverse-biased *pn*-junction has the *n*-side connected to the positive end of a voltage supply. The applied voltage is V_b, as shown in Fig. 2.26(c). Due to this biasing, the potential at the *n*-side is raised relative to that of the *p*-side, thus resulting in increase of the contact potential from V_o (the open-circuited case) to $V_o + V_b$ (the reverse-biased case). As a result of the increased barrier, the diffusion current is reduced, thereby resulting in the drift current exceeding the diffusion current and hence a net current in the direction from the *n*-side to the *p*-side. The drift current is small, due to the low concentration of minority carriers. Therefore, the net current in the reverse-biased case is close to zero.

The combination of the forward-biased behavior and the reverse-biased behavior results in the current-voltage relationship in Fig. 2.27. Current is substantial under forward bias, but is negligibly small under reverse bias. An electronic device that allows current to flow in only one direction is known as a diode.

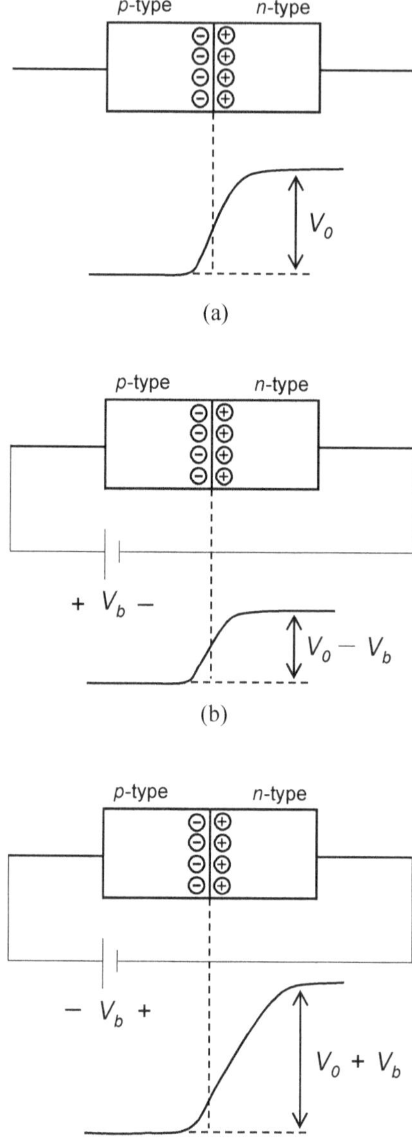

Fig. 2.26 A *pn*-junction. (a) Open-circuited condition. (b) Forward bias condition, with the potential barrier reduced from the open-circuited value by the bias voltage V_b. (c) Reverse bias condition, with the potential barrier increased from the open-circuited value by the bias voltage V_b.

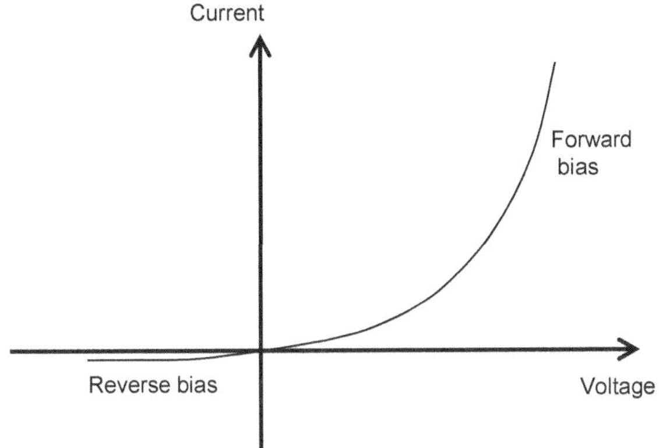

Fig. 2.27 Relationship of current with voltage for a *pn*-junction.

2.13 Pitfalls in electrical resistance measurement

The main pitfalls in resistance measurement relate to the neglect of the contact resistance (Sec. 2.13.1) and the electric polarization of the material (Sec. 2.13.2). This section describes thes pitfall and the associated circumvention.

2.13.1 *Pitfall due to the neglect of the contact resistance*

2.13.1.1 *Four-probe method vs. two-probe method*

There are in general two methods for measuring the electrical resistance of a material. They are the four-probe method and the two-probe method. The four-probe method involves four electrical contacts (electrodes), such that the outer two contacts are for passing a current, while the inner two contacts are for measuring the voltage. By separating the current and voltage contacts, the contact resistance is essentially eliminated from the measured resistance. On the other hand, the two-probe method involves only two electrical contacts, each of them being used for both current and voltage. As a consequence, the measured resistance using the two-probe method includes the contact resistance. Unless the contact resistance is negligible compared to the specimen resistance, the result obtained using the two-probe method is not reliable. A related pitfall is the assumption that the contact resistance is negligible when the contact material (e.g., silver) is highly conductive. Though the resistance may be small or negligible within the contact material, the resistance of the interface between the contact material and specimen can be substantial.

The two-probe method works if the distance l between the two electrical contacts is varied and the resistance is measured for each distance. The measured resistance R_m is given by

$$R_m = R + 2\,R_i, \tag{2.63}$$

where R is the specimen volume resistance and R_i is the contact resistance for each of the two contacts. The R is related to the resistivity ρ by the equation

$$R = \rho l/A, \tag{2.64}$$

where A is the cross-sectional area perpendicular to the resistance direction. The combination of Eq. (2.63) and (2.64) gives

$$R_m = \rho l/A + 2R_i. \tag{2.65}$$

The R_i is inversely related to A. Thus, a large value of A helps to reduce R_i. However, a large area cannot be used to support the assumption that R_i is negligible.

By plotting R_m versus l, a straight line is obtained, if all the electrodes used to provide the various values of l are similar in quality. The slope of the line equals ρ/A, according to Eq. (2.65). The intercept of this line with the vertical axis at $l = 0$ gives $2R_i$. Thus, R_i is decoupled from R. At least three values of l are needed for this method, which is known as the transmission line method. In contrast, in the four-probe method, only one value of l (the distance between the two inner contacts) is needed.

The use of a Wheatstone Bridge circuit does not remove the contact resistance. This is because the volume resistance and contact resistance are included in each leg of the bridge.

Current-voltage characteristic (curve) measurement is valuable, especially when the curve is not linear, as the curve gives more information than resistance measurement at a particular value of the current or voltage. However, the effect of the contact potential drop on the voltage in the curve remains a concern, which can be resolved by using the four-probe method.

There can be application-related material structures that make the implementation of the four-probe method difficult. Under this situation, a non-ideal four-probe method in which the current and voltage electrodes that are not in electrical contact are in the same plane (rather than in distinct planes) may be used (Fig. 2.10). This non-ideal four-probe method is more reliable than the two-probe method. In case that even the use of the non-ideal four-probe method is not feasible, the two-probe method may be used, provided the conclusions drawn from the study do not involve the assumption that the contact resistance is zero or negligible.

2.13.1.2 *Resistance response to a stimulus*

For both the science and applications, the response of the resistance to a certain stimulus is commonly studied. Examples of the stimulus include the voltage/current, strain/stress, temperature and damage. The response enables the sensing of the stimulus through the resistance measurement. In this situation, the material being tested serves as the sensor, as it senses itself without any sensor incorporation (i.e., self-sensing). Self-sensing is particularly valuable when the material is a structural material, so that the integrity of the structure may be monitored. For such studies, the two-probe method is not reliable, even if the contact resistance is small. This is because both the contact resistance and specimen volume resistance can change as the degree of the stimulus is varied. As a consequence, the observed change in resistance upon variation in the stimulus degree may be due to a combination of a change in contact resistance and a change in the volume resistance. In general, the contact resistance and volume fraction may change in the same direction or in opposite directions in response to the same stimulus. Even if the two contacts are embedded in the specimen, the contact resistance can change upon stimulus application. This is because the quality of the interface between the embedded contact and the material being studied can change upon experiencing the stimulus.

For example, upon the change in temperature, the thermal expansion of the contact material (e.g., silver paint after drying) and/or the change in the degree of conformability of the contact material with the specimen surface may occur, thus affecting the contact resistance. An increase in the fluidity and hence the conformability may occur upon heating and this tends to decrease the amount of air voids at the interface between the contact and specimen surface, thereby decreasing the contact resistance. As another example, upon compression, the interface between each of the two contacts and the specimen volume is tightened and hence improved. Thus, the contact resistance decreases with increasing compressive stress. This problem is particularly significant in case of contacts that are attached to the surface of the specimen. To alleviate this problem, the top contact (i.e., the contact on the surface receiving the load) may be limited to only a part of the area of the specimen, such that the contact does not experience the stress while the remaining area of the specimen does. For example, the loaded area is in the center part of the area, while the remaining area is large enough to allow the current used in the testing to go through essentially the entire area of the specimen. In case that the specimen is stiff enough, the force applied to a part of the specimen area is spread to the entire area. However, if the specimen is not stiff enough, the spreading of the force would not be adequate. A better solution, which does not require the material to be stiff, is to avoid using contacts on the surface, but to use embedded contacts.

2.13.1.3 *Types of electrical contacts*

The quality of attached contacts can vary greatly. Contacts (known as pressure contacts) in the form of a metal contact material (e.g., a conductor foil or tip) that touches the specimen surface through the application of pressure tend to be relatively low in quality, due to the fact that the proximate surfaces are never perfectly smooth in the microscopic level and, as a consequence, microscopic air pockets exist at the interface. Since air is an insulator, the contact quality is low. Furthermore, the contact resistance decreases with increasing pressure, which blunts or deforms the hillocks in the surface topography. Pressure contacts involving conductor tips (such as pointed metallic electrical probes) are particularly problematic, as the small area of the tip makes the contact resistance particularly large.

Contacts involving a conductive paint or paste (e.g., silver paint, nickel paint and graphite paint) that is present between a metal contact material (e.g., a metal foil) and the specimen surface tend to be superior to the pressure contacts, because the paint or paste displaces the air and is more conductive than air. For the purpose of displacing the air, the fluidity of the paint or paste is important. Between paints and pastes, paints tend to give superior contacts, due to their greater fluidity and hence stronger ability to enter the microscopic valleys in the surface topography of the specimen surface. For both paints and pastes, the fluidity decreases with increasing content of the conductive filler, while the conductivity increases. Therefore, for maximizing the effectiveness of the paints or pastes, the filler content should be kept sufficiently low. A paint or paste with high conductivity but low fluidity is not attractive. Obviously, the paints or pastes should be applied in the fluid (wet) state. As long as the paint or paste is more conductive than air, its presence helps the contact quality, unless the thickness of the paint or paste is excessive. The thickness should ideally be just enough for the paint or paste to fill the valleys in the topography of the proximate surfaces. The greater is the thickness, the higher is the contact resistance. In case that the paint or paste cannot fill the valleys completely, air pockets remain at the interface, thereby reducing the contact quality.

Contacts in the form of vacuum-deposited thin films should not be assumed to be associated with negligible contact resistance. The interface between the film and the specimen surface may still have substantial contact resistance, due to the contamination that may be present at the interface between the contact and specimen surface during the deposition. The amount of contamination depends on the quality of the vacuum used in the deposition. Furthermore, the typically small area of such contacts (e.g., in the form of dots) contributes to causing the contact resistance to be substantial. The possible non-Ohmic nature of the contacts is an additional concern.

Similar interface considerations apply to embedded contacts, though conductive paints or pastes are typically absent at the interface between an embedded contact and the material being tested. The application of pressure to the material containing embedded contacts tends to improve the quality of these contacts, as the pressure tends to tighten the interface. The processing of the material may induce pressure, as in the case that the material precursor shrinks as it is converted to the material during the process. However, the process-induced pressure is usually low compared to the pressure that is applied to the material.

2.13.2 *Pitfall due to the electric polarization of the material*

An electrically conductive material can exhibit a degree of dielectric behavior (Sec. 3.1). This means that the conductive material is associated with both resistance and capacitance, although the capacitance tends to be small.

The dielectric behavior refers to electric polarization behavior. The polarization means the separation between the positive charge center and negative charge center in the material. The polarization results in an electric dipole and hence an electric field that opposes the electric field applied to cause the conduction. As a consequence, the polarization increases the apparent DC resistance beyond the true resistance (Sec. 3.6.4). The build-up of the DC polarization takes time, so the increase of the apparent DC resistance occurs over a period of time. Typically, the higher is the conductivity, the faster is the polarization. For the purpose of measuring the true resistance, the resistance should be measured before the polarization becomes substantial. Since resistance measurement involves the meter putting out a small current to the specimen, polarization occurs as the measurement continues. Therefore, the resistance obtained at the very start of the resistance measurement is the value that is close to the true resistance. A pitfall is to wait till the resistance has become stable, i.e., essentially not changing with time. When the resistance becomes stable, the polarization has become saturated and the resistance corresponding to this state is higher than the true resistance.

An alternate method for obtaining the true resistance for a material that can be polarized is not to measure at the start of the polarization, but to reverse the polarity and take the average of the resistance immediately before the reversal (R_1) and that immediately after the reversal (R_2), as illustrated in Fig. 3.25. The average of R_1 and R_2 equals the true resistance in case of electrical symmetry. However, in general, R_1 and R_2 can be off from the true resistance by different amounts. The four-probe method should be used for the resistance measurement in accordance with this method. If the two-probe method is used, polarization that also occurs at the electrical contacts (i.e., at the interface between the contact and specimen surface) causes asymmetry upon polarity reversal, so that R_1 and R_2 are not equally far from the true resistance and the average of R_1 and R_2 does not equal the true resistance.

The phenomenon of dielectric relaxation differs greatly between conductors and nonconductors. For conductors, the polarization is due to the interaction of a fraction of the carriers (typically electrons) with the atoms. For nonconductors, the polarization is due to the asymmetry in the position of the ions or charged species in the material, e.g., asymmetry in the crystal structural unit cell. Due to the difference in polarization mechanism and thermodynamics, the dynamics of dielectric relaxation differ between conductors and nonconductors. In this regard, the case of conductors is still an emerging area of science.

Review questions

1. What is the unit of the mobility?
   Answer: The unit of the mobility μ is m^2/(V.s).

2. What is the unit of the contact electrical resistivity? What is the unit of the volume electrical resistivity?
   Answer: The unit of the contact electrical resistivity is Ω.m^2. The unit of the volume electrical resistivity is Ω.m.

3. Why is a ceramic typically a poor electrical conductor?
   Answer: A ceramic typically has a combination of covalent bonding and ionic bonding, both of which do not provide free electrons for electrical conduction.

4. Why does the electrical resistivity of a metal increase with increasing temperature?
   Answer: This is because the mobility of the electrons decreases with increasing temperature.

5. Why does the electrical resistivity of a semiconductor decrease with increasing temperature?
   Answer: This is because the carrier concentration increases with increasing temperature.

6. Why is a conductor with too high an electrical resistance not effective for resistance heating?
   Answer: With a high resistance, the current I in I^2R is too small.

7. Why is a conductor with too low an electrical resistance not effective for resistance heating?
   Answer: With a low resistance R, I^2R is too small.

8. Why is the four-probe method of electrical resistivity measurement superior to the two-probe method?

   Answer: The four-probe method essentially eliminates the contact resistance from the measured resistance. In contrast, the two-probe method includes the contact resistance in the measured resistance.

9. Why should the distance between the adjacent voltage and current contacts used in the four-probe method of electrical resistivity measurement be sufficiently large when each of the four electrical contacts is on the surface of the specimen around the entire perimeter of the specimen?

   Answer: A large enough distance between the adjacent voltage and current contacts enables the current to reach throughout the cross-sectional area in the region of the specimen between the two inner (voltage) contacts.

10. Under what situation should one use an ionic conductor (as opposed to an electronic conductor) as an electrical contact material for passing current?

    Answer: When the material under investigation is an ionic conductor, an ionic conductor is preferable for use as an electrical contact material for passing current.

11. What is meant by negative piezoresistivity?

    Answer: Negative piezoresistivity refers to the situation in which the fractional change in resistivity per unit strain is negative.

12. How is the gage factor associated with piezoresistivity related to the Poisson's ratio?

    Answer: The gage factor increases with increasing Poisson's ratio.

13. Why is the current very small when a *pn*-junction is under reverse bias?

    Answer: The current is very small under reverse bias because the contact potential drop at the junction is large and this drop is against conduction.

14. Why is the current substantial when a *pn*-junction is under forward bias?

    Answer: The current is substantial under forward bias because the contact potential drop at the junction is small.

15. What is the principle behind the ability of a thermocouple to sense temperature?

    Answer: The principle involves the Seebeck effect that occurs in each of the two dissimilar legs of the thermocouple that are electrically connected.

16. Why is a metal typically a poor thermoelectric material (i.e., one with a low value of ZT)?

    Answer: A metal has high thermal conductivity, which causes ZT to be low.

17. Why is a polymer typically a poor thermoelectric material (i.e., one with a low value of ZT)?

    Answer: A polymer has lower electrical conductivity, which causes ZT to be low.

Example problems

Problem 1 (Electrical resistivity)

A metal wire has diameter 0.25 mm, length 12.7 m and electrical resistivity 3.45×10^{-6} Ω.cm. Calculate the resistance of this wire.

Solution:

Resistivity $= 3.45 \times 10^{-6}$ Ω.cm $= 3.45 \times 10^{-8}$ Ω.m

Radius $= 0.25$ mm$/2 = 0.125$ mm $= 1.25 \times 10^{-4}$ m

$R = \rho l / A = (3.45 \times 10^{-8}$ Ω.m$)(12.7$ m$)/[\pi (1.25 \times 10^{-4}$ m$)^2] = 8.9$ Ω

Problem 2 (Conductivity equation)

Derive the equation

$$\sigma = qn\mu,$$

where σ is the electrical conductivity, q is the charge per carrier, and μ is the mobility of the carrier. Hint:

$$I = qnvA$$

$$v = \mu E$$

$$\sigma = \frac{\tilde{J}}{E}$$

Solution:

$$\tilde{J} = \frac{I}{A} = \frac{qnvA}{A} = qnv$$

$$\sigma = \frac{\tilde{J}}{E} = \frac{qnv}{E}$$

$$\sigma = qn\mu$$

Problem 3 (Electrical conductivity and resistivity)

A metal has electrical conductivity 2.57×10^7 $\Omega^{-1}.m^{-1}$ at 20°C. Its number of valence electrons per unit volume is 4.89×10^{13} cm^{-3}. Its temperature coefficient of electrical resistivity is 0.0039/°C. It is in the form of a wire of diameter 0.288 mm and length 759 mm. This wire is subjected to an electric field of 3.64×10^{-5} V/m along the length of the wire. The electronic charge q is 1.60×10^{-19} C.
Hint:

$R = \rho l/A$, where R is the resistance, ρ is the resistivity, l is the length, and A is the area.

$v = \mu E$, where v is the drift velocity, μ is the mobility, and E is the electric field.

$\sigma = qn\mu$, where σ is the conductivity, n is the free electron concentration (number of free electrons per unit volume) and μ is the mobility.

$\Delta\rho/\rho = \alpha\,\Delta T$, where $\Delta\rho/\rho$ is the fractional change in resistivity, α is the temperature coefficient of electrical resistivity, and ΔT is the change in temperature.

(a) Calculate the electrical resistance of this wire.

(b) Calculate the difference in voltage between the two ends of this wire.

(c) Calculate the mobility of the free electrons in this metal at 20°C.

(d) Calculate the drift velocity of the free electrons in this metal at 20°C.

(e) Calculate the electrical resistivity of this metal at 158°C.

Solution:

(a) Resistivity $\rho = 1/\text{conductivity}$

$$= 1/(2.57 \times 10^7\ \Omega^{-1}.\text{m}^{-1})$$

$$= 3.89 \times 10^{-8}\ \Omega.\text{m}$$

Resistance $R = \rho l/A$

$$= (3.89 \times 10^{-8}\ \Omega.\text{m})\,(0.759\ \text{m})/[\pi\,(1.44 \times 10^{-4}\ \text{m})^2]$$

$$= 0.453\ \Omega$$

(b) Voltage $=$ Electric field $\times$ length

$$= (3.64 \times 10^{-5}\ \text{V/m})\,(0.759\ \text{m})$$

$$= 2.76 \times 10^{-5}\ \text{V}$$

(c) $\mu = \sigma/(qn) = (2.57 \times 10^7 \ \Omega^{-1}.\text{m}^{-1})/[(1.60 \times 10^{-19} \ \text{C}) (4.89 \times 10^{19} \ \text{m}^{-3})]$

$= 3.28 \times 10^6 \ \text{m}^2/(\text{V.s})$

(d) $v = \mu E = [3.28 \times 10^6 \ \text{m}^2/(\text{V.s})] (3.64 \times 10^{-5} \ \text{V/m})$

$= 1.19 \times 10^2 \ \text{m/s}$

(e) (Resistivity at 158°C − resistivity at 20°C)/(resistivity at 20°C)

$= \alpha (158 - 20)°\text{C}$

Resistivity at 158°C − resistivity at 20°C

$= (\text{resistivity at } 20°\text{C}) \ \alpha (158 - 20)°\text{C}$

$= (3.89 \times 10^{-8} \ \Omega.\text{m}) (0.0039/°\text{C}) (138°\text{C}) = 2.09 \times 10^{-8} \ \Omega.\text{m}$

Resistivity at 158°C = resistivity at 20°C + $(2.09 \times 10^{-8} \ \Omega.\text{m})$

$= (3.89 \times 10^{-8} \ \Omega.\text{m}) + (2.09 \times 10^{-8} \ \Omega.\text{m})$

$= 5.98 \times 10^{-8} \ \Omega.\text{m}$

Problem 4 (Sheet resistance)

The sheet resistance of a sheet of material is defined as the resistance per square of the sheet. For a sheet having thickness 0.125 mm and volume electrical resistivity $7.43 \times 10^{-8} \ \Omega.\text{m}$, calculate the sheet resistance.

Solution:

Sheet resistance = Volume resistivity/thickness

$= (7.43 \times 10^{-8} \ \Omega.\text{m})/(1.25 \times 10^{-4} \ \text{m})$

$= 5.9 \times 10^{-4} \ \Omega/\text{square}$

Problem 5 (Current density)

An electric current of 278 mA flows through a metal wire of diameter 0.57 mm. What is the current density in the unit of A/m^2?

Solution:

Current density = current/area

$$= (278 \text{ mA})/[\pi \, (0.57 \text{ mm}/2)^2]$$

$$= 1090 \text{ mA/mm}^2$$

$$= 1.09 \times 10^6 \text{ A/m}^2.$$

Problem 6 (Drift velocity)

(a) Aluminum has electron mobility 34 $cm^2/(V.s)$. What is the drift velocity (in m/s) of the electrons under an electric field of 50 V/mm?

Solution:

Drift velocity = mobility × electric field

$$= (34 \text{ cm}^2/(V.s)) \, (50 \text{ V/mm})$$

$$= 170 \text{ m/s}$$

(b) The mobility of aluminum decreases with increasing temperature. Explain in terms of the scientific origin.

Solution:

The thermal vibration amplitude of the atoms increases with increase temperature, thus increasing the degree of scattering of the electrons.

Problem 7 (Free electron concentration)

A material has 5×10^{23} atoms per cm^3 and each atom contributes 2 free electrons for electrical conduction in the metal. What is the free electron concentration (i.e., the number of free electrons per unit volume)?

Solution:

$$\text{Free electron concentration} = (5 \times 10^{23})\, 2 \text{ cm}^{-3} = 1 \times 10^{24} \text{ cm}^{-3}$$

Problem 8 (Valence electron concentration)

A metal has 1 valence electron per atom. Its density is 5.87 g/cm^3. The atomic mass is 35 g/mol. What is the valence electron concentration in the unit cm^{-3}? That is, what is the number of valence electrons per cm^3? Note: Avogadro's number $= 6.02 \times 10^{23}$

Solution:

No. of atoms per $cm^3 = (5.87 \text{ g})/[(35 \text{ g})/(6.02 \times 10^{23})) = 1.01 \times 10^{23}$

No. of valence electrons per $cm^3 = 1.01 \times 10^{23}$

Valence electron concentration $= 1.01 \times 10^{23} \text{ cm}^{-3}$

Problem 9 (Conductivity of a metal)

Calculate the electrical conductivity [in $(\Omega.\text{cm})^{-1}$] of a metal that has a valence electron concentration of 2.7×10^{23} cm^{-3} and a valence electron mobility of 11 $m^2/(V.s)$.

Hint: $\sigma = q\, n\, \mu$

where $q = 1.6 \times 10^{-19}$ C

Solution:

$$\text{Conductivity} = (1.6 \times 10^{-19} \text{ C}) (2.7 \times 10^{23} \text{ cm}^{-3}) (11 \text{ m}^2/(\text{V.s}))$$

$$= 4.8 \times 10^9 \text{ } (\Omega.\text{cm})^{-1}$$

Problem 10 (Mobility and carrier concentration)

A metal has electrical resistivity 6.87×10^{-8} Ω.m. The valence electron concentration is 5.69×10^{23} m^{-3}. Calculate the mobility of the valence electrons.

Hint: The conductivity σ is given by $\sigma = qn\mu$

$$q = 1.6 \times 10^{-19} \text{ C}$$

Solution:

The mobility $\mu = \sigma/(qn) = 1/(\rho qn)$,

where ρ is the resistivity.

$$\mu = 1/[(6.87 \times 10^{-8} \text{ } \Omega.\text{m}) (1.6 \times 10^{-19} \text{ C}) (5.69 \times 10^{23} \text{ m}^{-3})]$$

$$= 1.6 \times 10^2 \text{ m}^2/(\text{V.s})$$

Note: $C = A.s$ and $\Omega = V/A$

Problem 11 (Mobility and carrier concentration)

Calculate the electrical resistivity of a material with electrons as the mobile charge carriers. The mobility of the carriers is 15.5 m^2/(V.s). The carrier concentration is 5.87×10^{23} m^{-3}. The magnitude of the charge of an electron is 1.6×10^{-19} C.

Solution:

The conductivity is given by

$$\sigma = qn\mu = (1.6 \times 10^{-19}\ \text{C})\,(5.87 \times 10^{23}\ \text{m}^{-3})\,(15.5\ \text{m}^2/(\text{V.s}))$$

$$= 146 \times 10^4\ \Omega^{-1}.\text{m}^{-1} = 1.46 \times 10^6\ \Omega^{-1}.\text{m}^{-1}$$

Note: $\text{C}/(\text{m.V.s}) = [(\text{C/s})/\text{V}]/\text{m} = (\text{A/V})/\text{m} = \Omega^{-1}.\text{m}^{-1}$

Resistivity $= 1/\text{conductivity} = 1/(1.46 \times 10^6\ \Omega^{-1}.\text{m}^{-1})$

$$= 0.68 \times 10^{-6}\ \Omega.\text{m}$$

$$= 6.8 \times 10^{-7}\ \Omega.\text{m}$$

Problem 12 (Semiconductor)

An intrinsic (undoped) semiconductor has hole concentration $= 3.47 \times 10^{17}\ \text{m}^{-3}$. What is the conduction electron concentration?

Solution: Conduction electron concentration $= 3.47 \times 10^{17}\ \text{m}^{-3}$

Problem 13 (Semiconductor)

What is meant by a semiconductor being n-type? An n-type semi-conductor has a donor atom concentration of $4.56 \times 10^{23}\ \text{m}^{-3}$. Assume every donor atom has released (donated) an electron. Assume that the hole concentration is negligible. What is the conduction electron concentration?

Solution:

An n-type semiconductor has the conduction electron concentration much greater than the hole concentration.

$$\text{Conduction electron concentration} = 4.56 \times 10^{23}\ \text{m}^{-3}$$

Problem 14 (Semiconductor)

A semiconductor has conduction electrons and holes. The current due to the conduction electrons is 28 mA. The current due to the holes is 17 mA. What is the total current?

Solution:

$$\text{Total current} = (28 + 17) \text{ mA} = 45 \text{ mA}$$

Problem 15 (Semiconductor)

An intrinsic semiconductor (silicon) has conduction electron mobility (0.14 m^2/(V.s)), hole mobility (0.038 m^2/(V.s)) and hole concentration 1.4×10^{16} m^{-3}. Calculate the conductivity in $\Omega^{-1}.m^{-1}$.

Solution:

For an intrinsic semiconductor,

$$n = p = 1.4 \times 10^{16} \text{ m}^{-3}$$

Conductivity $= qn\,(\mu_n + \mu_p)$

$$= (1.6 \times 10^{-19} \text{ C})\,(1.4 \times 10^{16} \text{ m}^{-3})\,(0.14 + 0.038) \text{ m}^2/(V.s)$$

$$= 0.40 \times 10^{-3} \ \Omega^{-1}.m^{-1}$$

$$= 4.0 \times 10^{-4} \ \Omega^{-1}.m^{-1}$$

Note: $C.m^{-3}.m^2/(V.s) = (C/s)/(V.m) = (A/V)/m = \Omega^{-1}.m^{-1}$

Problem 16 (Semiconductor)

An n-type semiconductor has conduction electron concentration 4×10^{19} cm^{-3}. Calculate the hole concentration in this n-type semiconductor. The intrinsic carrier concentration of this semiconductor is 5×10^{15} cm^{-3}.

Solution:

Mass-action law: $np = n_i^2$

$$p = n_i^2/n = (5 \times 10^{15})^2/(4 \times 10^{19}) = 6 \times 10^{11} \text{ cm}^{-3}$$

Problem 17 (Semiconductor)

Germanium (intrinsic, i.e., not doped) has carrier concentration $n = p = 2.3 \times 10^{13}$ cm^{-3}. The density of germanium is 5.323 g/cm³. The atomic mass of germanium is 72.61 g/mol. What is the fraction of valence electrons that are excited across the band gap at 300 K?

Solution:

No. Ge atoms per cm^3 = (Density of Ge)/(Mass of one Ge atom)

$$= (5.232 \text{ g/cm}^3)/[(72.61 \text{ g})/(6.02 \times 10^{23})]$$

$$= 4.34 \times 10^{22} \text{ cm}^{-3}$$

Since each Ge atom has 4 valence electrons, the concentration of valence electrons at 0 K is

$$4 \, (4.34 \times 10^{22} \text{ cm}^{-3}) = 1.74 \times 10^{23} \text{ cm}^{-3}$$

Therefore, the fraction of valence electrons that are excited across the energy band gap at 300 K

$$= (2.3 \times 10^{13} \text{ cm}^{-3})/(1.74 \times 10^{23} \text{ cm}^{-3}) = 1.3 \times 10^{-10}$$

Problem 18 (Effect of temperature on the electrical resistivity)

Copper exhibits electrical resistivity 1.72×10^{-6} Ω.cm at 20°C and temperature coefficient of resistivity 0.0039/°C. Calculate the resistivity at 100°C.

Solution:

$$\rho = \rho_o \, (1 + \alpha \, \Delta T),$$

where α is the temperature coefficient of resistivity and ρ_o is the resistivity at 20°C.

$$\Delta T = (100 - 20) \; °C$$

$$= 80°C$$

$$\rho = \rho_o \, (1 + \alpha \Delta T) = (1.72 \times 10^{-6} \; \Omega.cm) \, [1 + (0.0039/°C) \, (80°C)]$$

$$= (1.72 \times 10^{-6} \; \Omega.cm) \, (1.31)$$

$$= 2.26 \times 10^{-6} \; \Omega.cm$$

Problem 19 (Effect of temperature on the electrical resistivity)

Figure 2.27 shows a plot of $\log_{10}$ contact conductivity vs. 1/T (where T is in K). Calculate the activation energy per electron (in eV) for conduction

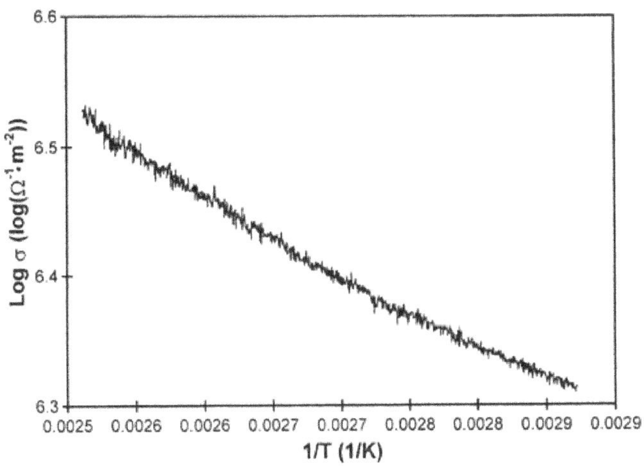

Fig. 2.27 Plot of log σ vs. 1/T, where T is in K.

in the direction perpendicular to the plane of the contact. Consider the approximately linear part of the curve below $1/T = 0.0027$ K^{-1}.

Solution:

$$\text{Slope} = -(6.53 - 6.40)/[(0.0027 - 0.0025)\text{ K}^{-1}]$$

$$= -(0.13/0.0002)\text{ K} = -650\text{ K} = -E/(2.3\text{ k}),$$

where E is the activation energy in eV per electron, and k is the Boltsmann's constant (8.63×10^{-5} eV/K).

$$E = (2.3)\,(8.63 \times 10^{-5}\text{ eV/K})\,(650\text{ K})$$

$$= 1.29 \times 10^4 \times 10^{-5}\text{ eV}$$

$$= 0.129\text{ eV}$$

Problem 20 (Composite material)

A polymer-matrix composite contains 56 vol.% unidirectional continuous carbon fibers and 44 vol.% matrix. The axial electrical conductivity of the fiber is 200 Ω^{-1}.cm^{-1}. The conductivity of the polymer is negligible. Calculate the conductivity of the composite in the fiber direction of the composite.

Hint:

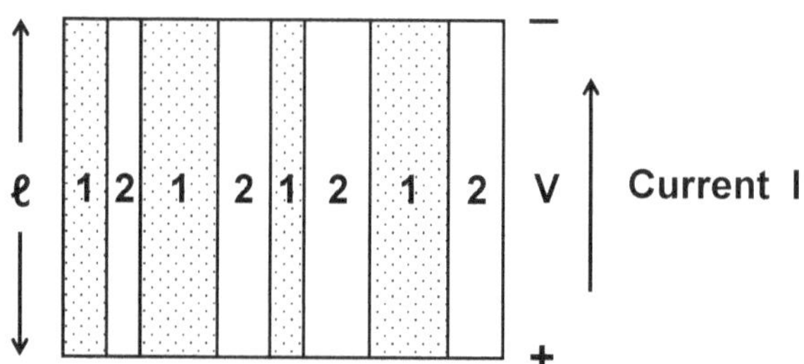

$$I_1 = V/R_1$$
$$I_2 = V/R_2$$
$$I = V\left[(1/R_1) + (1/R_2)\right]$$
$$R_1 = \rho_1\, l/A_1$$
$$R_2 = \rho_2\, l/A_2$$
$$I = (V/l)\left[(A_1/\rho_1) + (A_2/\rho_2)\right]$$

$$I = V/R$$
$$R = \rho l/A$$
$$I = VA/\rho l$$

$$VA/\rho l = (V/l)\left[(A_1/\rho_1) + (A_2/\rho_2)\right]$$
$$A/\rho = \left[(A_1/\rho_1) + (A_2/\rho_2)\right]$$
$$1/\rho = (A_1/A)(1/\rho_1) + (A_2/A)(1/\rho_2)$$
$$= v_1/\rho_1 + v_2/\rho_2$$
$$\sigma = v_1\sigma_1 + v_2\sigma_2$$

Solution:

Conductivity of composite $= (200\ \Omega^{-1}.\text{cm}^{-1})\,(0.56)$

$$= 110\ \Omega^{-1}.\text{cm}^{-1}$$

Problem 21 (Composite material)

A metal-matrix composite containing 56 vol.% unidirectional continuous fibers and 44 vol.% matrix. The transverse electrical conductivity of the fiber is 168 $\Omega^{-1}.\text{cm}^{-1}$. The conductivity of the matrix is $2.56 \times 10^5\ \Omega^{-1}.\text{cm}^{-1}$. Calculate the transverse conductivity of the composite.

Hint:

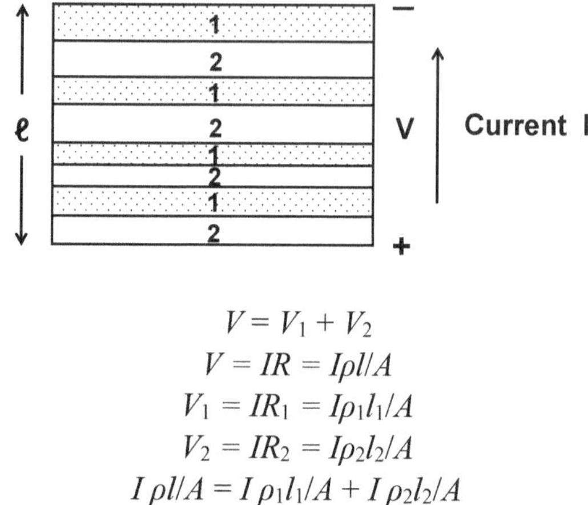

$$V = V_1 + V_2$$
$$V = IR = I\rho l/A$$
$$V_1 = IR_1 = I\rho_1 l_1/A$$
$$V_2 = IR_2 = I\rho_2 l_2/A$$
$$I\rho l/A = I\rho_1 l_1/A + I\rho_2 l_2/A$$
$$\rho l = \rho_1 l_1 + \rho_2 l_2$$
$$\rho = \rho_1 (l_1/l) + \rho_2 (l_2/l) = \rho_1 v_1 + \rho_2 v_2$$

Solution:

Transverse resistivity

$$= [1/(168\ \Omega^{-1}.\mathrm{cm}^{-1})]\ (0.56) + [1/(2.56 \times 10^5\ \Omega^{-1}.\mathrm{cm}^{-1})]\ (0.44)$$

$$= (3.3 \times 10^{-3} + 1.72 \times 10^{-6})\ \Omega.\mathrm{cm}$$

$$= 3.3 \times 10^{-3}\ \Omega.\mathrm{cm}$$

Transverse conductivity $= 1/(3.3 \times 10^{-3}\ \Omega.\mathrm{cm})$

$$= 300\ \Omega^{-1}.\mathrm{cm}^{-1}$$

Problem 22 (Composite material)

A composite in the form of parallel carbon continuous fibers in a nonconductive polymer matrix. The fiber volume fraction is 57%. The fiber resistivity is $3.46 \times 10^{-3}\ \Omega.\mathrm{cm}$. Calculate the resistivity of the composite in the fiber direction.

Solution:

For the parallel configuration,

$$\sigma = v_1\sigma_1 + v_2\sigma_2$$

$$\sigma = [(0.57)/(3.46 \times 100^{-3}\ \Omega.\text{cm})] + 0$$

$$= 1.65 \times 10^2\ \Omega^{-1}.\text{cm}^{-1}$$

$$\text{Resistivity} = 1/\sigma = 1/(1.65 \times 10^2\ \Omega^{-1}.\text{cm}^{-1})$$

$$= 0.61 \times 10^{-2}\ \Omega.\text{cm}$$

$$= 6.1 \times 10^{-3}\ \Omega.\text{cm}$$

Problem 23 (Resistors in parallel)

Two resistors with resistances $R_1 = 46\ \Omega$ and $R_2 = 98\ \Omega$ are in parallel. Calculate the resistance R of the combination of the two resistances.

Solution:

$$1/R = [(1/46) + (1/98)]\ \Omega^{-1}$$

$$= 0.0102\ \Omega^{-1}$$

$$R = 31\ \Omega$$

Problem 24 (Contact electrical resistivity)

The interface between A and B exhibits contact resistance $560\ \Omega$. The area of this interface is 1.35 cm^2. Calculate the contact electrical resistivity of this interface.

Solution:

Contact resistivity = contact resistance × contact area

$$= (560 \ \Omega) \ (1.35 \ cm^2)$$

$$= 756 \ \Omega \ cm^2$$

Problem 25 (Contact electrical resistance)

The contact resistivity of an interface is $6.97 \times 10^{-4} \ \Omega.m^2$. What is the minimum contact area needed to achieve a contact resistance of no more than 50 Ω?

Solution:

Area = contact resistivity/contact resistance

$$= (6.97 \times 10^{-4} \ \Omega.m^2)/(50 \ \Omega)$$

$$= 0.14 \times 10^{-4} \ m^2$$

$$= 1.4 \times 10^{-5} \ m^2$$

Problem 26 (Contact electrical resistivity)

A material of electrical resistivity $\rho = 2.47 \times 10^{-5} \ \Omega.cm$, area $A = 475 \ mm^2$ and thickness $l = 1.25 \ mm$ is sandwiched by two electrodes (i.e., two conductive sheets). The measured resistance R_m between the two electrodes is 17.9 mΩ. Calculate the contact electrical resistivity of the interface between this material and an electrode. Assume that the two interfaces are identical. Note: 1 mΩ = $10^{-3} \ \Omega$.

Hint: $R_m = R_v + 2 R_i$, where R_v is the volumetric resistance of the sandwiched material and is given by

$$R_v = \rho l / A$$

Solution:

$$R_v = (2.47 \times 10^{-5} \ \Omega.\text{cm}) \ (1.25 \ \text{mm})/(475 \ \text{mm}^2)$$

$$= (2.47 \times 10^{-4} \ \Omega.\text{mm}) \ (1.25 \ \text{mm})/(475 \ \text{mm}^2)$$

$$= 6.5 \ \text{m}\Omega$$

$$R_i = (R_m - R_v)/2$$

$$= (17.9 - 6.5) \ \text{m}\Omega/2$$

$$= 5.7 \ \text{m}\Omega$$

$$\text{Contact resistivity} = R_i A$$

$$= (5.7 \ \text{m}\Omega) \ (475 \ \text{mm}^2)$$

$$= 2.7 \ \Omega.\text{mm}^2$$

Problem 27 (Semiconductors)

Based on Fig. 2.28, calculate the energy bandgap (in eV) of GaAs. The activation energy is equal to the bandgap divided by 2. (The factor of 2 is not derived in this course, but it is because the conduction electron concentration and hole concentration are not independent.) Ignore the dashed curves in Fig. 2.28.

The Boltmann's constant $k = 1.38065 \times 10^{-23}$ J/K. 1 J $= 6.242 \times 10^{18}$ eV.

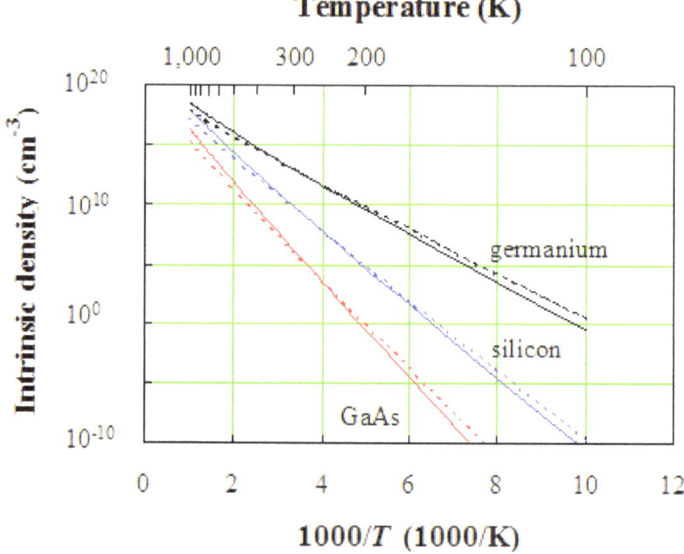

Fig. 2.28 Arrhenius plot of log carrier density vs. 1000/T, where T is in K.

Solution:

Slope $= -E/[(2.3)\,2k]$, where E is the energy band gap.

For GaAs, slope $= -25/[(7.3 - 1.2)/1000] = -4098$ K

$E = -$ slope $[(2.3)\,2k]$

$\quad = 4089$ K $[(2.3)\,2\,(1.38065 \times 10^{-23}$ J/K$)]$

$\quad = 2.60 \times 10^{-19}$ J

$\quad = (2.60 \times 10^{-19}$ J$)\,(6.242 \times 10^{18}$ eV/J$)$

$\quad = 1.6$ eV

Problem 28 (Semiconductor)

(a) A semiconductor's resistivity decreases with increasing temperature. Explain from the viewpoint of the scientific origin.

(b) A semiconductor with energy bandgap E_g = 1.42 eV exhibits electrical resistivity 2.78×10^{-3} Ω.cm at 25°C. Calculate the resistivity (in Ω.cm) at 75°C.

Hint: The conductivity is proportional to exp $[-E_g/(2kT)]$, where T is the temperature in K and k is the Boltzmann constant (8.6173×10^{-5} eV/K).

Hint: 25°C corresponds to $(25 + 273)$ K.

Solution:

(a) The higher is the temperature, the greater is the fraction of electrons that are excited from the valence band to the conduction band across the energy bandgap.

(b) At $T_1 = (25 + 273)$ K = 298 K, the conductivity is

$$\sigma_1 = Ae^{-E_g/(2kT_1)},$$

where A is a proportionality constant.

At $T_2 = (75 + 273)$ K = 348 K, the conductivity is

$$\sigma_2 = Ae^{-E_g/(2kT_2)}.$$

$$\sigma_2/\sigma_1 = e^{-[(E_g/(2kT_2)) - E_g/(2k\,T_1)]} = e^{-E_g/(2k)\,[(1/\,T_2 - 1/\,T_1)]}$$

$$= \exp\,[-1.42/(2 \times 8.6173 \times 10^{-5}) \times (1/348 - 1/298)]$$

$$= e^{3.972}$$

$$= 53.09$$

$$\sigma_2 = \sigma_1 \, (53.09)$$

$$= [1/(2.78 \times 10^{-3} \, \Omega.cm)] \, (53.09)$$

$$= 1.91 \times 10^4 \, \Omega^{-1}.cm^{-1}$$

Resistivity at $T_2 = 1/\sigma_2$

$$= 1/(1.91 \times 10^4 \, \Omega^{-1}.cm^{-1})$$

$$= 5.24 \times 10^{-5} \, \Omega.cm$$

Problem 29 (Semiconductor)

Gallium nitride (GaN) is a compound semiconductor with a direct band gap of 3.4 eV. What is the wavelength of the light that an LED made of GaP provides?

Planck's constant $h = 4.1357 \times 10^{-15}$ eV.s.

Speed of light $c = 3.008 \times 10^8$ m/s

Solution:

$$\text{Band gap} = \text{photon energy} = h f,$$

where f = frequency.

$$f = \text{photon energy}/h = \text{bandgap}/h$$

$$= (3.4 \text{ eV}) / (4.1357 \times 10^{-15} \text{ eV s})$$

$$= 8.22 \times 10^{14} \, s^{-1}$$

$$f = c/\lambda,$$

where λ = wavelength.

$$\lambda = c/f$$

$$= (3.008 \times 10^8 \text{ m/s})/(8.22 \times 10^{14} \text{ s}^{-1})$$

$$= 0.366 \text{ } \mu\text{m}$$

$$= 366 \text{ nm}$$

This is close to violet, which has wavelength ranging from 380–450 nm.

Problem 30 (Wurtzite structure)

The compound semiconductor gallium nitride (GaN) exhibits the Wurtzite structure (Fig. 2.29).

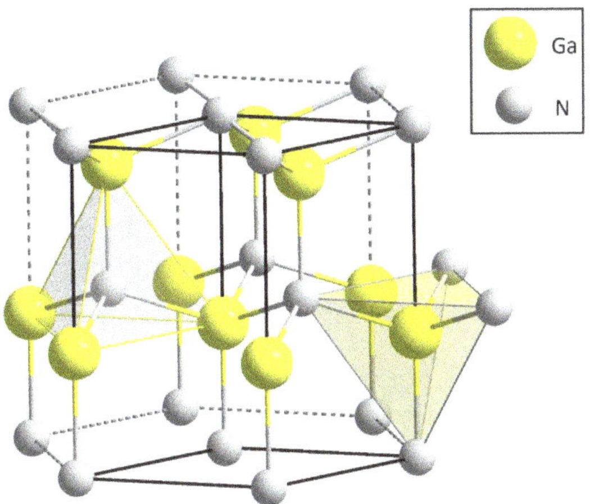

Fig. 2.29 The Wurtzite structure exhibited by GaN. The Ga atom is larger than the N atom.

(a) How many Ga atoms are there per unit cell (based on a hexagon)?

(b) Sketch the primitive unit cell of the Wurtzite structure and indicate the positions of all the atoms, with distinction between the two types of atoms.

Solution:

(a) 6

(b)

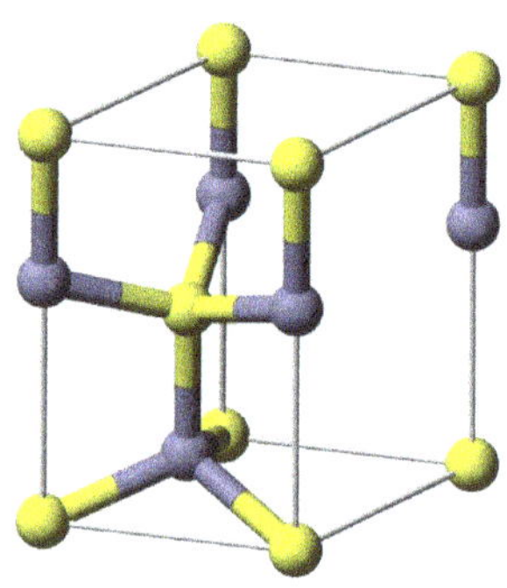

Problem 31 (Thermoelectric behavior)

(a) A thermocouple is made of material A and material B that are electrically connected to make a junction. The Seebeck coefficient of material A is +4.71 µV/K. The Seebeck coefficient of material B is +0.13 µV/K. What is the thermocouple voltage corresponding to a hot-end temperature of 578°C? Assume that the cold end is at room temperature (22°C).

(b) A thermoelectric energy generator consists of consists of 3 pieces of material A and 2 pieces of material B that are connected in series, as shown in Fig. 3. The Seebeck coefficients of material A and material B are +357 µV/K and −476 µV/K, respectively. Calculate the voltage provided by this generator when the temperature difference between the hot side and cold side is 389°C.

Hint: Consider Fig. 2.20.

Solution:

(a) Voltage = $(4.71 - 0.13)$ μV/K $(578 - 22)$ K

$$= 2547 \ \mu V$$

$$= 2.55 \ mV$$

(b) Voltage = $\{[3 \ (357) + 2 \ (476)] \ \mu V/K\} \ (389 \ K)$

$$= 787000 \ \mu V$$

$$= 0.787 \ V$$

Problem 32 (Effect of strain on the electrical resistance)

Derive the following equation, which relates to piezoresistivity:

$$\delta R/R = \delta \rho/\rho + (\delta \ell/\ell)(1 + 2v),$$

where R is the electrical resistance, ρ is the volume electrical resistivity, l is the length along the stress direction and v is the Poisson's ratio.

Hint: Lateral strain $= -v$ (axial strain)

Solution:

$$R = \rho l/A$$

Differentiation gives

$$\delta R/R = \delta \rho/\rho + (\delta \ell/\ell) - (\delta A/A)$$

$$A = w_1 \ w_2$$

$$\delta A/A = \delta w_1/w_1 + \delta w_2/w_2$$

If isotropic, $\delta w_1/w_1 = \delta w_2/w_2 = -v \ \delta \ell/\ell.$

Thus,

$$\delta A/A = -2v\ \delta\ell/\ell.$$

$$\delta R/R = \delta\rho/\rho + (\delta\ell/\ell) + 2v\ \delta\ell/\ell$$

$$= \delta\rho/\rho + (1 + 2v)\ \delta\ell/\ell$$

Problem 33 (Piezoresistivity)

The electrical resistance of a piezoresistive material decreases by 3.76% when a compressive strain of 1.22% is applied. What is the value of the gage factor?

Hint: The gage factor is the fractional change in resistance per unit strain.

Solution:

$$\text{Gage factor} = (-0.0376)/(-0.0122)$$

$$= 3.1$$

Problem 34 (Piezoresistivity)

A material exhibits piezoresistivity. The strain is 0.67%. The Poisson's ratio v is 0.31. The fractional increase in resistivity is 1.68%? What is the gage factor?

Hint: Gage factor = fractional change in resistance per unit strain
Fractional change in resistance
= (fractional change in resistivity) + [strain $(1 + 2v)$]

Solution:

Fractional change in resistance = 1.68% + [0.67% $(1 + 2\ (0.31))$]

$$= 0.0277$$

$$\text{Gage factor} = 0.0277/0.0067$$

$$= 4.1$$

Problem 35 (Piezoresistivity)

An isotropic piezoresistive material exhibits tensile modulus 247 GPa, Poisson's ratio v 0.347, resistivity (under zero stress) 3.46×10^{-3} $\Omega.cm$, and gage factor 258.

Hint: Gage factor = fractional change in resistance per unit strain

$$\delta R/R = \delta\rho/\rho + (\delta l/l)(1 + 2v), \text{ if isotropic.}$$

(a) Calculate the strain under a tensile stress of 22.87 MPa.

(b) Calculate the fractional change in resistivity per unit strain.

(c) Calculate the fractional change in resistance under a tensile stress of 22.87 MPa.

(d) Calculate the resistivity under a tensile stress of 22.87 MPa.

(e) In case that the material is not piezoresistive, calculate the gage factor.

Solution:

(a) Strain = stress/modulus = (22.87 MPa)/(247 GPa)

$$= 9.26 \times 10^{-5}$$

(b)
$$\delta R/R = \delta\rho/\rho + (\delta\ell/\ell)(1 + 2v)$$

$$\text{Gage factor} = [(\delta R/R)/(\delta\ell/\ell)]$$

$$= [(\delta\rho/\rho)/(\delta\ell/\ell)] + (1 + 2v)$$

Fractional change in resistivity per unit strain

$$= (\delta\rho/\rho)/(\delta\ell/\ell)$$

$$= \text{gage factor} - (1 + 2v)$$

$$= 258 - 1 - 2\,(0.347)$$

$$= 256$$

(c) Fractional change in resistance $= \delta R/R$

$$= \delta\rho/\rho + (\delta\ell/\ell)(1 + 2v)$$

$$= 256 + (9.26 \times 10^{-5})\,[1 + 2\,(0.347)]$$

$$= 256$$

(d) Fractional change in resistivity

$$= (\text{Fractional change in resistivity per unit strain})\,(\text{strain})$$

$$= 256\,(9.26 \times 10^{-5})$$

$$= 2.37 \times 10^{-2}$$

Resistivity $= (3.46 \times 10^{-3}\ \Omega.\text{cm})\,(1 + 2.37 \times 10^{-2})$

$$= (3.46 \times 10^{-3}\ \Omega.\text{cm})\,(1.0237)$$

$$= 3.54 \times 10^{-3}\ \Omega.\text{cm}$$

(e) $\delta\rho/\rho = 0$ (not piezoresistive)

Gage factor $= [(\delta R/R)/(\delta\ell/\ell)]$

$$= [(\delta\rho/\rho)/(\delta\ell/\ell)] + (1 + 2v)$$

$$= 1 + 2v$$

$$= 1 + 2\,(0.347)$$

$$= 1.69$$

Problem 36 (Resistive effect in the absence of piezoresistivity)

Derive an expression for the gage factor of a conductive material that does not exhibit piezoresistivity. The gage factor is defined as the fractional change in resistance per unit strain. The expression should be in terms of the Poisson's ratio, which should be assumed to be isotropic.

Solution:

$$R = \rho \frac{l}{A}$$

$$\frac{\Delta R}{R} = \frac{\Delta \rho}{\rho} + \frac{\Delta l}{l} - \frac{\Delta A}{A}$$

$$\frac{\Delta \rho}{\rho} = 0 \,(\text{no piezoresistivity})$$

$$\frac{\Delta R}{R} = \frac{\Delta l}{l} - \frac{\Delta A}{A}$$

$$= \frac{\Delta l}{l} - \frac{\Delta w_1}{w_1} - \frac{\Delta w_2}{w_2}$$

$$= \frac{\Delta l}{l} - \left(-v\frac{\Delta l}{l}\right) - \left(-v\frac{\Delta l}{l}\right)$$

$$= \frac{\Delta l}{l}(1 + 2v)$$

References

1. Chung DDL. Self-sensing concrete: from resistance-based sensing to capacitance-based sensing (Invited Paper). Int J Smart Nano Mater 2021;12(1):1-19.

2. Chung DDL. A critical review of piezoresistivity and its application in electrical-resistance-based strain sensing. J Mater Sci 2020; 55(32):15367-15396.

3. Chung DDL. Thermoelectric polymer-matrix structural and nonstructural composite materials. Adv Ind Eng Polymer Res 2018;1:61-65.

4. Chung DDL. Processing-structure-property relationships of continuous carbon fiber polymer-matrix composites. Mater Sci Eng R 2017; 113:1-29.

5. Snyder GJ, Toberer EC. Complex thermoelectric materials. Nature Mater 2008;7:105-114.

6. Tritt TM, Boettner H, Chen L. Thermoelectrics: direct solar thermal energy conversion. MRS Bull 2008;33:366-368.

7. Poudel B, Hao Q, Ma Y, Lan Y, Minnich A, Yu B, Yan X, Wang D, Muto A, Yashaee D, Chen X, Liu J, Dresselhaus MS, Chen G, Ren Z. High-thermoelectric performance of nanostructured bismuth antimony telluride bulk alloys. Science 2008;320:634-638.

8. Venkatasubramanian R, Siivola E, Colpitts T, O'Quinn B. Thin-film thermoelectric devices with high room-temperature figures of merit. Nature. 2001;413:597-602.

9. Ohta H, Kim S, Mune Y, Mizoguchi T, Nomura K, Ohta S, Nomura T, Nakanishi Y, Ikuhara Y, Hirano M, Hosono H, Koumoto K. Giant thermoelectric Seebeck coefficient of a two-dimensional electron gas in $SrTiO_3$. Nature Mater 2007;6:129-134.

Chapter 3

Dielectric Behavior

3.1 Introduction to the dielectric behavior

The dielectric behavior of a material relates to the material's response to an electric field (voltage gradient) inside the material, such that the response involves the separation of the positive charge center and negative charge center in the material, i.e., the formation of an electric dipole. A dielectric material is a material that exhibits dielectric behavior, regardless of the level of electrical conductivity in the material. This broad definition is in contrast to a narrower but traditional definition that limits these materials to those that are essentially nonconductive. In this book, the noun "dielectric" does not refer to an insulator. Rather, a conductive dielectric refers to a conductive material that exhibits dielectric behavior, whereas a nonconductive dielectric refers to a nonconductive material that exhibits dielectric behavior.

The electric field interacts electrostatically with some of the mobile charged species (e.g., electrons and ions) in the material. This interaction can cause the mobile charged species to move. This long-range movement is called drift, which results in an electric current (whether an electronic current or an ionic current) and is the main phenomenon in the electrical conduction behavior (Chapter 2). In case that both positive and negative mobile charges are present, these charges move in opposite directions in response to the same electric field, with both types of charge contributing to the overall current in an additive fashion.

In the dielectric behavior, the interaction of the electric field with the charged species causes asymmetry in the spatial distribution of the charged species, thereby resulting in the separation of positive and negative charge centers. This separation is known as electric polarization. In contrast, the conduction behavior (Chapter 2) does not involve this asymmetry.

In case of an ionic material (a ceramic material), the asymmetry behind the polarization may involve the asymmetry in the arrangement of the positive and negative ions in the crystal structure, so that the centers of positive and negative charges in the unit cell do not overlap; it may alternatively be due to the asymmetry in the ion distribution in the

181

material, as resulting from the difference in the concentration and/or mobility of the mobile ions of opposite charge. In case of a polymer, the asymmetry may involve asymmetry in the collection of electric dipoles associated with the polar bonds present in each molecule of the polymer. In case of an electronic conductor, the asymmetry involves the engagement of some of the mobile electrons with the atoms. Electron-atom interaction is required for a conductive dielectric, but it is not required for a conventional conductor or a nonconductive dielectric (Fig. 3.1).

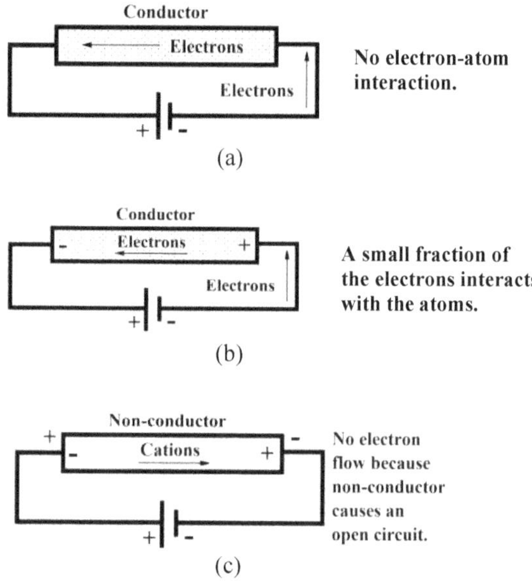

Fig. 3.1 Three types of electrical behavior. (a) Conventional conductor with no polarization due to the absence of electron-atom interaction. (b) Conductive dielectric with polarization, due to the interaction of a small fraction of the electrons with the atoms. (c) Nonconductive dielectric with polarization. Electron-atom (carrier-atom) interaction is required for (b), but not for (a) or (c).

If the material is nonconductive (e.g., typical ceramics and polymers), the electric field in the material can be high. In contrast, if the material is conductive (e.g., typical metals and carbons), the electric field in the material is low, since the electric potential (voltage) is almost constant throughout the material. On the other hand, the concentration of mobile charges is much higher in a conductive material than a nonconductive material. Thus, even if only a small fraction of the mobile charges in a conductive material engages with the atoms, the number of mobile charges that are engaged can be substantial.

Because of the high electric field that can be present in a nonconductive material, attention on the dielectric behavior of materials has long been dominated by the nonconductive materials. Obviously, dielectric materials that are nonconductive are important for electrical insulation, which is needed for electronics and electric cables.

An electric dipole is associated with an electric field. This electric field is inside the material, which is to be distinguished from the electric field applied from outside the material. The dipole also gives rise to a capacitance, which is an electrical quantity that is proportional to the magnitude of charge at each charge center. Usually, this magnitude is proportional to the voltage imposed to drive the charge separation. Thus, the capacitance is the charge magnitude divided by the voltage.

A capacitor is a device that provides a capacitance. In case that the device is a structure (e.g., an airframe), the capacitor is known as a structural capacitor [1]. Due to the Coulombic attraction between unlike charges, the separation of unlike charges costs energy. Hence, the separation of the charge centers embodies electrical energy and a capacitor is a device that can store electrical energy. The greater is the magnitude of charge at each charge center, the more is the amount of electrical energy stored.

Upon short circuiting the two charge centers, the capacitor discharges, with the magnitude of charge at each center decreasing. Upon subsequent application of a voltage (electric field) of a polarity that would cause the separation of the two charge centers, the capacitor charges, with the magnitude of charge at each center increasing. During discharge, a DC current flows, with the electrons flowing from the negative charge center to the positive charge center through an outer circuit that connects the two charge centers until the charge is zero at either center. In other words, the DC current is a pulse. Thus, electrical energy is provided during discharge. On the other hand, during charge, electrical energy is input to the capacitor. The charged state is not thermodynamically stable, since the charged state is at a higher potential energy than the discharged state. Thus, discharge occurs spontaneously once the two charge centers are short circuited.

If the voltage that drives the polarization is AC rather than DC, the polarization changes polarity as the voltage changes polarity. This means that discharge and charge alternately occur as the voltage changes polarity. The current flows in the outer circuit in opposite directions during discharge and charge, thus resulting in an AC current. However, when the AC frequency is too high, the polarization may not be able to change its polarity fast enough, thus resulting in the weakening of this AC phenomenon.

A nonconductive dielectric primarily behaves as a capacitor, whereas a conductive dielectric primarily behaves as a resistor (Fig. 3.2). Under DC condition, a nonconductive dielectric allows a current pulse to flow in the outer circuit, whereas a conductive dielectric allows a continuous current to flow in the material.

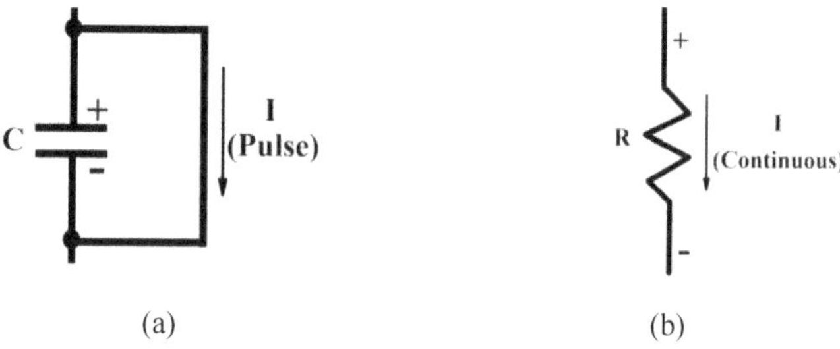

(a)   (b)

Fig. 3.2 Electrical models of (a) a nonconductive dielectric, and (b) a conductive dielectric. In both (a) and (b), I refers to a current.

It is possible for a material to have inherent polarization in the absence of an applied electric field. This occurs when the arrangement of the charged species in the material is asymmetric, as in the case of a single crystal with a crystal structure that does not have a center of symmetry in its unit cell.

Whether the polarization is inherent in the material or induced by an applied electric field, it can be valuable for numerous applications. These applications include the sensing of strain, temperature or damage by measuring the capacitance, electrical energy storage (capacitors), conversion between electrical energy and mechanical energy (e.g., the piezoelectric effect), and conversion between thermal energy and electrical energy (e.g., the pyroelectric effect).

The direct piezoelectric effect is associated with the conversion of mechanical energy to electrical energy, whereas the converse (or reverse) piezoelectric effect is associated with the conversion of electrical energy to mechanical energy. The direct effect is useful for strain/stress sensing, as the electricity generated serves as the output of the sensor. In addition, the direct effect is useful for the generation of electricity without pollution, in contrast to the generation of electricity from coal and the associated pollution. The converse effect is useful for actuation, i.e., the use of an electrical energy input to provide stress or strain, as needed for motors, micromanipulators and movements in general.

3.2 Electric permittivity

A traditional capacitor, known as a parallel-plate capacitor, is in the form of two parallel conductor plates sandwiching a dielectric material (i.e., an electrical insulator). Figure 3.3 illustrates a hypothetical situation in which the dielectric material is simply air (or vacuum). The separation between the plates is l. The area of each plate is A. The two plates are connected to a voltage supply (voltage V), which causes one plate to be positively charged (with charge $+Q$), while the other plate is negatively charged (with charge $-Q$). The magnitude of charge on each plate is Q. The magnitude of charge per unit area is abbreviated D_o (the subscript indicating that this is for the case of just air between the two plates)

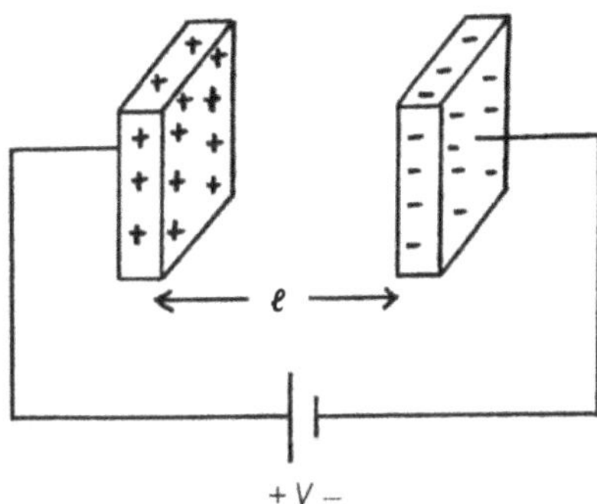

Fig. 3.3 A parallel-plate dielectric capacitor in the form of air separated by two conductor plates that are connected to a voltage supply, which makes one plate positively charged, while the other plate is negatively charged.

which is equal to Q/A. The quantities Q and V are linearly related, as illustrated in Fig. 3.4. The slope of the plot of Q vs. V is the capacitance C_o (the subscript indicating that this is for the case of just air between the two plates), which thus has the unit Coulomb/Volt (the same as Farad, abbreviated F).

$$C_o = \varepsilon_o \, A/l, \qquad (3.1)$$

where ε_o is the permittivity of free space (a universal constant equal to 8.85×10^{-12} F/m). In other words, the capacitance is proportional to A and is inversely proportional to l.

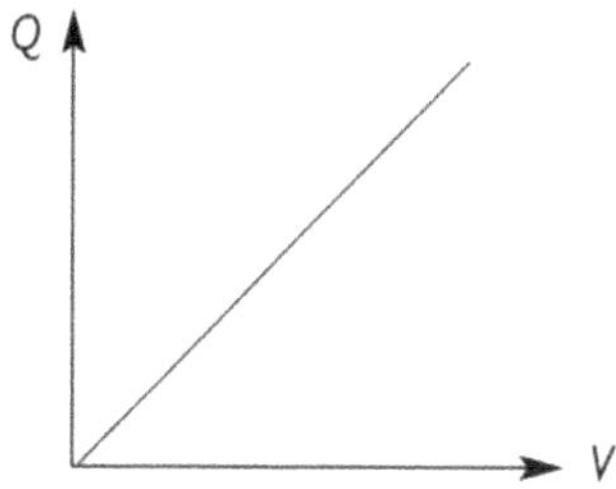

Fig. 3.4 The magnitude (Q) of electrical charge on each plate of the capacitor of Fig. 3.3 versus the applied voltage (V) between the two plates. The curve is linear through the origin, with the slope equal to the capacitance (C).

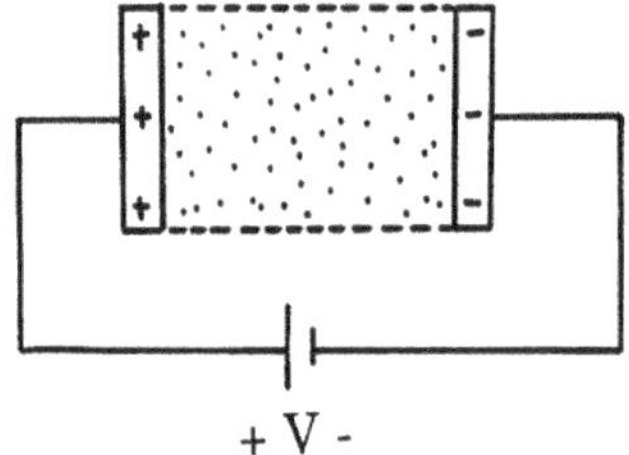

Fig. 3.5 A parallel-plate capacitor like that in Fig. 3.3, except that the air between the two plates is replaced by a dielectric material (i.e., an electrical insulator).

When the air is replaced by a dielectric material with relative permittivity κ, as illustrated in Fig. 3.5, the magntitude of charge on each plate is κQ instead of Q and, as a consequence, the capacitance C is given by

$$C = (\kappa Q)/V = \kappa\, C_o = \varepsilon_o\, \kappa\, A/l. \tag{3.2}$$

The relative permittivity κ is 1 for vacuum (by definition), 1.00054 for air, and is greater than 1 for any material. The permittivity of a material depends on the microstructure [2] and the temperature [3].

The permittivity (dielectric constant) ε is defined as

$$\varepsilon = \varepsilon_o\, \kappa. \tag{3.3}$$

Hence, κ is dimensionless, whereas ε has the same unit as ε_o. Since κ exceeds 1 for a dielectric material, C is greater than C_o. The higher is κ, the greater is C. Therefore, for capacitors, a dielectric material with a very high value of κ is preferred. On the other hand, for the dielectric material between conductor lines in an electric circuit, a low value of κ is preferred, so as to reduce the signal propagation delay associated with the capacitor that stems from the presence of a dielectric material between two conductors (a geometry that is akin to that of a parallel-plate capacitor).

The separation of the positive and negative charge centers of a material due to an applied electric field is known as polarization. The greater is the tendency for polarization, the higher is κ. Polarization can involve a variety of mechanisms, which include (i) the skewing of an electron cloud toward the positive end of the applied electric field, (ii) the shifting of ions (e.g, positive ions shifting toward the negative end of the electric field), (iii) the rotation of molecular segments that contain charges in the form of functional groups, such that the rotation brings the negative charge in the segment closer to the positive end of the electric field, (iv) the reorientation of polarized molecules, such that the reoriented molecule has its negative end closer to the positive end of the electric field, and (v) the gathering of ions of a certain charge at an interface (e.g., the interface between two conductive fibers that almost touch one another) that has a relatively high local electric field.

Table 3.1 shows the values of κ for various materials, including polymers and ceramics. Polymers tend to have lower κ values than ceramics. Teflon has a very low value of κ (2.1). The value is lower than those of polyethylene (2.25) and polystyrene (2.4–2.7), because of teflon's low degree of branching and the symmetry in the mer structure of teflon. Assymetry promotes polarization. Branching provides molecular segments that also promote polarization. Polyethylene tends to have a high degree of branching, although its mer structure is symmetric. Polystyrene has a low degree of branching, but its mer structure is asymmetric. Rubber has a much higher value of κ (7), due to the easy movement of the molecules in rubber (an elastomer).

Table 3.1 Relative permittivity κ (at 1 kHz and room temperature) and the dielectric strength for various materials.

Material	κ	Dielectric strength (MV/m)
Vacuum	1 (by definition)	/
Air	1.00054	3
Teflon	2.1	60
Polyethylene	2.25	25
Polystyrene	2.4–2.7	24
Fused silica	3.91	25–40
Concrete	4.5	/
Cordierite	4.1–5.3	2.32
Pyrex glass	4.7	14
Beryllia BeO (99.5%)	6.7	10–14
Rubber	7	12
Diamond	5.5–10	1,000
Alumina (99.5%)	9.8	16.7
Graphite	10–15	/
Silicon	11.68	/
Water	80.1	/
Titanium dioxide	86–173	/
Strontium titanate ($SrTiO_3$)	310	8

Silicon dioxide (silica) has a rather high value of κ (3.9), due to the ionic character of this solid. Pyrex glass (borosilicate) has a higher value of κ (4.7) than silicon dioxide, due to the availability of more mobile ions than silicon dioxide.

Silicon and diamond have the same crystal structure, but silicon has a higher value of κ (11.68) than diamond (5.5–10). This is because of the small energy band gap of silicon compared to diamond and the consequent greater ease to skew electron clouds. Graphite also has a high value of κ (10–15), due to the ease of skewing the delocalized π electrons, which are not engaged in any covalent bond.

Water has a high κ value (80.1), because of the mobility of the ions (e.g., H^+ ions) in water. Titanium dioxide has a high κ value (86–173), due to the involvement of both ions and electrons. Strontium titanate has a very high κ value (310), because it is ferroelectric (Sec. 3.7.4).

Example Problems 1–6 at the end of the chapter are relevant.

3.3 Calculation of the electric permittivity of a composite

3.3.1 *Parallel configuration*

Consider a capacitor that has two dielectric components in a parallel configuration, as illustrated in Fig. 3.6(a). Component 1 has relative

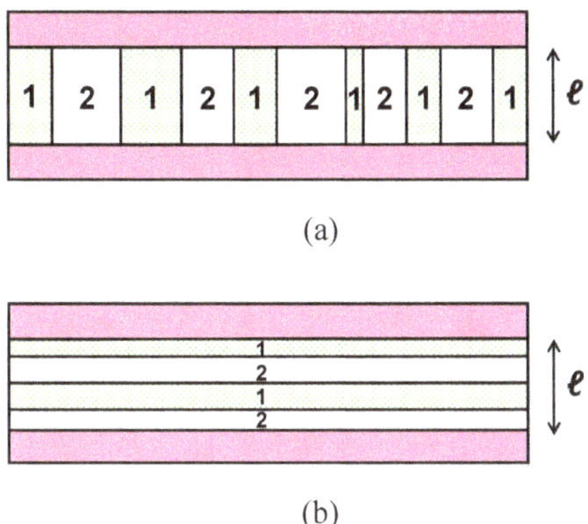

(a)

(b)

Fig. 3.6 A composite material as the dielectric material in a parallel-plate capacitor. The two electrode plates are the pink regions. The composite consists of Component 1 (green regions) and Component 2 (white regions). (a) Components 1 and 2 in parallel. (b) Components 1 and 2 in series.

permittivity κ_1, whereas Component 2 has relative permittivity κ_2. The area is A_1 and A_2 for Components 1 and 2 respectively. For this geometry, the volume fraction is the same as the area fraction of the same component. Hence, the volume fraction of Component 1 is $v_1 = A_1/A$, and the volume fraction of Component 2 is $v_2 = A_2/A$, where A is the total area.

The charge Q is given by

$$Q = Q_1 + Q_2, \tag{3.4}$$

where Q_1 is the charge due to Component 1 and Q_2 is the charge due to Component 2. Since $Q = CV$, $Q_1 = C_1V$ and $Q_2 = C_2V$ (due to the fact that both components are subjected to the same voltage V), Eq. (3.4) gives

$$CV = C_1V + C_2V. \tag{3.5}$$

Hence,

$$C = C_1 + C_2. \tag{3.6}$$

Therefore, with l being the same for the two components, Eq. (3.6) becomes

$$C = (\kappa_1 \, \varepsilon_o \, A_1/l) + (\kappa_2 \, \varepsilon_o \, A_2/l) \tag{3.7}$$

However, C is related to the relative permittivity κ of the composite by

$$C = \kappa \, \varepsilon_o \, A/l. \tag{3.8}$$

From Eq. (3.7) and (3.8),

$$\kappa \, \varepsilon_o \, A/l = (\kappa_1 \, \varepsilon_o \, A_1/l) + (\kappa_2 \, \varepsilon_o \, A_2/l). \tag{3.9}$$

Hence,

$$\kappa A = \kappa_1 A_1 + \kappa_2 A_2, \tag{3.10}$$

or

$$\kappa = (\kappa_1 A_1/A) + (\kappa_2 A_2/A) \tag{3.11}$$

or

$$\kappa = v_1 \kappa_1 + v_2 \kappa_2. \tag{3.12}$$

Equation (3.12) means that, for the parallel configuration, the relative permittivity of the composite is the weighted average of the values of the two components, such that the weighting factors are the volume fractions of the two components. This equation is a manifestation of the Rule of Mixtures (ROM).

3.3.2 Series configuration

Consider a capacitor that has two dielectric components in a series configuration, as illustrated in Fig. 3.6(b). Component 1 has relative permittivity κ_1, whereas Component 2 has relative permittivity κ_2. The area of both components are the same, namely A. The thickness of Component 1 (i.e., all the layers of Component 1 together) is l_1; the thickness of Component 2 (i.e., all the layers of Component 2 together) is l_2. For this geometry, the volume fraction is the same as the thickness fraction of the same component. Hence, the volume fraction of Component 1 is $v_1 = l_1/l$ and the volume fraction of Component 2 is $v_2 = l_2/l$, where l is the total thickness.

The voltage V is shared by the two components, so

$$V = V_1 + V_2, \tag{3.13}$$

where V_1 is the voltage share of Component 1 and V_2 is the voltage share of Component 2. Since $V = Q/C$, $V_1 = Q/C_1$ and $V_2 = Q/C_2$ (due to the fact that Q is the same for the two components), Eq. (3.13) becomes

$$Q/C = Q/C_1 + Q/C_2. \tag{3.14}$$

Hence

$$1/C = 1/C_1 + 1/C_2. \tag{3.15}$$

Therefore,

$$1/(\kappa\,\varepsilon_o A/l) = 1/(\kappa_1\,\varepsilon_o A/l_1) + 1/(\kappa_2\,\varepsilon_o A/l_2), \qquad (3.16)$$

or

$$l/\kappa = l_1/\kappa_1 + l_2/\kappa_2, \qquad (3.17)$$

or

$$1/\kappa = (l_1/l)\,(1/\kappa_1) + (l_2/l)\,(1/\kappa_2), \qquad (3.18)$$

or

$$1/\kappa = v_1/\kappa_1 + v_2/\kappa_2. \qquad (3.19)$$

Equation (3.19) means that, for the series configuration, $1/\kappa$ is the weighted average of $1/\kappa_1$ and $1/\kappa_2$, such that the weighting factors are the volume fractions of the two components. This equation is also a manifestation of the Rule of Mixtures (ROM).

For a dielectric composite that consists of two dielectric components that are in neither the series configuration nor the parallel configuration, as in the case of Component 2 in the form of particles in Component 1, which is the matrix, the κ value of the composite may be estimated using the equation for the parallel configuration, i.e., Eq. (3.12). Since air has a low value of 1 for the relative permittivity, by using air, say, in the form of bubbles, as Component 2, a composite with a low value of the relative permittivity can be obtained. The air bubbles may be provided by hollow glass spheres, which act as a filler in the composite.

Example Problems 7–13 at the end of the chapter are relevant.

3.4 Origin of the dielectric behavior

Table 3.1 lists the values of κ for various dielectric materials, including polymers and ceramics. At a given frequency, κ is higher for polyvinyl chloride (PVC) than polyethylene (PE). This is because of the presence of a polar C-Cl bond in each mer (repeating unit) of PVC. That this bond is polar (partially ionic in character, although it is a covalent bond) is due to the strong electronegativity of Cl. In contrast, PE does not have any electronegative atom, as it only has C and H atoms. The presence of polar bonds in PVC is the cause of the high κ of PVC, as the polar bonds interact with the applied electric field. Alumina (Al_2O_3) is an ionic solid, with Al^{3+} cations and O^{2-} anions, which, due to their charge, interact strongly with the applied electric field, thus resulting in an even higher κ for alumina compared to PVC. In general, ceramics tend to have higher values of κ than polymers, due to the ionic character in ceramics.

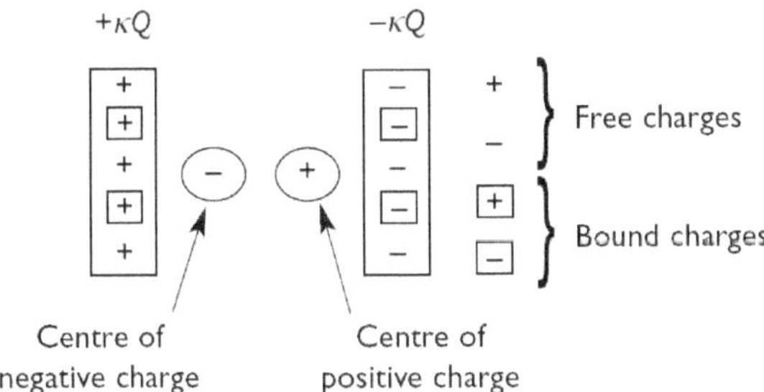

Fig. 3.7 The magnitude of charge on each plate of the capacitor of Fig. 3.3 is κQ, which consists of the free charges present in case of only air between the plates (case of Fig. 3.1) and the bound charges that are present due to the insulator between the plates. The centers of negative and positive charges are within the dielectric material between the two plates.

In the presence of the dielectric material between the two plates, the magnitude of charge on each plate (κQ) is greater than that when the dielectric material is absent (Q). This is because the dielectric material responds to the applied voltage by having its center of negative charge displaced toward the positive plate and its center of positive charge displaced toward the negative plate. This charge displacement can involve various mechanisms, such as ion movement (in case that the dielectric material contains ions) and alignment of polarized molecules or polarized molecular segments with the applied electric field in the direction perpendicular to the plates (in case that the dielectric material contains polarized molecules or polarized molecular segments). As illustrated in Fig. 3.7, the separation of the positive and negative charge centers in the dielectric material induces additional charges on either plate (beyond the amount when the dielectric material is absent). The additional charges are known as bound charges, whereas the charges present in the absence of the dielectric material are known as free charges. Hence,

$$\text{the magnitude of free charges} = Q$$

and

$$\text{the magnitude of bound charges} = \kappa Q - Q = (\kappa - 1)\,Q \qquad (3.20)$$

Hence, the quantity $\kappa - 1$ is of interest and is known as the electric susceptibility χ.

The relative permittivity κ depends on the material. For a given material, it decreases with the frequency, due to the difficulty of certain mechanisms of charge separation to keep up with fast electric field variation. Figure 3.8 illustrates the decrease in κ at each frequency where a certain mechanism starts to fail to catch up. The orientation of polarized bonds tends to start not catching up at rather low frequencies. The skewing of electron clouds is a mechanism that can persist to higher frequencies. When all the mechanisms fail, $\kappa = 1$, which is the value for vacuum and is the value at the highest frequency in Fig. 3.8.

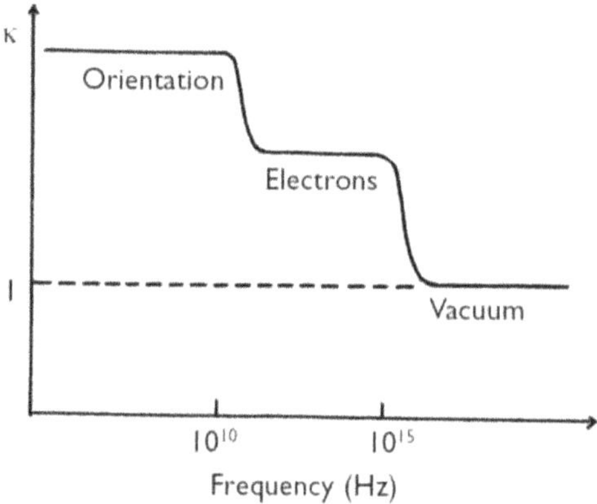

Fig. 3.8 Variation of the relative permittivity κ with the frequency.

An electric dipole refers to a pair of positive and negative charges that are separate. The dipole moment is defined as the magnitude of charge at each of the two points multiplied by the distance between the two points. The polarization (P) in the dielectric material is defined as the dipole moment per unit volume. There is a one-to-one correspondence between the surface charges on the dielectric material (due to the separation between the positive and negative charge centers) and the bound charges on the proximate plate. This is because the interior charges of opposite signs cancel one another, as shown in Fig. 3.9, where there are three unit cells stacked on top of one another. Therefore, the dipole moment in the dielectric material is $(\kappa - 1)\ Ql$ (based on Eq. (3.20)). Thus,

$$P = \text{the dipole moment per unit volume}$$

$$= [(\kappa - 1)\ Ql]/(Al) = (\kappa - 1)\ Q/A \tag{3.21}$$

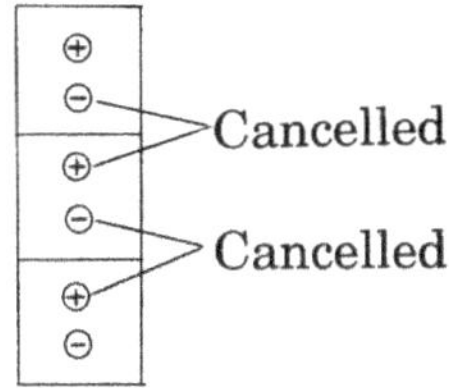

Fig. 3.9 Three crystal structural unit cells of a dielectric material, showing cancellation of interior charges of opposite signs so that the surface charges are responsible for the polarization.

The applied electric field E is V/l, which is the voltage gradient. Rearrangement of Eq. (3.2) gives

$$E = V/l = Q/\varepsilon_o A \tag{3.22}$$

Hence, Eq. (3.21) becomes

$$P = \varepsilon_o\ (\kappa - 1)\ E. \tag{3.23}$$

Equation (3.23) means that the polarization is linearly related to the electric field, provided that κ is independent of the electric field. The slope of the curve is $\varepsilon_o\ (\kappa - 1)$. This curve is known as the polarization curve, as illustrated in Fig. 3.8(a).

Rearrangement of Eq. (3.22) gives

$$Q/A = \varepsilon_o E, \tag{3.24}$$

which means that Q/A is proportional to E, with the proportionality constant being ε_o. Similarly, in case of the presence of the dielectric material,

$$\kappa Q/A = \kappa \varepsilon_o E. \tag{3.25}$$

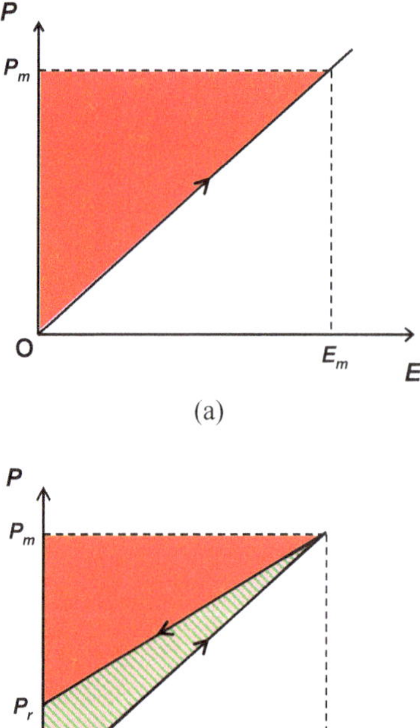

(a)

(b)

Fig. 3.10 Polarization P vs. electric field E. (a) Curve of P vs. E as E is increased from zero to a maximum value E_m (i.e., the polarization curve). The red area is the energy input per unit volume for polarizing to $P = 0$ to $P = P_m$ while E is increased from 0 to E_m. (b) Curve of P vs. E as E is increased from 0 to E_m and subsequently decreased from E_m back to zero. The red area is the energy output per unit volume for depolarizing from the maximum polarization P_m to the remanent polarization P_r. The area between the two curves in (b) is the energy loss per unit volume per cycle of E change.

Hence, the charge per unit area on a plate is given by

$$D = \kappa Q/A = \kappa \varepsilon_o E. \qquad (3.26)$$

An electric dipole is a pair of positive and negative charges that are separated from one another. It is associated with potential energy, which is the electrical energy. Thus, polarization requires energy input. This energy per unit volume for polarizing from $P = 0$ to $P = P_m$ while E is increased from zero to E_m is given by the shaded area in Fig. 3.10(a), i.e.,

$$\text{Energy per unit volume for polarization} = \int E \,\Delta P \qquad (3.27)$$

The unit of E is V/m. The unit of P is $C.m/m^3 = C/m^2$, since the P is dipole moment per unit volume and the unit of dipole moment is C.m. Hence, the unit of $E \,\Delta P$ is (V/m) $(C/m^2) = V.C/m^3 = J/m^3$, which is the unit for energy per unit volume.

After polarization, reduction of E causes decrease of P. This depolarization is associated with a release in energy, due to the decrease in the dipole moment per unit volume. This energy release is responsible for the spontaneous tendency to depolarize once E is removed. The energy release per unit volume is given by the dotted area in Fig. 3.10(b), where the depolarization is only partial, resulting in a nonzero polarization (called remanent polarization P_r) when E is returned to 0. In other words, there is hysteresis. In general, the degree of completeness of the depolarization depends on the microstructure of the material.

The entire process of polarization and subsequent partial depolarization involves an energy input that is larger than the energy output. The difference between the energy input and the energy output is the energy loss. The energy loss per unit volume is given by the area between the two curves in Fig. 3.10(b). The energy lost typically becomes heat. If the two curves in Fig. 3.10(b) overlap, there is no energy loss. The more severe is the hysteresis, the greater is the energy loss. Thus, this energy loss is also called hysteresis energy loss.

A high value of P_m is desirable for a computer memory that uses the remanent polarization in the material to store information after the electric field has been switched off. By using a reverse electric field, polarization is in the reverse direction, causing remanent polarization in the reverse direction. The two opposite directions of remanent polarization allow the storage of digital data as 0 and 1. Thus, for computer memories using this phenomenon, a large hysteresis is desirable.

3.5 Lossy capacitor

In reality, the dielectric material in a capacitor may not be perfect in two ways. In one way, it may not be an ideal insulator. In the other way, it may not be able to have its electric dipole flip quickly enough in response to a high-frequency AC electric field, so that it is unable to respond to the electric field. The sluggishness of the response is also described in Sec. 3.4 in relation to the incomplete depolarization that follows polarization and the consequent hysteresis in the plot of polarization vs. electric field during the increase and subsequent decrease of the applied DC electric field (Fig. 3.10(b)). Both issues result in energy loss (known as dielectric loss), because the input electrical energy is not perfectly received by the material. The energy is usually lost as heat. In other words, there can be two contributions to the energy loss. Due to the loss, the capacitor is said to be lossy.

In case that the loss is due to the material being not an ideal insulator, the loss is known as conduction loss. Under this situation, the capacitor is not associated with a pure capacitance, but is also associated with a finite resistance. The current going through the resistance would cause resistance heating (Sec. 2.7) and hence energy loss as heat.

Two equivalent circuits are commonly used for modeling a lossy capacitor involving conduction loss. One model involves the capacitor and the resistor in parallel. The other model involves the capacitor and the resistor in series.

3.5.1 *Lossy capacitor modeled as a capacitor and a resistor in parallel*

The lossy capacitor is modeled by a capacitor (C) and a resistor (R) in parallel, as shown in Fig. 3.11. If the resistor is infinite in resistance (i.e., the case of an ideal capacitor), the resistor in the circuit will be replaced by an open circuit, which means that the resistor branch of the circuit disappears. Analysis of this circuit is provided below.

Consider that the applied electric field E is a sinusoidal function of time t, with angular frequency ω (in radians per s), which is equal to $2\pi\upsilon$, where υ is the frequency in Hz. Using complex notation,

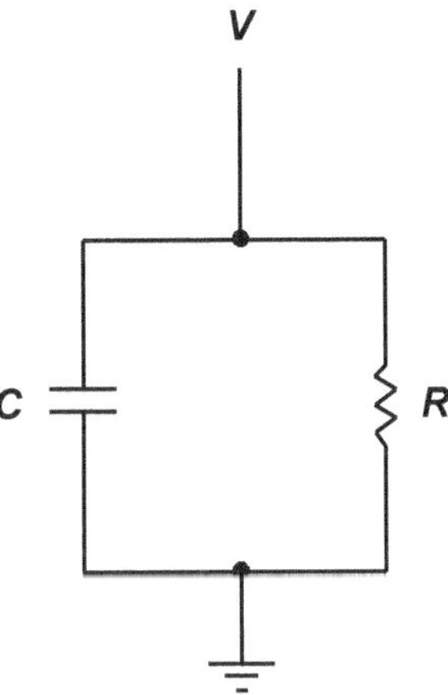

Fig. 3.11 Equivalent circuit model of a lossy capacitor, i.e., a capacitor that has a finite resistance R associated with its dielectric material. This circuit has the capacitance C and the resistance R in parallel. The voltage across either circuit component is V.

$$E = E_o\, e^{i\omega t} = E_o\,(\cos \omega t + i \sin \omega t), \tag{3.28}$$

where E_o is the amplitude of the electric field wave. Due to the presence of R in the circuit, D lags behind E by a phase angle δ (in radians). Hence,

$$D = D_o\, e^{i(\omega t-\delta)} = D_o\,[\cos (\omega t-\delta) + i \sin (\omega t-\delta)], \tag{3.29}$$

where D_o is the amplitude of the D wave. From Eq. (3.26), (3.28) and (3.29),

$$D_o\, e^{i(\omega t-\delta)} = \kappa\varepsilon_o E_o\, e^{i\omega t}. \tag{3.30}$$

Rearrangement of Eq. (3.30) gives

$$\kappa\varepsilon_o = (D_o/E_o)e^{-i\delta} = (D_o/E_o)\,(\cos \delta - i \sin \delta). \tag{3.31}$$

From Eq. (3.31),

$$\tan \delta = (\sin \delta)/(\cos \delta) = -(\text{imaginary part of } \kappa\varepsilon_o)/(\text{real part of } \kappa\varepsilon_o)$$

$$= -(\text{imaginary part of } \kappa)/(\text{real part of } \kappa). \tag{3.32}$$

Equation (3.32) means that $\delta = 0$ when the imaginary part of κ is zero. In other words, the real part of κ describes the behavior of an ideal capacitor; since there is no resistance branch for the equivalent circuit of an ideal capacitor, $\delta = 0$. On the other hand, the imaginary part of κ is due to the presence of the resistance branch. Figure 3.12 shows the complex plane of κ, i.e., the imaginary part of κ versus the real part of κ. It shows that κ lags the real part of κ by the angle δ. This means that κ lags the κ for the case of an ideal capacitor by the angle δ. This is consistent with the lag of D from E by the angle δ, as shown by Eq. (3.29).

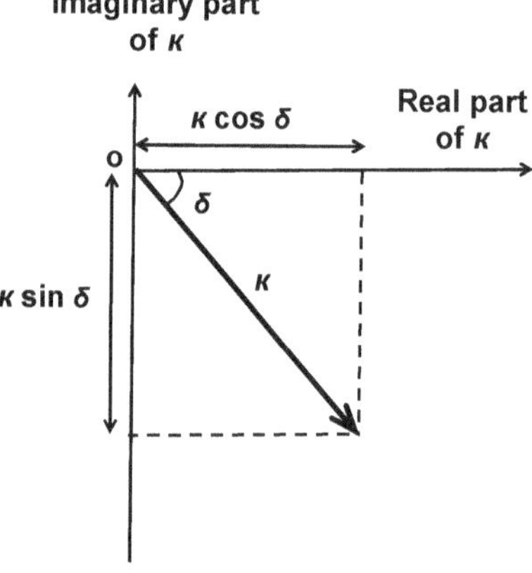

Fig. 3.12 Complex plane of the relative permittivity κ, showing the imaginary part of κ versus the real part of κ.

The current I_C through the capacitance branch of the circuit is given by

$$I_C = \Delta Q/\Delta t, \tag{3.33}$$

since current is defined as charge per unit time. From Fig. 3.4,

$$C = \Delta Q/\Delta V, \tag{3.34}$$

where V is the voltage across the capacitance branch as well as that across the resistance branch. Rearrangement of Eq. (3.34) and differentiation with respect to t gives

$$\Delta Q / \Delta t = C\,(\Delta V / \Delta t). \tag{3.35}$$

Combination of Eq. (3.29) and (3.31) gives

$$I_C = C\,(\Delta V / \Delta t). \tag{3.36}$$

Since E is the voltage gradient, E and V are in phase. Hence, from Eq. (3.28),

$$V = V_o\, e^{i\omega t} = V_o\,(\cos \omega t + i \sin \omega t), \tag{3.37}$$

where V_o is the amplitude of the voltage wave. Combination of Eq. (3.36) and (3.37) gives

$$I_C = C\, V_o\, i\omega\, e^{i\omega t} \tag{3.38}$$

Since i is 90° from the real axis of a complex plane, Eq. (3.38) means that I_C leads V in phase by 90°.

The impedance Z, which is a complex quantity, is defined as

$$Z = V/I. \tag{3.39}$$

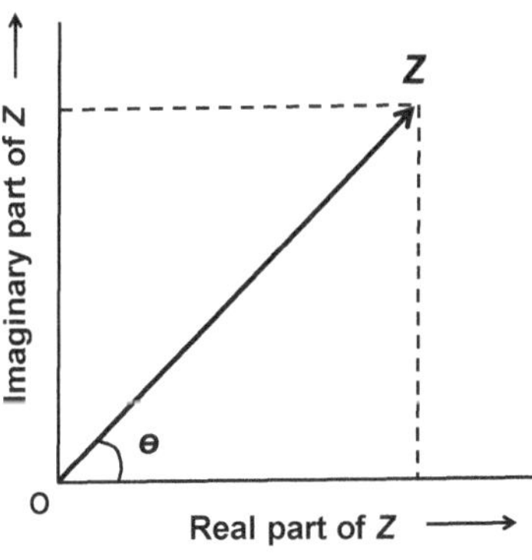

Fig. 3.13 Complex plane of the impedance Z, showing the imaginary part X (the reactance) versus the real part R (the resistance).

The real part of Z is known as the resistance, which is not equal to R in the equivalent circuit of Fig. 3.11, as shown below. The imaginary part of Z is known as the reactance. The complex plane of Z is shown in Fig. 3.13. The magnitude of Z, i.e., $|Z|$, is given by

$$|Z| = \sqrt{[(\text{imaginary part})^2 + (\text{real part})^2]}, \qquad (3.40)$$

where

$$\text{real part of } Z = |Z| \cos \theta \qquad (3.41)$$

and

$$\text{imaginary part of } Z = |Z| \sin \theta, \qquad (3.42)$$

with θ being the angle between Z and the axis for the real part of Z, as defined in Fig. 3.13.

The contribution of the capacitance branch to the impedance, i.e., Z_C, is given by

$$Z_C = V/I_C \qquad (3.43)$$

Using Eq. (3.40), (3.38) and (3.37),

$$Z_C = (V_o \, e^{i\omega t})/(C \, V_o \, i\omega \, e^{i\omega t})$$

$$= 1/(i\omega C)$$

$$= -i/(\omega C) \qquad (3.44)$$

Equation (3.44) means that Z_C is purely a reactance, i.e., $X = -1/(\omega C)$. Since the unit of ω is s^{-1} and the unit of C is Farad = Coulomb/Volt (i.e., C/V), the unit of X is $1/(s^{-1}.C/V) = (V.s)/C = V/A = \Omega$. Note that C/s = Ampere (i.e., A). Thus, the units of X and R are both Ω.

The current I_R through the resistance branch is in phase with the voltage, so

$$I_R = V/R \qquad (3.45)$$

Equation (3.45) means that I_R and V are in phase. Thus, I_C leads I_R in phase by 90°. Using Eq. (3.39) and (3.37), we have

$$I_R = (V_o/R) \, e^{i\omega t} \tag{3.46}$$

The contribution of the resistance branch to the impedance, i.e., Z_R, is given by

$$Z_R = V/I_R \tag{3.47}$$

Using Eq. (3.45), (3.46) and (3.47), we have

$$Z_R = R \tag{3.48}$$

Equation (3.48) means that Z_R is purely a resistance, as expected.
For the equivalent circuit of Fig. 3.9, Z is given by

$$(1/Z) = (1/Z_R) + (1/Z_C), \tag{3.49}$$

since the capacitance and resistance in the circuit are in parallel. Hence, from Eq. (3.48) and (3.44),

$$(1/Z) = (1/R) + i\omega C$$

$$= (1 + i\omega RC)/R \tag{3.50}$$

Upon reciprocating both sides of the equation,

$$Z = R/(1 + i\omega RC) \tag{3.51}$$

Upon multiplying both the numerator and the denominator by $(1 - i\omega RC)$,

$$Z = [R\,(1 - i\omega RC)]/\,[(1 + i\omega RC)\,(1 - i\omega RC)]$$

$$= [R\,(1 - i\omega RC)]/(1 + \omega^2 R^2 C^2)$$

$$= [R/(1 + \omega^2 R^2 C^2)] - [i\omega R^2 C/(1 + \omega^2 R^2 C^2)] \tag{3.52}$$

Hence,

$$\text{real part of } Z = R/(1 + \omega^2 R^2 C^2) \tag{3.53}$$

and

$$\text{imaginary part of } Z = -\omega R^2 C/(1 + \omega^2 R^2 C^2) \tag{3.54}$$

The complex plane of Z corresponding to Eq. (3.53) and (3.54) is shown in Fig. 3.14. It is the same as Fig. 3.11, except that the imaginary part is negative. From Fig. 3.14,

$$\tan \theta = \text{magnitude of the imaginary part of } Z/\text{real part of } Z$$

$$= (\omega R^2 C)/R = \omega RC \tag{3.55}$$

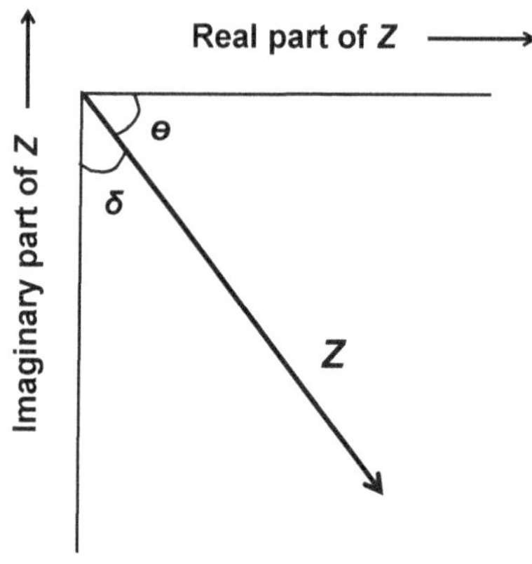

Fig. 3.14 The complex plane of the impedance Z for the equivalent circuit of Fig. 3.11 or 3.17. The imaginary part X (the reactance) is negative, but the real part R (the resistance) is positive.

The angle δ, as defined in Eq. (3.29), is the angle between Z and the imaginary part of Z, since the case of $\delta = 0$ corresponds to the case of $R = \infty$, i.e., the absence of the resistance branch in Fig. 3.11. In other words, when R is absent, Z is purely imaginary. Thus, as shown in Fig. 3.14,

$$\delta = 90° - \theta \tag{3.56}$$

and

$$\tan \delta = \text{real part of } Z / \text{magnitude of the imaginary part of } Z$$

$$= 1/(\omega RC) \tag{3.57}$$

Based on Eq. (3.38) and (3.45), the total current I is given by

$$I = I_C + I_R = C\,V_o\,i\omega\,e^{i\omega t} + (V_o/R)\,e^{i\omega t}$$

$$= V_o\,[i\omega C + (1/R)]\,e^{i\omega t} \tag{3.58}$$

The phasor diagram (phase relationship diagram) of I, as shown in Fig. 3.15, indicates that I_C is ahead of I_R in phase by 90°. In addition, it shows that the total current I is behind I_C in phase by the angle δ, such that $\tan \delta$ is given by

$$\tan \delta = (V_o/R)/(\omega C V_o) = 1/(\omega RC) \tag{3.59}$$

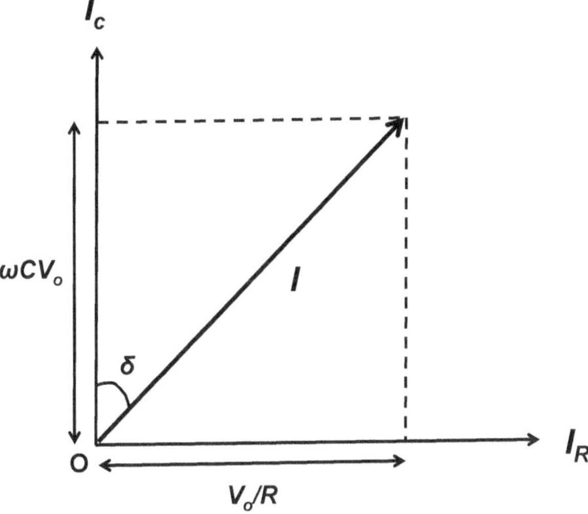

Fig. 3.15 Phasor diagram showing the complex plane of the current I, with the vertical axis being I_c and the horizontal axis being I_R.

Equation (3.59) is consistent with Eq. (3.57). That the current I_c is ahead of I by the angle δ is consistent with Fig. 3.12, since the case of $I = I_c$ in Fig. 3.15 corresponds to (i) the case of κ = the real part of κ in Fig. 3.12, where the real part of κ is ahead of κ by the angle δ, and (ii) the case of $\delta = 0$ in Fig. 3.14.

Equation (3.59) means that δ decreases with increasing ω, C or R. This trend is consistent with the notion that a higher R corresponds to a less lossy capacitor in the equivalent circuit of Fig. 3.11.

The product of current (in Ampere, abbreviated A) and voltage (in Volt, abbreviated V) is power (in Watts, abbreviated W, which equals A.V). Energy (in Watts.second, which equals Joules, abbreviated J) is the integration of power over time. Thus,

the energy stored in the capacitance branch

$$= \int VI_C \, dt = \int V \left[C \left(dV/dt \right) \right] dt = C \int V \, dV = CV^2/2 = (CV_o^2/2) \, e^{i2\omega t},$$

$$(3.60)$$

where Eq. (3.36) and (3.37) have been used. From Eq. (3.60), the maximum energy stored is $CV_o^2/2$, which occurs at $2\omega t = \pi$, or $\omega t = \pi/2$.

The energy loss (to heat) in the resistance branch

$$= \int VI_R \, dt = \int V \, (V/R) \, dt$$

$$= \int (V^2/R) \, dt = (1/R) \int V^2 \, dt = (1/R) \int V_o^2 \, e^{i2\omega t} \, dt$$

$$= (V_o^2/R) \int e^{i2\omega t} \, dt, \qquad (3.61)$$

where Eq. (3.37) and (3.45) have been used. Integration of Eq. (3.61) from $t = 0$ to $t = t$ gives

energy loss in the resistance branch

$$= [V_o^2/(i2\omega R] \, (e^{i2\omega t} - 1)$$

$$= [V_o^2/(i2\omega R] \, (\cos 2\omega t - 1 + i \sin 2\omega t) \qquad (3.62)$$

Taking the real part,

$$\text{energy loss in the resistance branch} = [V_o^2/(2\omega R)]\,(\sin 2\omega t) \quad (3.63)$$

The maximum energy lost in the resistance branch is $V_o^2/(2\omega R)$, which occurs at $2\omega t = \pi/2$, or $\omega t = \pi/4$.

Therefore, the ratio of the maximum energy loss in the resistance branch to the maximum energy stored in the capacitance branch = $[CV_o^2/2]/[V_o^2/(2\omega R)] = 1/(\omega CR) = \tan \delta$, where Eq. (3.59) has been used. Thus, a physical meaning of $\tan \delta$ is this ratio. In other words, $\tan \delta$ (known as the dielectric loss factor) relates to the energy loss.

Table 3.2 lists the values of the $\tan \delta$ for various dielectric materials, including polymers and ceramics. PVC has a higher $\tan \delta$ value than PE, due to the partial ionic character in PVC causing the electrical resistivity to be lower, as shown in Table 3.2. As indicated by Eq. (3.59), which applies to the equivalent circuit of Fig. 3.11, $\tan \delta$ is inversely related to R. Similarly, alumina has a higher $\tan \delta$ value than PE, due to the ionic character of alumina causing the electrical resistivity to be lower. The resistivity of PVC and alumina are comparable, but $\tan \delta$ is higher for PVC than alumina. This is because of the higher κ value of alumina causing higher C. As indicated by Eq. (3.59), $\tan \delta$ is inversely related to C. In general, polymers tend to have higher resistivity than ceramics, due to the greater ionic character in ceramics.

Table 3.2 Relative permittivity κ, dielectric loss factor $\tan \delta$, dielectric strength and DC electrical resistivity of selected dielectric materials. 1 MHz = 10^6 Hz. 1 MV = 10^6 V.

Material	κ (60 Hz)	κ (1 MHz)	$\tan \delta$ (1 MHz)	Resistivity (Ω.cm)	Dielectric strength (MV/m)
Polyethylene	2.3	2.3	0.00010	$>10^{16}$	20
Polyvinyl chloride	3.5	3.2	0.05000	10^{12}	40
Alumina	9.0	6.5	0.00100	10^{11}–10^{13}	6

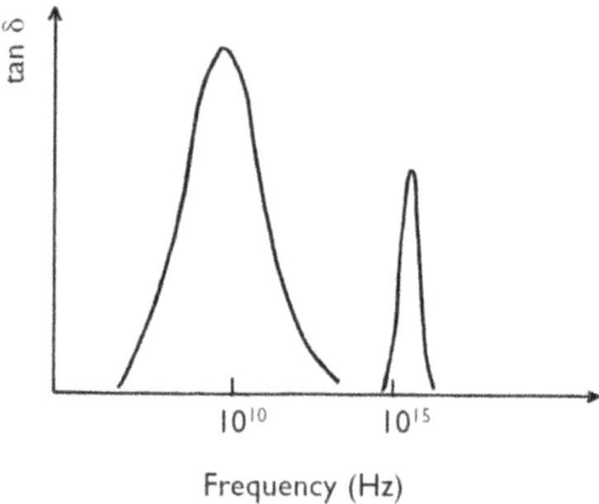

Fig. 3.16 Dependence of tan δ on frequency. The value is highest at frequencies where a certain polarization mechanism starts to not catch up. The peaks correspond to the steps in Fig. 3.8.

As shown in Table 3.2, for the same material, κ tends to decrease with increasing frequency from 60 Hz to 1 MHz. In general, for a given material, the energy loss depends on the frequency. When a certain polarization mechanism starts to not catch up with the fast electric field variation, there is a peak in the energy loss, as illustrated in Fig. 3.16. This is because at such a frequency, a particular polarization mechanism tries to respond to the electric field, but fails. The difficulty of the electric dipoles in responding to the varying applied electric field is known as dipole friction. The peaks in Fig. 3.16 correspond to the steps in Fig. 3.8.

3.5.2 *Lossy capacitor modeled as a capacitor and a resistor in series*

If the alternate equivalent circuit of Fig. 3.17 is used instead of that of Fig. 3.11,

$$Z = Z_R + Z_C, \tag{3.64}$$

since the resistance and capacitance are in series in Fig. 3.17. Using again Eq. (3.48) and (3.44), we have

$$Z = R - i/(\omega C). \tag{3.65}$$

Hence,

$$\text{real part of } Z = R, \tag{3.66}$$

and

$$\text{imaginary part of } Z = -1/(\omega C). \tag{3.67}$$

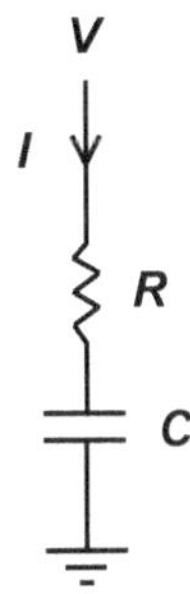

Fig. 3.17 Equivalent circuit model of a lossy capacitor, i.e., a capacitor that has a finite resistance R associated with its dielectric material. This circuit has the capacitance C and the resistance R in series. The voltage across the R and C together is V. The current through R and C is the same.

Both equivalent circuits of Fig. 3.11 and 3.17 are commonly used to model a lossy capacitor. In order to determine the values of C and R from Z, an equivalent circuit needs to be assumed. The values of C and R obtained by assuming the parallel configuration in Fig. 3.11 are known as C_p and R_p respectively. The values of C and R obtained by assuming the series configuration in Fig. 3.17 are known as C_s and R_s respectively.

In case of the equivalent circuit of Fig. 3.17, Eq. (3.65) applies and means that tan θ, as shown in Fig. 3.14, is given by

$$\tan \theta = 1/(\omega RC) \tag{3.68}$$

and that tan δ, as shown in Fig. 3.14, is given by

$$\tan \delta = \omega RC \tag{3.69}$$

When $R = 0$, the resistor in Fig. 3.17 does not exist, so the circuit is a pure capacitor and $\delta = 0$, as also shown by Eq. (3.69). The greater is R, the more lossy is the capacitor and the higher is δ, as also shown by Eq. (3.69).

Note that Eq. (3.69), which applies to the equivalent circuit in Fig. 3.17, is very different from Eq. (3.59), which applies for the equivalent circuit in Fig. 3.11. The equivalent circuit of Fig. 3.11 is usually closer to reality than that of Fig. 3.17. Unless noted otherwise, the equivalent circuit may be assumed to be Fig. 3.11.

To analyze the energy, consider the current I through the entire circuit of Fig. 3.17. The current through the capacitor and the resistor in the circuit is the same. Let

$$I = I_o\, e^{i\omega t}, \tag{3.70}$$

where I_o is the amplitude of the current wave of angular frequency ω. Using Eq. (3.36),

$$I = I_C = C\,(\Delta V_C/\Delta t), \tag{3.71}$$

where V_C is the voltage across the capacitor in the circuit. Using Eq. (3.71) and (3.70),

$$\Delta V_C/\Delta t = I/C = (I_o/C)\, e^{i\omega t} \tag{3.72}$$

Upon integrating both sides of the equation,

$$V_C = \int (I_o/C)\, e^{i\omega t}\, \Delta t$$

$$= (I_o/C)\int e^{i\omega t}\, \Delta t \tag{3.73}$$

Integration from $t = 0$ to $t = t$ gives

$$V_C = [I_o/(i\omega C)]\, e^{i\omega t} \tag{3.74}$$

The voltage V_R across the resistor in the circuit is given by

$$V_R = IR = R\,I_o\, e^{i\omega t}, \tag{3.75}$$

where Eq. (3.70) is used.

The energy stored in the capacitor in the circuit is given by

$$\text{energy stored} = \int IV_C \, \Delta t \tag{3.76}$$

Using Eq. (3.70) and (3.76),

$$\text{energy stored} = \int IV_C \, \Delta t = [I_o{}^2/(i\omega C)] \int e^{i2\omega t} \, \Delta t \tag{3.77}$$

Integration from $t = 0$ to $t = t$ gives

$$\text{energy stored} = [-I_o{}^2/(2\omega^2 C)] \, e^{i2\omega t} \tag{3.78}$$

Taking the maximum,

$$\text{maximum energy stored} = I_o{}^2/(2\omega^2 C), \tag{3.79}$$

which occurs at $2\omega t = \pi$, i.e., $\omega t = \pi/2$, as in the case of the equivalent circuit of Fig. 3.11.

The energy loss in the resistor in the circuit is given by

$$\text{energy loss} = \int IV_R \, \Delta t \tag{3.80}$$

Using Eq. (3.80) and (3.75),

$$\text{energy loss} = R\, I_o{}^2 \int e^{i2\omega t} \, \Delta t \tag{3.81}$$

Integration from $t = 0$ to $t = t$ gives

$$\text{energy loss} = [(R\, I_o{}^2)/(i2\omega)] \, e^{i2\omega t}$$

$$= -i\,[(R\, I_o{}^2)/(2\omega)] \, e^{i2\omega t}$$

$$= -i\,[(R\, I_o{}^2)/(2\omega)] \, (\cos 2\omega t + i \sin 2\omega t) \tag{3.82}$$

Taking the real part,

$$\text{energy loss} = [(R\, I_o{}^2)/(2\omega)] \sin 2\omega t \tag{3.83}$$

Taking the maximum,

$$\text{maximum energy loss} = (R\,I_o^2)/(2\omega). \qquad (3.84)$$

This maximum occurs at $2\omega t = \pi/2$, i.e., $\omega t = \pi/4$, as in the case of the equivalent circuit of Fig. 3.11. From Eq. (3.83) and (3.79),

$$\text{the ratio of the maximum energy loss to the}$$

$$\text{maximum energy stored} = \omega RC, \qquad (3.85)$$

which is equal to tan δ (Eq. (3.69)). This means that a physical meaning of tan δ is this ratio. This meaning of tan δ applies to both equivalent circuits of Fig. 3.11 and 3.17.

Example Problems 14–16 are relevant.

3.6 Dielectric material evaluation

3.6.1 *Measurement of the electric permittivity*

Unless stated otherwise, the term "relative permittivity" refers to the real part of κ. At normal frequencies, which include the utility frequency (60 Hz in the U.S.) and are much below the frequencies in the radio wave, microwave and optical regimes, the measurement of κ of a material involves capacitance measurement. This measurement typically uses a parallel-plate capacitor geometry, in which the specimen is the dielectric material sandwiched by two electrodes (e.g., copper sheets), which are electrically conductive and serve as electrical contacts for imposing an AC electric field in the direction perpendicular to the plane of the parallel-plate capacitor, as illustrated in Fig. 3.18. For high frequencies in the radio wave, microwave or optical regime, the electrical input is the form of electromagnetic radiation and the measurement of κ does not involve measuring the capacitance, but involves measuring the degree of absorption or reflection of the incident radiation by the material (Sec. 4.4). This section pertains to the normal (low) frequency range only.

The capacitance is typically measured by using an LCR meter, which is not designed for measuring the capacitance of a low-resistance material. In case that the material under evaluation is substantially conductive, an insulating film (e.g., double-sided adhesive tape) should be positioned between the specimen and electrode, so that the overall material system has a high enough resistance for the LCR meter to function properly. This is illustrated in Fig. 3.18.

The variables in the test include the electric field magnitude (i.e., the applied AC voltage divided by the distance over which the voltage is applied) and the AC frequency. The appropriate electric field depends on the particular material, as different materials respond to an electric field to different degrees, due to the difference in polarizability. The appropriate frequency also depends on the particular material, as different mechanisms of polarization respond to the electric field of various frequencies with different degrees of sluggishness.

Whether the insulator film is present or not, the configuration of Fig. 3.18 suffers from the contribution of the capacitance (C_i) associated with the interface between the specimen and each of the electrical contacts to the measured capacitance (C_m). These three capacitances are in series. Using the well-known equation for capacitances in series, the measured capacitance is given by

$$1/C_m = 1/C + 2/C_i, \tag{3.86}$$

where C is the capacitance due to the specimen and the factor of 2 is due to the presence of two interfaces. Using Eq. (3.2) for C, Eq. (3.86) becomes

$$1/C_m = l/(\varepsilon_o \kappa A) + 2/C_i. \tag{3.87}$$

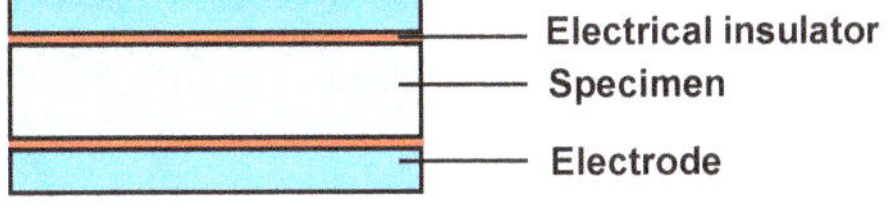

Fig. 3.18 Configuration for measurement of the relative permittivity of a specimen.

By conducting the capacitance measurement for three (or more) specimens of different thicknesses using the same configuration and plotting the measured $1/C_m$ against l as in Fig. 3.19(a), the value of κ can be obtained from the slope and the value of C_i can be obtained by the intercept at the $1/C_m$ axis at $l = 0$. The slope is equal to $1/\varepsilon_o \kappa A$. The intercept is equal to $2/C_i$. In this way, the volumetric and interfacial capacitances are decoupled and the effect of the insulating film is included in the interfacial term. This decoupling should be performed whether the thin insulating film is used at the electrode-specimen interface or not. This is because one should not assume that the interface is perfect, with $C_i = \infty$, even if the electrode material is an excellent conductor.

The greater is C_i, the less is the influence of C_i on C_m and the more superior is the contact. For a perfect contact, C_i is infinite. A pitfall is the unsubstantiated assumption that C_i is infinite, so that $1/C_i$ equals 0 in Eq. (3.86). When C_i is not infinite, C_m is less than C, as indicated by Eq. (3.86). The assumption that C_i is infinite causes C to be underestimated. Even if the contact material is highly conductive, the interface between the contact and specimen surface may not be perfect and, as a result, may be associated with a finite value of C_i.

The thickness of the insulating film is ideally just enough to provide adequate resistance to the material system. This is because the larger is the film thickness, the lower is C_i, and, based on Eq. (3.86), the lower is C_m (i.e., the less sensitive is the capacitance measurement). The presence of the film affects the capacitance, as it affects C_i (Eq. (3.86)). However, without the film, the measured capacitance may be very unreliable, due to the conductive nature of the specimen and the limitation of the meter. By testing the material at three or more different values of l and using Eq. (3.87), C_i can be decoupled from C. Such decoupling is needed to remove the effect of the electrode-specimen interface, which includes the film.

The imperfect interface has been referred to as a "dead layer". The interface is a region with a non-zero thickness. For example, microscopic pores can exist in this region, due to the roughness of the proximate surfaces and the inadequate conformability of the electrode material with the surface topography of the specimen.

Another issue is the possible accumulation of charge at the interface between the contact (electrode) and specimen surface. Charge accumulation may be due to space charge polarization, which is particularly common if the material has mobile ions (e.g., a cement-based material, which has mobile ions in the pore solution). It may also be due to the difference in the charge carrier relaxation time between the materials on the two sides of the interface (i.e., the Maxwell–Wagner effect, which commonly occurs when one material is more conductive than the other).

In the presence of charge accumulation at the interface, the interfacial polarization can be significant, even possibly overshadowing the volumetric polarization in the material under investigation. The charge accumulation at the interface may increase or decrease the measured capacitance, depending on the effect of the charge accumulation on C_i. In other words, the charge accumulation may tighten or loosen the interface. Under this situation, the abovementioned decoupling of the interfacial and volumetric contributions to the capacitance is critically important.

The providing of multiple values of l is typically feasible in a materials research setting. However, in an application-related material system, it may be infeasible to provide multiple thicknesses of the material for testing. Under this situation, the electrode-specimen interface quality needs to be as high as possible, and the conclusions drawn from the study should not involve the assumption that the interface is perfect or nearly perfect.

The linearity in Fig. 3.19(a) would not occur adequately if the specimen is not uniform in quality or dimensions, or the electrical contacts are not uniform in quality. Therefore, care in the preparation of the specimen and electrical contacts is important.

An alternate method of determining κ involves using different values of A (with l fixed) instead of the abovementioned use of different values of l (with A fixed). Rearrangement of Eq. (3.87) gives

$$C_m = \varepsilon_o \kappa A C_i / (l C_i + 2\varepsilon_o \kappa A). \tag{3.88}$$

In case that $C_i \gg 2\varepsilon_o \kappa A / l$ (i.e., C_i is sufficiently large due to the adequate quality of the electrical contact), Eq. (3.88) becomes

$$C_m = \varepsilon_o \kappa A / l, \tag{3.89}$$

and a plot of C_m vs. A is a straight line with slope $\varepsilon_o \kappa / l$. Hence, from the slope, κ is obtained. Although decoupling by varying A is less clean than decoupling by varying, it may be experimentally more convenient to vary A instead of l.

The above discussion is for a parallel-plate capacitor configuration. In case of coplanar electrodes (i.e., the electrodes being on the same plane, such as being on the same surface of the specimen), varying the electrode area is relatively simple. For example, the two coplanar electrodes for capacitance measurement are in the form of two parallel strips of the same length that are separated by a fixed small distance. In order to vary the electrode area, different lengths of the parallel strips are used.

In this configuration, the electrode area is not the same as the cross-sectional area A of the capacitor. Since the capacitance is in the plane of the coplanar electrodes, in the direction from one electrode to the other, the cross-sectional area is perpendicular to this plane. Nevertheless, the greater is the electrode area (i.e., the longer is each electrode strip), the larger is the cross-sectional area. Therefore, Eq. (3.88) and (3.89) are still applicable, with the understanding that A is proportional to the length of the electrode strip. It should be noted that the depth of penetration of the electric field below the coplanar electrodes is not uniform in the direction from one electrode to the other, since the depth increases nonlinearly with the distance away from either electrode. However, provided that the electrode strips are sufficiently long, the electric field lines are essentially entirely in the direction perpendicular to the electrode strips and the electric field non-uniformity in this perpendicular direction is the same for the different lengths of the electrodes. As a consequence, under this condition, the proportionality mentioned above is reliable.

The vast majority of reported work concerns the parallel-plate capacitor configuration rather than the coplanar electrodes configuration. As a result, the vast majority of reported work was performed on specimens with small thicknesses, with the thickness direction being the direction of the capacitance measurement. However, the material behavior may differ between the through-thickness direction and in-plane direction, particularly if the material is inherently anisotropic. An example of an anisotropic material is a thick film with preferred orientation of the functional particles of the film in the plane of the film. Another example of an anisotropic material is a thin film with preferred crystallographic orientation. Furthermore, the material dimension is much larger in an in-plane direction than the through-thickness direction and the dimension affects the charged species excursion distance under an AC electric field, thereby affecting the permittivity and capacitance. In other words, the permittivity can be larger in the in-plane longitudinal direction than the through-thickness direction because of the dimensional difference between these directions. For measuring the permittivity or capacitance in an in-plane direction, the coplanar electrode configuration is suitable, whereas the parallel-plate capacitor configuration is not suitable.

For a nonconductive material, the inter-electrode distance l (whether in the parallel-plate capacitor configuration or the coplanar electrodes configuration) needs to be relatively small, in order to alleviate the issue related to the fringing electric field. The parallel-plate configuration is more amenable to a small value of l than the coplanar electrodes geometry. Thus, the parallel-plate capacitor geometry involving sandwiching electrodes is commonly used.

For a conductive material, the fringing field is negligible, so l can be quite large and coplanar electrodes can be suitable. Due to the feasibility of a large value of l, the coplanar electrodes configuration is more suitable for implementation to multifunctional structures (such as a

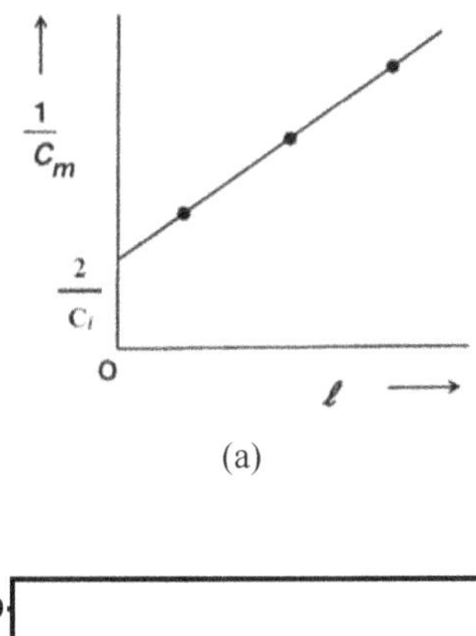

(a)

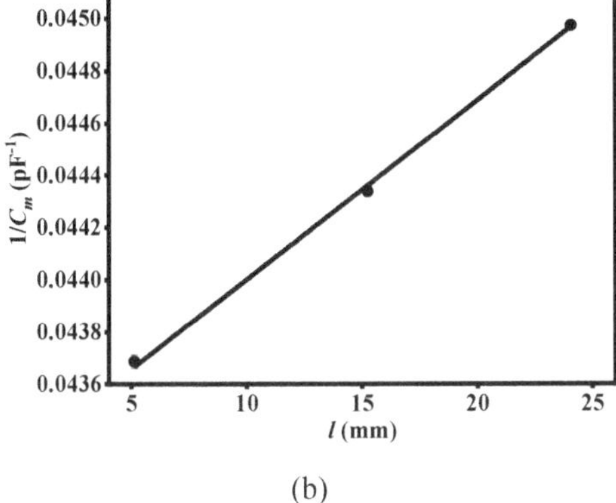

(b)

Fig. 3.19 Plot of $1/C_m$ vs. l for the determination of C_i and κ, using Eq. (3.87). (a) Schematic plot. (b) Experimental results for polycrystalline graphite of grain size 1 μm. The error bars are too short to be legible for two of the three data points. [2].

bridge) than the parallel-plate capacitor configuration. The coplanar electrodes can be applied, such that they are spaced along the length of the material or structure.

In case that the material is not conductive, the fringing field causes the capacitance measured using the coplanar electrodes to be inaccurate. However, the change of the measured capacitance in response to a certain stimulus (e.g., stress) can still be measured, thus providing a means to sense the stimulus.

The specimen in the parallel-plate capacitor configuration is commonly in the form of a thin disc (not necessarily circular), i.e., it is thin in the direction perpendicular to the plane of the capacitor and is wide in the lateral direction. This shape helps to provide a measured capacitance that is not too small, since the capacitance is proportional to the area and is inversely proportional to the thickness (Eq. (3.2)). Furthermore, this shape limits the extent of the fringing electrical field, which is associated with electric flux lines that go from one electrode to the other through the surrounding at the edge of the specimen, rather than going through the specimen, as illustrated in Fig. 3.20. The fringing field is more severe in Fig. 3.20(b) than Fig. 3.20(a), due to the relatively large thickness of the specimen in Fig. 3.20(b). The issue related to the fringing electric field occurs whether the insulating films of Fig. 3.18 are present or not.

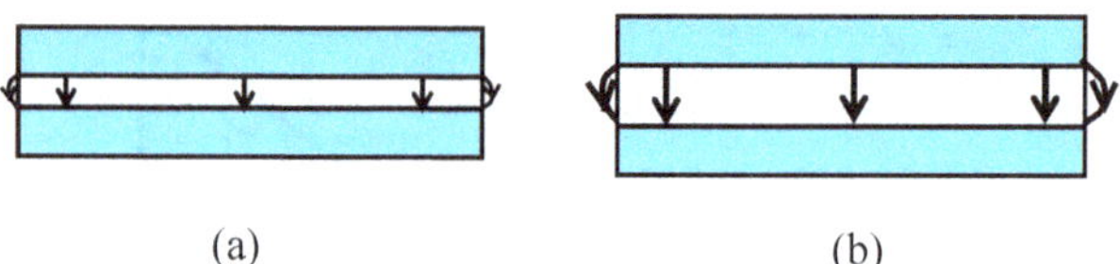

(a)   (b)

Fig. 3.20 Electric field fringing at the edges of a specimen (dotted area) during dielectric testing. The specimen is sandwiched by two electrodes (sold blue areas). The arrows indicate the electric flux lines. (a) A thin specimen, with a small amount of electric field fringing. (b) A thick specimen, with a large amount of electric field fringing.

In case that the material is a poor conductor, fringing electric field may be significant at the exposed surface of the specimen, thus causing inaccuracy in the capacitance measurement. In case of a parallel-plate capacitor, the fringing field at the specimen edge causes the area of the capacitor to be effectively larger than the true area. As a consequence, the capacitance is higher than the value without the fringing field. In case of coplanar electrodes, the fringing field is even more significant, due to the relatively large amount of exposed area.

An effective method to avoid the fringing electric field in the parallel-plate capacitor configuration is to use a guard electrode, as illustrated in Fig. 3.21. The guard electrode is in the plane of one of the two electrodes and is around the edge of this electrode, but it is separate from this electrode by a small gap. For example, the guard electrode is an annular ring around the circular electrode (labeled electrode 1 in Fig. 3.21) in the same plane. The other electrode (labeled electrode 2 in Fig. 3.21) is larger in area than the first electrode, as it includes the area of the guard electrode. This configuration is known as the three-electrode configuration, since the guard electrode may be counted as the third electrode. The electric field is applied between electrodes 1 and 2, such that electrode 1 and the guard electrode are at the same potential. In other words, electrode 1 and the guard electrode are electrically connected, in spite of the physical gap between them. The area of the specimen is the same as that of electrode 1. Due to the electric field between the guard electrode and electrode 2, fringing field cannot occur at the edge of electrode 1. With the presence of the guard electrode, the specimen thickness can be substantial.

In case of the parallel-plate capacitor configuration, the use of a large-area electrode, a small specimen thickness and the use of a guard electrode to surround one of the two electrodes two-dimensionally would alleviate this problem (Fig. 3.21). In case of coplanar electrodes, the use of a short distance between the electrodes helps reduce the fringing field.

In case that the specimen is sufficiently conductive, the electric field lines are trapped inside the specimen, with no fringing field, regardless of the distance between the two electrodes. Under this situation, the specimen does not need to be in the shape of a thin disc (for measuring κ in the direction perpendicular to the disc), but it can be in the shape of a wire or bar (for measuring κ along the length of the wire or bar). By using a bar-shaped specimen of polycrystalline graphite and three inter-electrode distances along the length of the bar, the results in Fig. 3.19(b) are obtained, giving a κ value of $(5.1 \pm 0.2) \times 10^3$ at 2 kHz.

When the capacitance is studied during the application of a stimulus (e.g., voltage/current, stress/strain, temperature and damage), both C (which relates to κ) and C_i can change when the degree of the stimulus

 Functional Materials

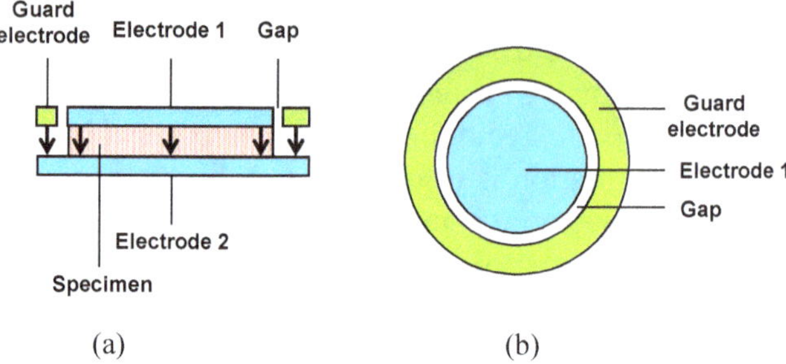

Fig. 3.21 The three-electrode configuration for dielectric testing of a specimen. The specimen (red area) is sandwiched by two electrodes (blue areas labeled electrode 1 and electrode 2). The arrows indicate the electric flux lines. There is a gap between electrode 1 and the guard electrode (green area). (a) Side view. (b) Top view.

changes. Moreover, the changes in C and C_i can be in the same direction or in different directions under the same stimulus. For example, upon compression, C tends to increase due to the decrease in the specimen dimension in the direction of the capacitance measurement, but C_i may increase (in case that the electrode-specimen interface is improved by the compression) or decrease (in case that the electrode-specimen interface is degraded by the compression). Whether the electrode-specimen interface is improved or degraded by the compression depends on the properties of the materials involved, such as the brittleness of the electrode material and the quality of the interface prior to the compression. If the quality of the interface prior to compression is low, compression tends to improve the interface. If the material at the interface between the electrode and specimen surface is brittle, as in the case of silver paint after drying, the compression may degrade the interface. Therefore, the assumption that C_i is infinite throughout the variation in voltage/current, stress/strain, temperature or damage degree can cause unreliable determination of the effect of the stimulus on C.

The effect of the temperature on the permittivity or capacitance is considered below. Based on Eq. (3.2), the fractional change in capacitance per unit temperature change is given by

$$\xi = (\delta C/C)/\delta T = (\delta\kappa/\kappa)/\delta T + (\delta A/A)/\delta T - (\delta l/l)/\delta T. \qquad (3.90)$$

Let the first term on the right side of Eq. (3.90) be ς, i.e., the fractional change in κ per unit change in temperature. Based on Eq. (2.38),

Eq. (2.32) and Eq. (3.90), the fractional change in A per unit change in temperature (i.e., the second term on the right side of Eq. (3.90)) is given by

$$(\delta A/A)/\delta T = -2v\,(\delta l/l)/\delta T = -2v\gamma, \qquad (3.91)$$

where v is the Poisson's ratio (for the isotropic case) and γ is the coefficient of thermal expansion. By definition of the coefficient of thermal expansion, the last term on the right side of Eq. (3.90) is given by

$$-(\delta l/l)/\delta T = -\gamma. \qquad (3.92)$$

The combination of Eq, (3.90), (3.91) and (3.92) gives

$$\xi = \varsigma - 2v\gamma - \gamma = \varsigma - (1 + 2v)\,\gamma. \qquad (3.93)$$

Equation (3.93) provides the relationship between ς and ξ.

3.6.2 *Measurement of the impedance*

For measuring the impedance Z, which relates to both the real and imaginary parts of κ, a current needs to flow through the specimen. Thus, the electrical insulators in Fig. 3.18 should be absent. However, the interfacial impedance at the electrode-specimen interface should not be ignored. The three-electrode configuration of Fig. 3.21 may be used to avoid the fringing field.

Impedance measurement suffers from the same pitfalls as those for resistance and capacitance measurements, respectively. The real part of the complex impedance is the resistance; the imaginary part relates to the capacitance and inductance. The fact that the impedance is for AC rather than DC does not alleviate the issues. The analysis of the frequency dependence of the real and imaginary parts of the impedance using plots such as the Nyquist Plot also does not remove the issues.

Unfortunately, impedance measurements reported in the literature commonly involve the two-probe method, with the assumption that the two electrodes are perfect or nearly perfect at all frequencies, i.e., with zero (or negligible) resistance and infinite (or very high) capacitance at the electrode-specimen interface at all frequencies, and with the assumption that the same measurement method can be the used, regardless of the resistance of the material. The specifications of the impedance meter need to be consulted to obtain the range of the combination of resistance and capacitance that the meter can handle properly. This range varies with the frequency. If the specimen is conductive, this limitation should be carefully considered for the purpose of measuring the capacitance correctly.

The frequency may affect both the electrode-specimen interface characteristics and specimen characteristics, and these effects may be qualitatively or quantitatively different. The excursion distance of the charged species decreases with increasing frequency. The smaller is this distance, the less is the effect of the microstructure, such as the grain boundaries. The available distance is small at the interface compared to that in the volumetric bulk. Furthermore, the degree of the dipole friction may differ between the volumetric bulk and the interface.

To alleviate the problems mentioned above concerning impedance measurement, one should perform decoupling of the volumetric and interfacial contributions to the impedance. For the decoupling, the impedance should be measured for various specimen dimensions. In case of the measurement of the impedance as a function of the frequency, the decoupling should be performed at each frequency, because the frequency may affect the volumetric impedance and interfacial impedance differently. A change in frequency is a type of stimulus, akin to a change in temperature being a stimulus.

In case that the capacitance part of Z is important, the specimen should be shaped like a disc, as in Fig. 3.18, so that the capacitance is not too small. However, in case that the capacitance part of Z is not important while the resistance part of Z is important, the thickness of the specimen can be substantial. The larger the thickness, the greater is the resistance and the smaller is the capacitance. On the other hand, the smaller the cross-sectional area, the larger is the resistance and the smaller is the capacitance. Thus, both the thickness and the area need to be considered in arriving at an appropriate specimen shape.

3.6.3 *Frequency dependence of the impedance*

The impedance Z may be measured as the frequency is scanned. The effect of frequency on Z gives valuable information concerning the equivalent circuit model, which may be more complicated than those in Fig. 3.11 and 3.17. In other words, the frequency dependence of Z

provides information that indicates the appropriate equivalent circuit model. This is because resistance and capacitance are different in their frequency dependence. According to Eq. (3.48), Z_R is independent of frequency. However, according to Eq. (3.44), Z_C is inversely related to the angular frequency ω.

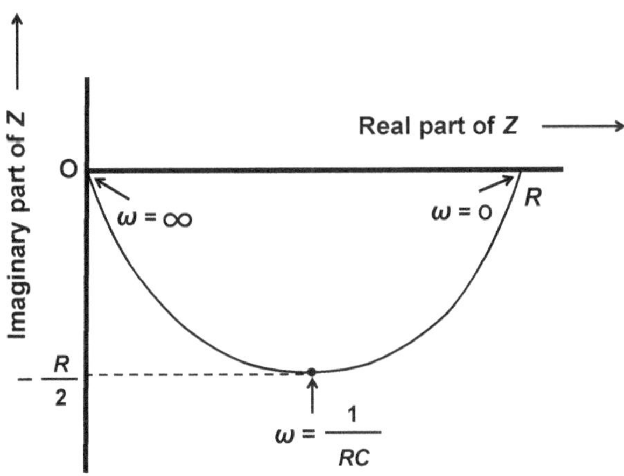

Fig. 3.22 Nyquist plot for the equivalent circuit of Fig. 3.11.

Consider the effect of frequency on Z for the equivalent circuit of Fig. 3.11 (i.e., the case of the resistor and capacitor in parallel). According to Eq. (3.53), the real part of Z depends on ω, such that it is equal to R when $\omega = 0$ and is equal to 0 when $\omega = \infty$. According to Eq. (3.54), the imaginary part of Z depends on ω, such that it is equal to 0 when $\omega = 0$ and is also equal to 0 when $\omega = \infty$. The plot of the imaginary part of Z versus the real part of Z is in general called the Nyquist plot. This plot can be experimentally obtained by measuring Z while the frequency is scanned. It is shown in Fig. 3.22, as calculated for the case of the equivalent circuit of Fig. 3.11. In this case, it is a semicircle of radius $R/2$. The imaginary part of Z is negative, while the real part of Z is positive, as shown by Eq. (3.54) and (3.53) respectively. The maximum in the magnitude of the imaginary part of Z is equal to $R/2$ and this maximum occurs at $\omega = 1/(RC)$, as shown in Fig. 3.22 and confirmed by the following conventional differentiation method for maximum determination. In the following equations, the imaginary part of Z is

                    Functional Materials

abbreviated Im Z and the real part of Z is abbreviated Re Z. Using Eq. (3.54), the differentiation of the magnitude of the imaginary part of Z with respect to ω is given by

$$(\Delta|\text{Im } Z|)/\Delta\omega = (\Delta/\Delta\omega)\,[(\omega R^2 C)/(1 + \omega^2 R^2 C^2)]$$

$$= R^2 C\,(1 - \omega^2 R^2 C^2) \tag{3.94}$$

At the maximum, the differential is equal to zero. From Eq. (3.94), (d $|\text{Im } Z|$)/dω = 0 when $\omega = 1/(RC)$. To confirm that this is the maximum rather than the minimum, take the second derivative of $|\text{Im } Z|$. Using Eq. (3.94),

$$(\Delta^2\,|\text{Im } Z|)/(\Delta\,\omega^2) = R^4 C^3\,(-2\omega) < 0. \tag{3.95}$$

Since the second differential is negative, $\omega = 1/(RC)$ indeeds correspond to the maximum of the magnitude of the imaginary part of Z, rather than the minimum. At $\omega = 1/(RC)$, Eq. (3.54) gives

$$\text{Im } Z = -R/2, \tag{3.96}$$

which is as shown in Fig. 3.22. Equation (3.96) is consistent with the fact that the unit of Im Z is Ω.

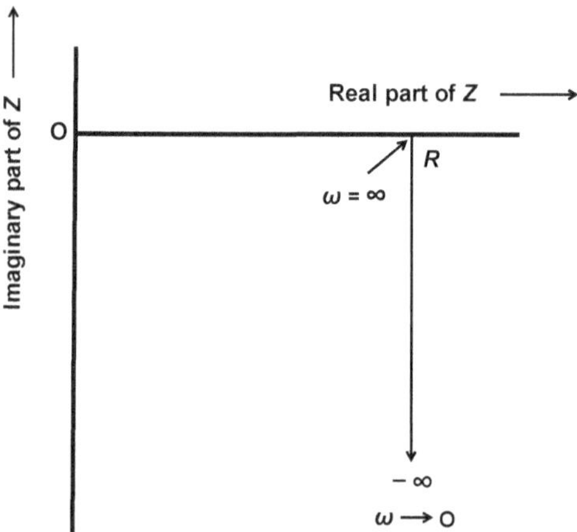

Fig. 3.23 Nyquist plot for the equivalent circuit of Fig. 3.17.

In case of the equivalent circuit of Fig. 3.17 (i.e., the series configuration), Eq. (3.66) shows that the real part of Z is R, i.e., it is independent of ω, while Eq. (3.67) shows that the imaginary part of Z is $-1/(\omega C)$, i.e., it is inversely related to ω, being $-\infty$ when $\omega = 0$ and being 0 when $\omega = \infty$. The corresponding Nyquist plot is shown in Fig. 3.23.

The large difference between Fig. 3.22 and 3.23 indicates that the Nyquist plot strongly depends on the equivalent circuit model. Thus, by experimentally determining the Nyquist plot, the equivalent circuit model can be derived through fitting the measured Nyquist plot to plots calculated by assuming various equivalent circuit models. However, the circuit model is typically not unique (as the plot is typically not simple in shape, unlike the simplicity of Fig. 3.22 and 3 23), making the values of the circuit elements in the model not very meaningful from the viewpoint of the basic science.

The equivalent circuit model commonly derived from analysis of the Nyquist plot by curve fitting assumes that the circuit elements in the model are independent of the frequency. In fact, both the resistance and capacitance can vary with the frequency, particularly when the frequency is varied over a wide range. A wide frequency range is required for the analysis based on the Nyquist plot. In addition, the circuit model is typically not unique (as the plot is typically not simple in shape), making the values of the circuit elements in the model not very meaningful from the viewpoint of the basic science (as opposed to the empirical science).

Example Problem 31 at the end of the chapter is relevant.

3.6.4 *Effect of polarization on the electrical resistivity*

A highly conductive material cannot support a substantial electric field, since it itself is close to being a short circuit. However, a material that is sufficiently high in electrical resistivity can support a substantial electric field, so that an applied electric field can cause the charged species in the material to move. For example, the movement involves the positive ions moving toward the negative end of the voltage gradient that corresponds to the applied electric field, while the negative ions move toward the positive end. This movement results in polarization (i.e., separation between the positive and negative charge centers in the material), thus giving rise to an electric field of the opposite polarity inside the material, as illustrated in Fig. 3.24(a).

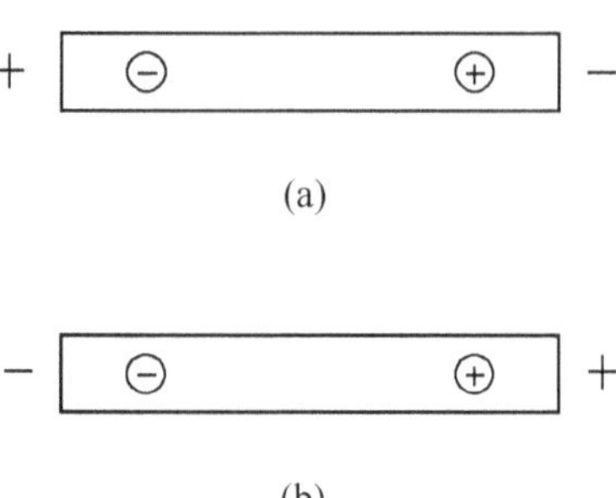

(a)

(b)

Fig. 3.24 Electric polarization that occurs during electric field application. (a) Before field polarity reversal. (b) After field polarity reversal.

In measuring the electrical resistivity of a material, an electric field needs to be applied. If this field is DC, polarization occurs. As a consequence, the electric field in the material is reduced from the applied electric field by an amount equal to the opposing electric field. Due to Ohm's Law, this causes the current to be reduced from the level in the absence of polarization. In resistance measurement, the resistance is given by the applied voltage divided by the current. Since the current is reduced, the measured resistance is higher than the value in the absence of polarization. This increased resistance is not the true resistance, but is

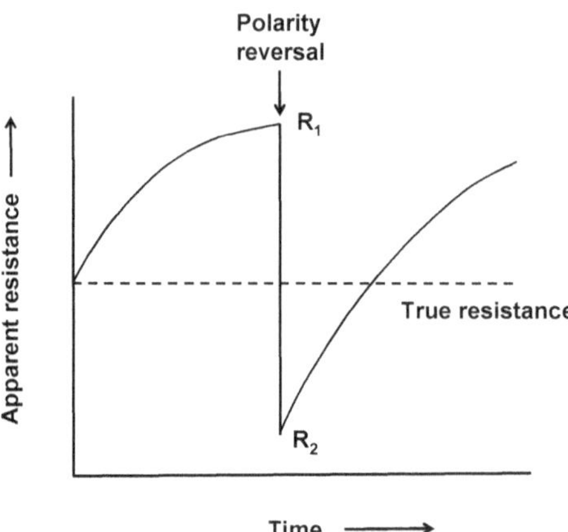

Fig. 3.25 Effects of an applied electric field and its polarity reversal on the apparent volume electrical resistivity [4].

the apparent resistance in the presence of polarization. The more is the polarization, the higher is the apparent resistance. During the resistance measurement at a constant DC field, polarization gradually builds up, so that the measured resistance gradually increases from the true value to the apparent value, which is higher than the true value, as illustrated in Fig. 3.25.

The increase in the apparent resistance means that the polarization impedes conduction and is thus undesirable for conduction applications in general. Furthermore, the capacitance associated with the polarization causes signal propagation delay, due to the RC time constant (with the product of R and C having the unit of seconds) that is exhibited by a circuit that involves a resistance R and a capacitance C (Fig. 3.26). The delay is particularly undesirable for high-speed electronics such as high-speed computers.

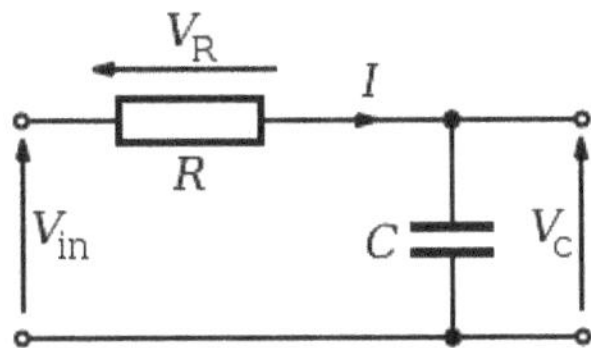

Fig. 3.26 A series RC circuit assiciated with the RC time constant. (https://en.wikipedia.org/wiki/RC_time_constant, public domain)

Upon reversal of the polarity of the applied electric field, the situation changes from that of Fig. 3.24(a) to that of Fig. 3.24(b). In Fig. 3.24(b), the applied field has the same polarity as the field resulting from polarization, so the effective field encountered by the material is higher than the applied field by an amount equal to the field due to polarization. As a consequence, the measured (apparent) resistance is lower than the true resistance, as illustrated in Fig. 3.25, where R_1 is the apparent resistance immediately before polarity reversal and R_2 is the apparent resistance immediately after polarity reversal. The average of R_1 and R_2 equals the true resistance. The method involving polarity reversal and taking the average can be used to determine the true resistance of a material that has tendency to polarize.

After polarity reversal, the depolarization gradually occurs, thereby bringing the apparent resistance to the true resistance. As the field is continuously present, this is immediately followed by polarization in the opposite direction, so that the apparent resistance becomes greater than the true resistance, as illustrated in Fig. 3.25.

In case of an AC field, polarity reversal occurs frequently, so that there is not much time for polarization to build up at any polarity. Polarization takes time because the movement of ions takes time. Therefore, the effect in Fig. 3.25 does not occur in the AC case.

Figure 3.27 shows the results for carbon fiber (short) reinforced cement paste. The two-probe method (Fig. 3.27(b)) gives a slightly larger fractional change in resistivity than the four-probe method (Fig. 3.27(a)), indicating slightly more polarization. For the two-probe case, the resistivity does not drop as much at the time of switching and the resistivity rise after the switching is faster and more complete (reaching the value just before the switching). As the two-probe method includes the contact resistance in the measured resistance, it includes the polarization at the contact-specimen interface in addition to the polarization within the specimen in the overall polarization. The polarization reversal associated with the contact-specimen interface is faster than that associated with the specimen itself, thus causing the shorter rise time for the case of the two-probe method. The observation that the resistivity drop at the time of switching is less for the two-probe method is also due to the faster polarization reversal for the contact-specimen interface. The polarization reversal associated with the contact-specimen interface is also more complete than that associated with the specimen itself. Due to the fast response of the contact-specimen interface, it dominates the observed effect of polarity switching in Fig. 3.27(b). In contrast, the polarization of the specimen itself dominates the observed effect of polarity switching in Fig. 3.27(a).

Polarization also occurs for a cement-based material in the absence of conductive admixtures like carbon fiber. In this case, sand addition slows down polarization saturation, but enhances the polarization. Silica fume addition does not slow down polarization saturation, but it diminishes the polarization slightly. An increase in temperature enhances the polarization due to increase in ionic mobility. (J. Cao, D.D.L. Chung. Electric polarization and depolarization in cement-based materials, studied by apparent electrical resistance measurement. Cem. Concr. Res. 34(3), 481–485 (2004).) This relates to the increase in permittivity with increasing temperature, a phenomenon known as pyropermittivity, which is to be distinguished from pyroelectricity.

The polarization and depolarization under a constant DC current are faster and more significant when the current is higher, as shown in Fig. 3.26(a) for polycrystalline graphite [4]. The polarization/depolarization rate decreases as polarization/depolarization progresses, as indicated by the decreasing slope of the curve of apparent resistivity vs. time of polarization/depolarization. In relation to polarization, this trend is because the opposing electric field resulting from the polarization hinders further polarization. In relation to depolarization, this trend is because the forward electric field resulting the depolarization hinders further depolarization.

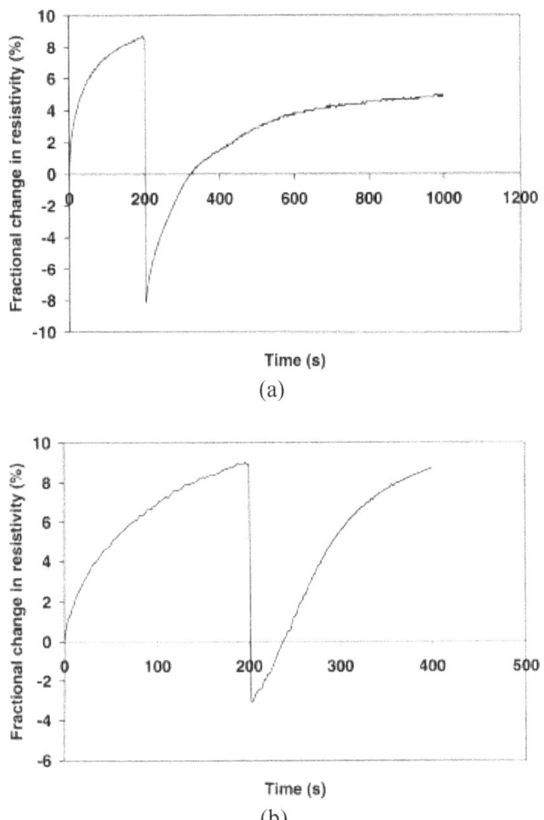

Fig. 3.27 Variation of the measured (apparent) volume electrical resistivity with time before and after voltage polarity switching for cement paste containing carbon fiber (0.5% by weight of cement) and silica fume. (a) Four-probe method. (b) Two-probe method. [5]

In case of the charged species responsible for the polarization being carriers such as electrons, the polarization stems from the interaction between a fraction of the carriers and the atoms [2]. The depolarization is slower and less significant than the prior polarization (Fig. 3.28(a)). The relative sluggishness of depolarization is due to the disengagement of the carrier-atom interaction during depolarization being more difficult than engagement of the interaction during polarization. This difference between polarization and depolarization results in asymmetry in the curve upon polarity reversal. The degree of asymmetry increases with increasing degree of polarization prior to the polarity reversal. The asymmetry in the curve is a reflectin of the electrical asymmetry associated with the presence of an electret, which refers to a permanent electric dipole in the material.

The DC polarization is associated with DC capacitance (Fig. 3.28(b)), which is obtained with consideration of the separately measured AC permittivity of the material. The AC permittivity is based on the measured AC capacitance. Meters such as the LCR meter measure the AC capacitance, but not the DC capacitance. However, the DC capacitance can be assessed through DC polarization consideration.

For graphite, the carriers include electrons and holes, since graphite is a semi-metal. Let n be the number of carriers that participate in the DC polarization. Only a small fraction of the carriers participates in the DC polarization. This fraction increases with the number of injected electrons (N_I) to a point, but it decreases beyond this point (Fig. 3.28(c)). The N_I relates to the product of the current and time, as this product is the charge and the charge divided by the electronic charge is the number of electrons. This nonmonotonic trend is explained below.

For a material at a fixed temperature, the total number of carriers N_c is fixed, while N_I increases continuously as the polarization current application occurs. At the end of the period of current application, N_c is below N_I. On the other hand, at the start of the polarization current application (i.e., time $= 0$ s), $N_I = 0$ and hence $N_I + N_c = N_c$. As the polarization current application time increases from 0 to the maximum value, $n/(N_I + N_c)$ changes from n/N_c to roughly n/N_I, while N_I increases continuously to values much exceeding N_c. The non-monotonic trend of $n/(N_I + N_c)$ vs. time is because n increases with increasing time relatively sharply at early times, thus causing $n/(N_I + N_c)$ to increase with time, whereas n increases with increasing time less sharply at later times (due

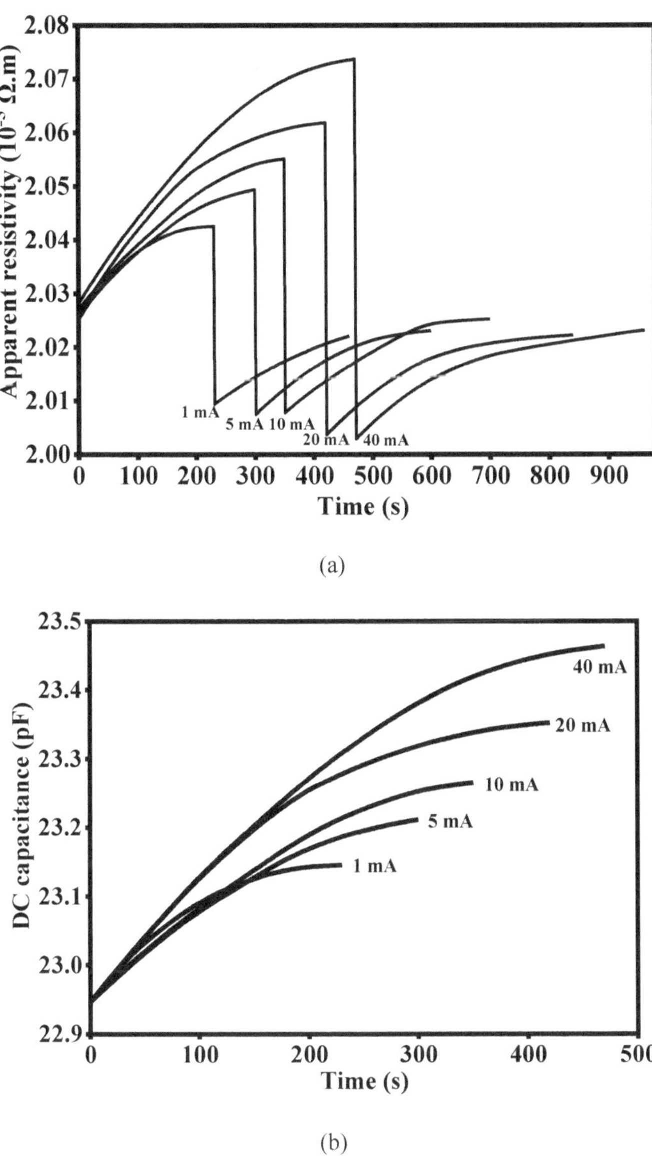

Fig. 3.28 DC polarization and depolarization of polycrystalline graphite at various values of the constant DC current. (a) The behavior shown by the DC apparent resistivity, which increases upon polarization. (b) The behavior shown by the DC capacitance caused by the polarization. (c) The fraction of carriers that participate in the DC polarization vs. the number of injected electrons. [4]

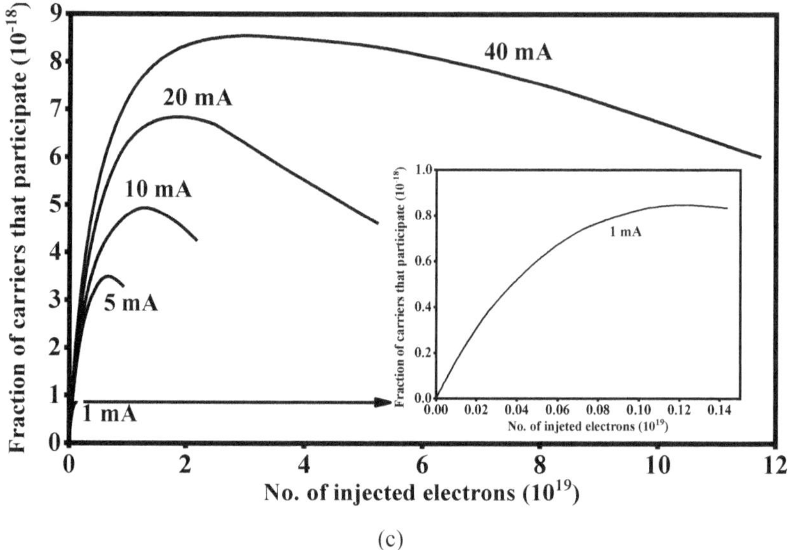

(c)

Fig. 3.28 (*Continued*).

to the slowing down of the polarization), thus causing $n/(N_l + N_c)$ to decrease with time.

3.7 Electrical insulation

An electrical insulator has no mobile carrier to conduct electricity. When the electric field is sufficiently high, charged species (e.g., electrons and ions) may be pulled out of chemical bonds by the electrical force (a phenomenon known as dielectric breakdown), thus resulting in mobile carriers, which cause the material to conduct electricity. The dielectric strength is the electric field above which an electrical insulator becomes conductive.

Table 3.1 shows the values of the dielectric strength for various materials. A material that has a low value of the relative permittivity κ tends to have a high value of the dielectric strength. This is because a low κ value is associated with a low tendency for polarization, which involves interaction of the charged species with the electric field. Thus, Teflon has a low κ value of 2.1 and a high dielectric strength of 60 MV/m; rubber has a high κ value of 7 and a low dielectric strength of

12 MV/m; strontium titanate has an even higher κ value of 310 and an even lower dielectric strength of 8 MV/m. However, the correlation between κ and the dielectric strength is limited, due to the strong effect of impurities (particularly ionic impurities) on the dielectric strength. It is due to impurities that air has a low dielectric strength of 3 MV/m, in spite of its low κ value of 1.00054.

3.8 Conversion between mechanical energy and electrical energy

The conversion of mechanical energy to electrical energy (i.e., the generation of electricity by an applied stress, as needed for stress/strain sensing and clean energy production) and the conversion of electrical energy to mechanical energy (i.e., the generation of stress or strain by an applied electric field, as needed for motors and actuators that are electrically controlled) can be obtained by the use of dielectric materials. The mechanism behind the conversion from mechanical energy to electrical energy involves the change in polarization upon stress application. The mechanism behind the conversion from electrical energy to mechanical energy involves the interaction of the applied electric field with the dielectric material and the consequent change in dimension.

In case that the mechanism specifically involves the change of the separation of the positive and negative charge centers in the dielectric material during stress or electric field application, the phenomenon is known as the piezoelectric effect. The direct piezoelectric effect converts mechanical energy to electrical energy, whereas the converse (or reverse) piezoelectric effect converts electrical energy to mechanical energy. The word "piezoelectric" is derived from the Greek word piezein, which means squeezing or pressing.

3.8.1 *Conversion of mechanical energy to electrical energy*

3.8.1.1 *Scientific basis*

Electricity can be generated upon stress application, thus allowing conversion of mechanical energy to electrical energy. This effect stems from the effect of an applied stress/strain on the polarization (i.e., the electric dipole moment per unit volume) in a dielectric material, which is sandwiched by electrodes in the form of an electrical conductor that

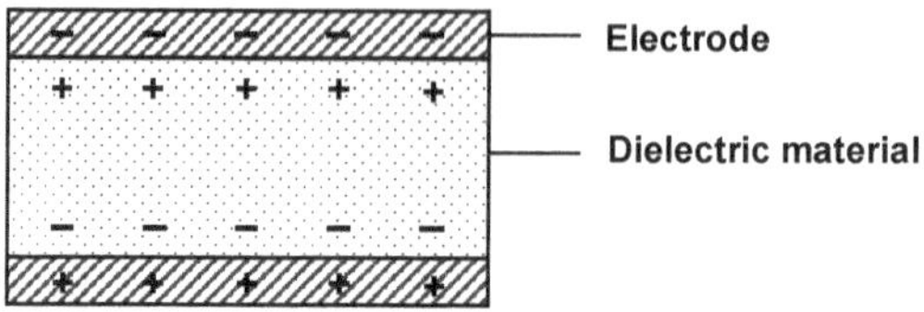

Fig. 3.29 A polarized dielectric material with induced charges at the two electrodes sandwiching it.

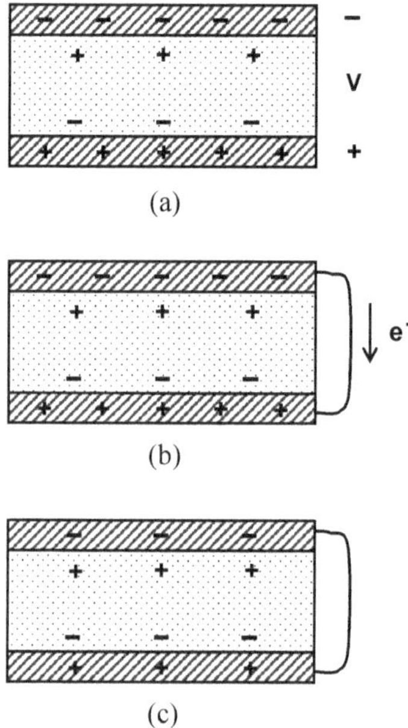

Fig. 3.30 A dielectric material (dotted region) with its polarization reduced by an applied stress. This causes a voltage between the two electrodes in case that the two electrodes are open-circuited, as shown in (a). In case that the two electrodes are short-circuited, this causes a pulse of current, with electrons flowing from the negative electrode to the positive electrode, as shown in (b). The current persists until the magnitude of charge at each electrode equals the magnitude of charge at the proximate surface of the dielectric material, as shown in (c), where there is no more current.

conducts by the flow of electrons. Polarization in the dielectric material results in charges of opposite sign in the two electrodes (Fig. 3.29). A decrease in the polarization in the material (a situation that commonly occurs upon compression in the direction of polarization, due to the decrease in the dielectric material thickness and the consequent decrease in the separation between the centers of positive and negative charges) causes an imbalance in the amount of charge on the dielectric material surface and the amount of bound charge in the electrode facing it. As a consequence, a voltage appears between the two electrodes when the two electrodes are open-circuited, with the negative end of the voltage at the negative electrode, as shown in Fig. 3.30(a). This is known as the open-circuit voltage. If the two electrodes are, instead, short-circuited, a pulse of electron flow occurs from the negative electrode to the positive electrode (Fig. 3.30(b)). The pulse occurs until the charge imbalance no longer exists (Fig. 3.30(c)). The voltage and current outputs in response to the applied stress are illustrated in Fig. 3.31. The open-circuit voltage

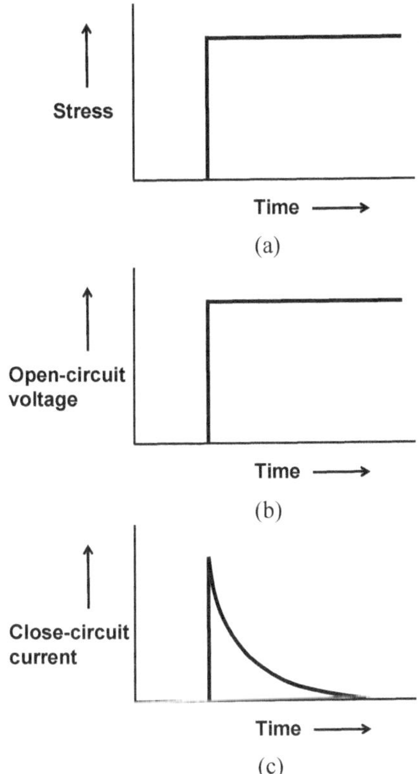

Fig. 3.31 An applied stress shown in (a) resulting in the open-circuit voltage shown in (b) and the close-circuit current pulse shown in (c).

is constant as the stress is maintained constant. However, the close-circuit current is a pulse.

Figure 3.32 illustrates the opposite scenario, where the applied stress causes the polarization in the dielectric material to increase. This scenario commonly occurs when upon tension in the direction of polarization, due to the increase in the dielectric material thickness and the consequent increase in the separation between the centers of positive and negative charges. This change in polarization causes an open-circuit voltage with the negative end of the voltage at the positive electrode, as shown in Fig. 3.32(a) and causes a close-circuit current pulse with the electrons flowing from the positive electrode to the negative electrode (Fig. 3.32(b)) until the charge imbalance no longer exists (Fig. 3.32(c)). The electrical response is as illustrated in Fig. 3.31.

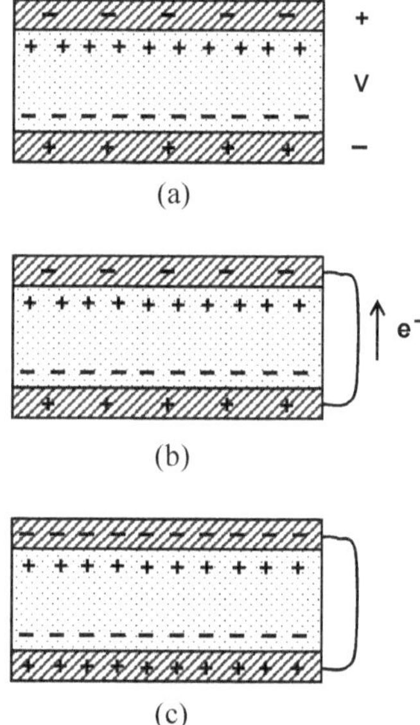

Fig. 3.32 A dielectric material (dotted region) with its polarization increased by an applied stress. This causes a voltage between the two electrodes in case that the two electrodes are open-circuited, as shown in (a). In case that the two electrodes are short-circuited, this causes a pulse of current, with electrons flowing from the positive electrode to the negative electrode, as shown in (b). The current persists until the magnitude of charge at each electrode equals the magnitude of charge at the proximate surface of the dielectric material, as shown in (c), where there is no more current.

If the applied stress is a wave, the open-circuit voltage and the close-circuit current are also in the form of waves, as illustrated in Fig. 3.33. That the close-circuit current is not a pulse in Fig. 3.33(c) is because the stress keeps changing and the superposition of many current pulses results in a current wave. Because the measurement of a current pulse is not as simple as that of a current wave, testing involving current measurement is commonly conducted using a stress wave.

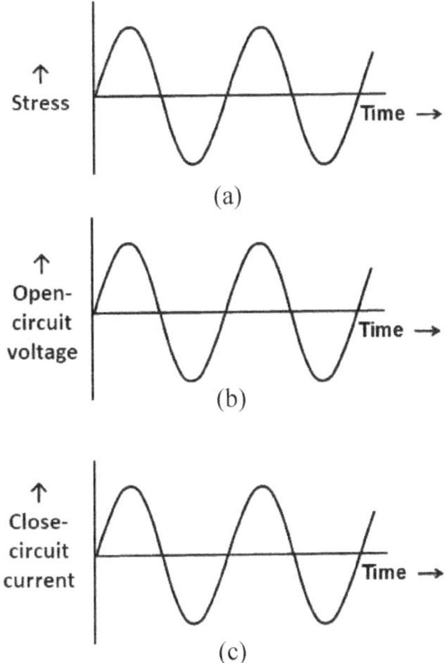

Fig. 3.33 An applied stress wave, as shown in (a), applied on a dielectric material and resulting in an open-circuit voltage wave, as shown in (b), and a close-circuit current wave, as shown in (c). The possible time lag between the voltage and current waves is not shown.

When the two electrodes are electrically connected with a series resistor, they are neither open-circuited nor open-circuited. Under this situation, the voltage is less than the open-circuit voltage and the current is less than the close-circuit current, as illustrated in Fig. 3.34. The higher is the operating current, the lower is the operating voltage.

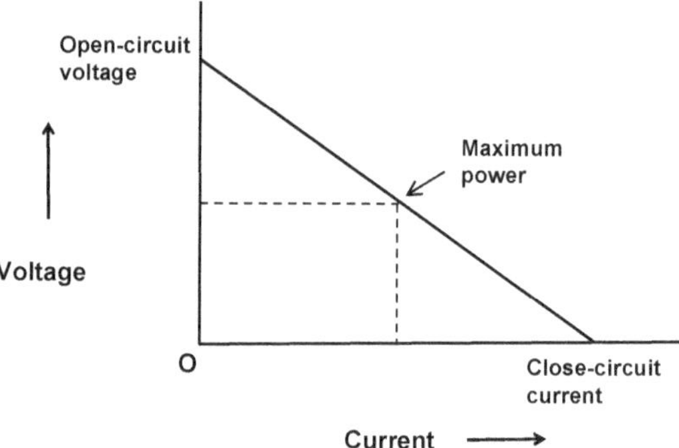

Fig. 3.34 Relationship of the open-circuit voltage with the close-circuit current. The voltage and current result from the application of a stress on a dielectric material.

The electric power output P (in Watts, abbreviated W, with Watts = Volts x Amperes) is given by

$$P = VI, \tag{3.97}$$

where V is the operating voltage and I is the operating current. Due to the linear relationship between the voltage and the current (Fig. 3.33), the power is maximum when the operating voltage is equal to half of the open-circuit voltage and when the operating current is equal to half of the close-circuit current, as shown in Fig. 3.34. In other words, the maximum power P_{max} is given by

$$P_{max} = \tfrac{1}{4} \text{ (open-circuit voltage) (close-circuit current)} \tag{3.98}$$

Since $I = V/R$ (Ohm's Law), Eq. (3.97) can be rewritten as

$$P = V^2/R. \tag{3.99}$$

Equation (3.99) means that P is higher when R is lower. Thus, the dielectric material should have a reasonably low resistivity in order to provide a high power. As shown in Eq. (2.10), R is inversely proportional to the area A. Thus, a large area of the dielectric material is advantageous. As also shown in Eq. (2.10), R is proportional to the length of the conductor, i.e., the thickness of the dielectric material. Hence, a small thickness of the dielectric material is also advantageous.

The mechanical energy input is given by

$$X = \Delta F \, \Delta l, \tag{3.100}$$

where ΔF is the change in force normal to the area A of the dielectric material and Δl is the change in thickness of the dielectric material.

Based on Eq. (3.97), the electrical energy output (Y) is given by

$$Y = V I t, \tag{3.101}$$

where t is the time and I is the current. Note that, by definition, power is energy per unit time.

If the mechanical energy input (X) equals the electrical energy output (Y), the combination of Eq. (3.100) and (3.101) gives

$$\Delta F \, \Delta l = V I t. \tag{3.102}$$

Dividing by l gives

$$\Delta F \, \Delta l / l = (V/l) \, I t = E I t, \tag{3.103}$$

where $E = V/l$ is the electric field. Dividing by ΔF gives

$$\Delta l / l = E I t / \Delta F. \tag{3.104}$$

Equation (3.104) means that the strain $\Delta l / l$ is proportional to the electric field E. This equation may be expressed as

$$\Delta l / l = d \, E, \tag{3.105}$$

where d, called the piezoelectric coupling coefficient (although this term is strictly used for describing the piezoelectric effect only), is defined as

$$d = I t / \Delta F. \tag{3.106}$$

This coefficient describes the strain per unit electric field. Its values for the case of the strain and the electric field in the same direction are shown in Table 6.3 for selected piezoelectric materials.

Since the current, by definition, is charge per unit time, we have the equation

$$It = (\kappa - 1)\Delta Q, \tag{3.107}$$

where $(\kappa - 1)\Delta Q$ is the imbalance in charge between the dielectric material surface and the electrode facing it. This charge imbalance equals the change in the amount of charge on the surface of the dielectric material (Sec. 3.3, Eq. (3.20)). In addition, by definition, the change in stress, abbreviated ΔU, is given by

$$\Delta U = \Delta F/A. \tag{3.108}$$

Hence, Eq. (3.106) can be rewritten as

$$d = (\kappa - 1)\Delta Q/[(\Delta U)\, A]. \tag{3.109}$$

From Eq. (3.21), $P = (\kappa - 1)\, Q/A$. Let ΔP be the change in the polarization due to the change in stress and let $\Delta \kappa$ be the change in the relative permittivity due to the change in stress. Hence,

$$P + \Delta P = (\kappa + \Delta \kappa - 1)\, (Q + \Delta Q)/A$$

$$P + \Delta P = (\kappa - 1)\, Q/A + (\kappa - 1)\, \Delta Q/A + \Delta \kappa\, (Q/A) + \Delta \kappa\, \Delta Q/A,$$

$$\tag{3.110}$$

where κ is the value of the relative permittivity in the absence of the change in stress. Subtraction of P from Eq. (3.110) gives

$$\Delta P = (\kappa - 1)\, (\Delta Q/A) + (\Delta \kappa)\, (Q/A) + \Delta \kappa\, \Delta Q/A. \tag{3.111}$$

In case that $\Delta \kappa$ is negligible, while ΔQ is not negligible, Eq. (3.111) becomes

$$\Delta P = (\kappa - 1)\, (\Delta Q/A). \tag{3.112}$$

The ΔQ gives rise to an imbalance between the charge on the dielectric material surface and the charge in the electrode, thus resulting in a voltage or current output, i.e., the direct piezoelectric effect.

In case that ΔQ is negligible, while $\Delta\kappa$ is not negligible, Eq. (3.111) becomes

$$\Delta P = (\Delta\kappa)\,(Q/A). \tag{3.113}$$

The $\Delta\kappa$ gives rise to a change in capacitance ΔC (Eq. (3.2)), which in turn causes a change in the amount of charge on the dielectric material surface (Fig. 3.4). The resulting imbalance between the charge on the dielectric material surface and the charge in the electrode gives rise to a voltage or current output. This effect, which stems from the fact that the dielectric material provides a capacitance that is stress dependent, is not considered the direct piezoelectric effect. Nevertheless, it converts mechanical energy to electrical energy, and the phenomenon is known as piezopermittivity (i.e., the effect of stress on the permittivity) [6]. Piezoelectricity does not require the permittivity to change with strain, whereas piezopermittivity does.

The mechanism of the direct piezoelectric effect (i.e., the effect that involves ΔQ) involves, for example, the slight squishing of the electric dipole (associated with the polarization) upon compression in the direction of the dipole (with at least a component of the applied compressive stress in this direction). In other words, there is a change in the distance between the center of positive charge and the center of negative charge. A compressive stress in this direction may alternatively be obtained by applying tension in a perpendicular direction, due to the Poisson effect. On the other hand, the effect of stress on κ can be due to the stress affecting the material's microstructure (e.g., the preferred orientation) and/or composition distribution (e.g., the distribution of certain ions). This is piezopermittivity, which is to be distinguished from piezoelectricity.

The effect on Q tends to be highly dependent on the direction, since it typically involves change in the separation between the positive and negative charge centers and the effect on the separation is greatest when the stress is in the same direction as the polarization. On the other hand, the effect on κ tends to be associated with microstructural effects that may not be highly dependent on the direction.

In spite of the difference in mechanism, the ΔQ and $\Delta\kappa$ effects cannot be distinguished by measurement of the electrical output, since both effects convert mechanical energy to electrical energy. Distinction requires measurement of κ at various stresses.

If $\Delta\kappa$ is small compared to κ and ΔQ is small compared to Q (a case which is practically not very important), the third term on the right side of Eq. (3.111) may be neglected, so that

$$\Delta P = (\kappa - 1)(\Delta Q/A) + (\Delta\kappa)(Q/A). \qquad (3.114)$$

However, if both $\Delta\kappa$ and ΔQ are large (a case which is rare), the third term dominates in Eq, (3.111), so that

$$\Delta P = \Delta\kappa\, \Delta Q/A. \qquad (3.115)$$

Equation (3.114) and (3.115) describe effects that involve both ΔQ and $\Delta\kappa$.

At the same stress, ΔQ and $\Delta\kappa$ may have the same or opposite signs. In case that they have the same sign, with $\Delta\kappa$ being small compared to κ and ΔQ being small compared to Q, the ΔQ and $\Delta\kappa$ effects combine to result in a larger ΔP (Eq. (3.114)); in case that they have opposite signs, the ΔQ and $\Delta\kappa$ effects combine to result in a smaller ΔP (Eq. (3.114)).

In case that both $\Delta\kappa$ and ΔQ are large, ΔP is negative when either $\Delta\kappa$ or ΔQ is negative, and is positive when both $\Delta\kappa$ and ΔQ are either positive or negative (Eq. (3.115)).

3.8.1.2 *Effect of stress on the charge center separation*

Based on Eq. (3.112), the direct piezoelectric effect, which is associated with the effect of stress on Q when the effect of stress on κ is negligible, is described by the equation

$$\Delta P = (\kappa - 1)\,\Delta Q/A, \qquad (3.116)$$

Thus, Eq. (3.109) can be rewritten as

$$d = \Delta P/\Delta U, \qquad (3.117)$$

or

$$\Delta P = d\, \Delta U. \tag{3.118}$$

Equation (3.118) is an equation that is commonly used to describe the direct piezoelectric effect. It means that the change in polarization ΔP that results from the change in stress ΔU is proportional to ΔU, with the proportionality constant being d. From Eq. (3.106), the unit of d is C/N (which is the same as V/m), since the unit of It is A.s = C. Equation (3.116) and Eq. (3.117) also mean that d is relatively high when κ is high.

From Eq. (3.105) and the fact that $E = V/l$,

$$\Delta l/l = d\, V/l. \tag{3.119}$$

Rearrangment gives

$$V = \Delta l/d. \tag{3.120}$$

Using Hooke's Law, the stress is given by

$$\Delta U = M\, \Delta l/l, \tag{3.121}$$

where M is the elastic modulus. Combination of Eq. (6.121) and Eq. (3.120) gives

$$V = l\, (\Delta U)/(Md). \tag{3.122}$$

Equation (3.122) can be rewritten as

$$V = g\, l\Delta U, \tag{3.123}$$

where g, called the piezoelectric voltage coefficient (although this term is strictly used for describing the piezoelectric effect only), is defined as

$$g = 1/(Md). \tag{3.124}$$

The unit of g is V^2/C, which is the same as V.m/N. Equation (3.123) means that the voltage output is proportional to the product of ΔU and l, with the proportionality constant being g. In other words, g is equal to the electric field (V/l) divided by ΔU. Equation (3.124) means that g and

d are inversely related through the modulus M. Hence, g can be simply calculated from M and d. Equation (3.123) and Eq. (3.124) also mean that the voltage output is relatively high when the modulus M is low. Thus, a structural material that has a high modulus tends to be weak in the voltage output.

Example Problems 20–27 at the end of the chapter are relevant.

3.8.1.3 *Effect of stress on the electric permittivity*

The effect of stress on the permittivity is known as piezopermittivity [6]. More exactly, piezopermittivity refers to the reversible change of the electric permittivity with elastic strain. It is useful for capacitance-based strain/stress sensing.

The capacitance C relates to relative permittivity κ of the specimen by Eq. (3.2). This equation is well-known for a parallel-plate capacitor. However, it applies to a capacitor with coplanar electrodes also, provided that the cross-sectional area of the capacitor in the direction perpendicular to the capacitance direction is well-defined. Based on Eq. (3.2), the fractional change in C upon the change in l is given by

$$\frac{\Delta C}{C} = \frac{\Delta \kappa}{\kappa} + \frac{\Delta A}{A} - \frac{\Delta l}{l}$$

$$= \frac{\Delta \kappa}{\kappa} - \frac{\Delta l}{l}(1 + 2v), \tag{3.125}$$

where v is the Poisson's ratio. In Eq. (3.125), it is assumed that the material is isotropic in the two directions that are transverse to the l direction. If this assumption is not made, the term $2v$ would be replaced by $v_1 + v_2$, where v_1 and v_2 are the values of the Poisson's ratio in the two transverse directions. According to Eq. (3.125), both the change in κ (i.e., $\Delta \kappa$) and the change in l (i.e., Δl) contribute to the change in C (i.e., ΔC).

Let us consider three cases. In Case 1, $(\Delta \kappa)/\kappa \gg (\Delta l)/l$. In Case 2, $(\Delta \kappa)/\kappa \sim (\Delta l)/l$. In Case 3, $(\Delta \kappa)/\kappa \ll (\Delta l)/l$.

In Case 1, $(\Delta \kappa)/\kappa \gg (\Delta l)/l$, Eq. (3.125) becomes

$$\frac{\Delta C}{C} = \frac{\Delta \kappa}{\kappa}, \tag{3.126}$$

and the capacitance-based strain sensing effectiveness (defined as the fractional change in capacitance per unit strain) is given by

$$\frac{\frac{\Delta C}{C}}{\frac{\Delta l}{l}} = \frac{\frac{\Delta \kappa}{\kappa}}{\frac{\Delta l}{l}}. \tag{3.127}$$

Tensile strain is positive, whereas compressive strain is negative. If $(\Delta\kappa)/\kappa$ is positive when $(\Delta l)/l$ is positive, or if $(\Delta\kappa)/\kappa$ is negative when $(\Delta l)/l$ is negative, the piezopermittivity is said to be positive. If $(\Delta\kappa)/\kappa$ is negative when $(\Delta l)/l$ is positive, or if $(\Delta\kappa)/\kappa$ is positive when $(\Delta l)/l$ is negative, the piezopermittivity is said to be negative.

In Case 2, $(\Delta\kappa)/\kappa \sim (\Delta l)/l$. For the specific situation under Case 2 in which $(\Delta\kappa)/\kappa = (\Delta l)/l$, Eq. (3.125) becomes

$$\frac{\Delta C}{C} = -\frac{\Delta l}{l}(2v), \tag{3.128}$$

and the sensing effectiveness is given by

$$\frac{\frac{\Delta C}{C}}{\frac{\Delta l}{l}} = -2v. \tag{3.129}$$

In Case 3, $(\Delta\kappa)/\kappa \ll (\Delta l)/l$. Then Eq. (3.125) becomes

$$\frac{\Delta C}{C} = -\frac{\Delta l}{l}(1 + 2v), \tag{3.130}$$

and the sensing effectiveness is given by

$$\frac{\frac{\Delta C}{C}}{\frac{\Delta l}{l}} = -(1 + 2v). \tag{3.131}$$

With $(\Delta\kappa)/\kappa$ being negligible in Case 3, the capacitance change is not a manifestation of piezopermittivity, but is merely due to the dimensional changes. Comparison of Eq. (3.125) and Eq. (3.131) shows that the fractional change in capacitance has a term missing in Case 3, thus tending to reduce the strain sensing effectiveness. Nevertheless, Case 3 commonly occurs in practical sensing, partly due to the common assumption that κ is not affected by the strain.

Since v is positive, Eq. (3.129) and (3.131) indicate that the sensing effectiveness is negative for both Case 2 and Case 3. In other words, $(\Delta C)/C$ and $(\Delta l)/l$ are opposite in sign. Since $v < 0.5$, the maximum magnitude of the sensing effectiveness is -1 for the specific equality condition under Case 2, and is -2 for Case 3. This means that the sensing effectiveness is limited in both Case 2 and Case 3. In contrast, in Case 1, Eq. (3.127) indicates that the sensing effectiveness is dictated by $(\Delta \kappa)/\kappa$, which has a magnitude that is not limited. Thus, for achieving high sensing effectiveness, Case 1 is preferred. Although Case 3 (with $(\Delta \kappa)/\kappa$ being negligible) does not correspond to piezopermittivity, it allows a typically less sensitive form of capacitance-based strain sensing. In Case 3, the sensitivity can be high if $(\Delta l)/l$ is high, as in the case of an elastomeric material.

The capacitance C in the above theory is the volumetric capacitance of the material, with the capacitance associated with the specimen-electrode interface excluded. The poorer is the interface, the smaller is the interfacial capacitance and the lower is the measured capacitance, which is the series capacitance of the volumetric capacitance and two interfacial capacitances Eq. (3.86).

The piezopermittivity stems from the effects of strain on the dielectric continuity and conduction continuity. The dielectric continuity relates to the effectiveness of dipole-dipole interaction in the direction of the permittivity. In other words, if a chain of dipoles has the dipole-dipole interaction broken at one or more points in the chain, the dielectric continuity is lost within the chain. The conduction continuity relates to the effectiveness for carrier movement in the direction of the conductivity. The microstructure, which can change upon stress application, affects both types of continuity.

The DC conductivity is more sensitive to the microstructure than the AC capacitance, due to the larger-distance excursion of the carriers under DC condition. Tension tends to decrease both types of continuity, while compression tends to increase them. The degree of continuity tends to change reversibly in the elastic regime of deformation, whereas it changes with increasing irreversibility when the elastic limit is increasingly exceeded.

The scientific origin of the electric polarization that underlies the permittivity differs between conductive materials (e.g., carbons and metals) and nonconductive materials (e.g., ceramics and polymers). The latter has long been studied, but the former has only been studied recently. The dielectric behavior of a nonconductive material typically stems from the asymmetry in the crystal structure and the movement of ions or polar bonds in response to an applied electric field. However, the dielectric behavior of a conductive material typically stems from the interaction of a small fraction of the charge carriers with the atoms. This interaction causes some carriers to engage with some of the atoms, thus causing polarization.

Equation (3.113) describes the effect of stress on κ when the effect of stress on Q negligible. The electrical energy per unit volume associated with the change in κ is given by the product of the polarization change ΔP and the electric field E. Using Eq. (3.113), this energy is given by

$$\text{energy per unit volume associated with the change in } \kappa$$

$$= (\Delta P)\, E$$

$$= (\Delta\kappa)\,(Q/A)\, E. \tag{3.132}$$

Since the volume is Al,

$$\text{energy associated with the change in } \kappa$$

$$= (\Delta\kappa)\,(Q/A)\, E\, A\, l = (\Delta\kappa)\, Q\, E\, l. \tag{3.133}$$

Since $E = V/l$,

$$\text{energy associated with the change in } \kappa$$

$$= (\Delta\kappa)\, Q\,(V/l)\, l = (\Delta\kappa)QV \tag{3.134}$$

This electrical energy is responsible for the voltage (V) or current (I) output. Hence,

$$(\Delta\kappa)QV = VIt. \qquad\qquad (3.135)$$

Rearrangement gives

$$I = (\Delta\kappa)Q/t. \qquad\qquad (3.136)$$

Assuming that the mechanical energy input equals the electrical energy output,

$$\Delta F \,\Delta l = (\Delta\kappa)QV = VIt, \qquad\qquad (3.137)$$

where Eq. (3.128) and (3.102) have been used.
Example Problem 19 at the end of the chapter is relevant.

3.8.2 *Conversion of electrical energy to mechanical energy*

In case of the converse piezoelectric effect, the conversion of electrical energy to mechanical energy involves the effect of an applied electric field on the separation of the center of positive charge and the center of negative charge, thus changing the polarization, as described in Eq. (3.112). The change in this separation in turn results in a strain. Another mechanism of this energy conversion (not the converse piezoelectric effect) involves the effect of an applied electric field on κ, thus changing the polarization, as described in Eq. (3.113). The change in κ in turn results in a strain.

If the specimen is not constrained at all, the energy conversion gives a strain, with stress being zero. However, if the specimen is constrained from any expansion and expansion rather than contraction is the consequence of the energy conversion, the energy conversion gives a stress, with zero strain. In general, the specimen may be partially constrained, so it gives a nonzero strain that is below the unconstrained value and gives a nonzero stress that is below the fully constrained value, as illustrated in Fig. 3.35. In the totally constrained state, the strain output is zero, but the stress output is the highest. A nonzero power output requires both a nonzero stress and a nonzero strain, since work is given by the product of force and distance. Therefore, maximum work occurs at the midpoint of the curve in Fig. 3.35.

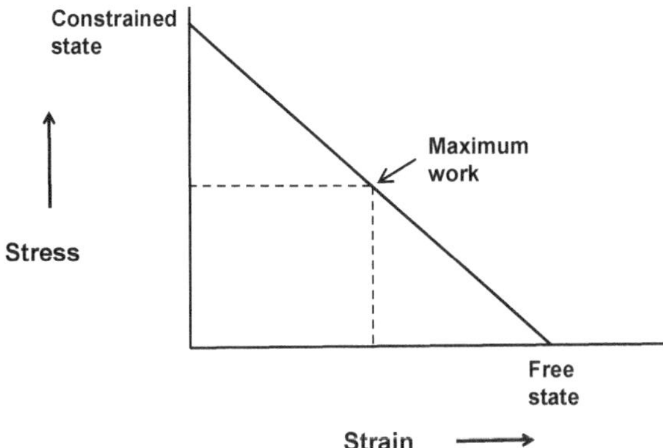

Fig. 3.35 Relationship of the stress with the strain. The stress and strain result from the application of an electric field on a dielectric material.

3.8.2.1 *Effect of electric field on the charge center separation*

In the converse piezoelectric effect (also known as the reverse piezoelectric effect), an applied electric field results in a strain or stress, thereby allowing the conversion of electrical energy to mechanical energy. This effect stems from the applied electric field changing the distance separation between with the positive and negative charge centers in the dielectric material, thereby causing change in thickness (i.e., strain) in the dielectric material. In case that the applied electric field is such that the negative end of the associated voltage is at the negative electrode (Fig. 3.36(b)), the separation of the centers of positive and negative charges is increased, thereby resulting in an increase in thickness of the dielectric material, i.e., a positive strain, as shown by comparing Fig. 3.36(b) with Fig. 3.36(a). In case that the applied electric field is such that the negative end of the associated voltage is at the positive electrode (Fig. 3.36(c)), the separation of the centers of positive and negative charges is decreased, thereby resulting in a decrease in thickness of the dielectric material, i.e., a negative strain, as shown by comparing Fig. 3.36(c) with Fig. 3.36(a).

An example of converse piezoelectric behavior is shown in Fig. 3.36(d) [7]. It is a plot of the strain output vs. the applied DC electric field for three-dimensionally (3D) printed polymer in the direction perpendicular to the printed layers. In the range of low electric field, the strain is linearly related to the electric field, with the curve going through the origin. The slope of the plot is equal to the piezoelectric coupling coefficient d (0.2 pC/N in this case), according to Eq. (3.119). However, in the high range of the electric field, the slope is diminished, due to the saturation of the strain. The effect stems from a degree of alignment of the molecules in a printed layer. The alignment is a consequence of the shear stress experienced by the resin (pre-polymer) during the printing.

Figure 3.36 illustrates the case in which the material is not constrained at all, so that it is totally free to change its dimension. Hence, the output is the highest strain possible, but it gives zero stress. This corresponds to the free state in Fig. 3.35, which shows the variation of the stress with strain as the degree of constraint is varied.

As in Sec. 3.8.1.2, assume that the electrical energy input (Y) equals the mechanical energy output (X). Thus, the equations in Sec. 3.8.1.2 for the direct piezoelectric effect apply. In particular, Eq. (3.105) is an

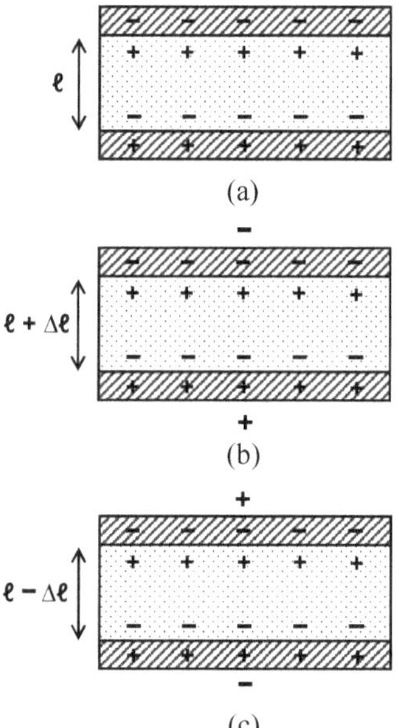

Fig. 3.36 The converse piezoelectric effect. (a) Without an applied electric field. (b) The thickness increased by Δl in the presence of an applied electric field with the polarity shown. (c) The thickness decreased by Δl in the presence of an applied electric field with the polarity shown. (d) Plot of the strain output vs. the applied DC electric field for three-dimensionally (3D) printed polymer in the direction perpendicular to the printed layers. [7]

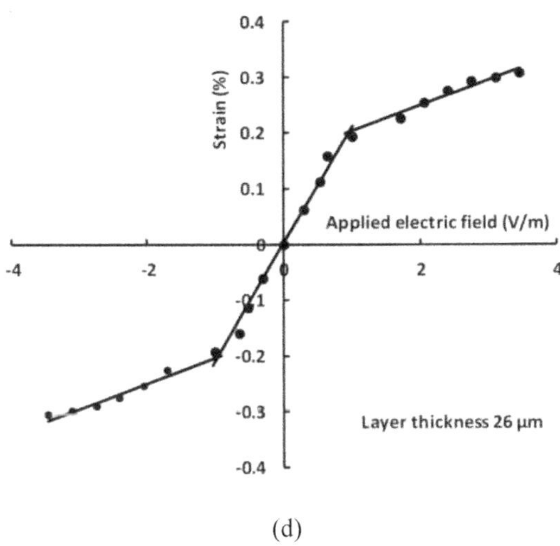

(d)

Fig. 3.36 (*Continued*).

equation that is commonly used to describe the converse piezoelectric effect. This equation means that the strain is proportional to the electric field, with the proportionality constant being d.

That the equations are the same for the direct and converse effects means that a dielectric material that exhibits the direct effect should also exhibit the converse effect, and vice versa. From Eq. (3.121), a high modulus results in a low strain for the same energy. Therefore, a structural material with a high modulus tends to be weak in the strain/stress output.

3.8.2.2 *Effect of electric field on the permittivity*

In general, the relative permittivity κ of a material depends on the electric field. In other words, the plot of polarization versus electric field is not linear, so its slope depends on the electric field. The dependence of κ on the electric field stems from the requirement of various levels of electric field for interaction with various species (e.g., molecules that are attached to various components) in the material. The higher is the electric field, the more are the types of species that can respond to the field. The interaction of the field with the material results in microstructural effects, which give rise to strain.

3.8.3 *Electromechanical coupling factor*

The electromechanical coupling factor (also known as the electromechanical coupling coefficient) is abbreviated k and is defined as

$$k^2 = \text{output mechanical energy/input electrical energy} \qquad (3.138)$$

or

$$k^2 = \text{output electrical energy/input mechanical energy} \qquad (3.139)$$

Equation (3.138) applies when the conversion is from electrical energy to mechanical energy. Equation (3.139) applies when the conversion is from mechanical energy to electrical energy. Due to non-ideal operating conditions, the value of k^2 in the two directions of energy conversion may be different. The factor k^2 describes the efficiency of energy conversion. For a piezoelectric material, k is typically less than 0.1. For a conventional piezoelectric material (Sec. 3.8.4), k is typically between 0.4 and 0.7 (e.g., 0.38 for $BaTiO_3$ and 0.66 for $PbZrO_3$-$PbTiO_3$ solid solution).

3.8.4 *Piezoelectric materials*

The piezoelectric effect is well-known for ceramic materials that exhibit the distorted Perovskite structure (the slightly tetragonal form of the cubic crystal structure shown in Fig. 3.37). These ceramic materials are expensive and are not rugged mechanically, so they are not usually used as structural materials. Devices made from these ceramic materials are commonly embedded in a structure or attached on a structure.

Another method of attaining a structural composite that is itself piezoelectric is to use piezoelectric fibers as the reinforcement in the composite. However, fine piezoelectric fibers with acceptable mechanical properties are not well developed. Their high cost is another issue.

Yet another method to attain a structural composite that is itself piezoelectric is to use conventional reinforcing fibers (e.g., carbon fibers) while exploiting the polymer matrix for the reverse piezoelectric effect. Due to some ionic character in the covalent bonds, some polymers are

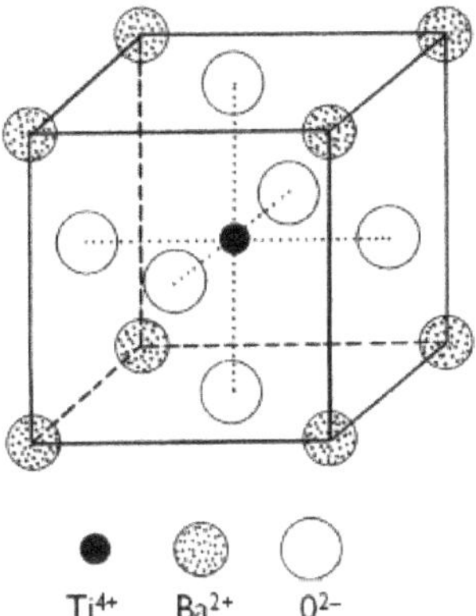

Fig. 3.37 The perovskite crystal structure. The general formula is ABX_3, where A and B are two cations that are very different in size and X is an anion that bonds to both. The A atoms are larger than the B atoms. In the undistorted cubic unit cell, as illustrated, the B cation is surrounded by six cations (octahedral coordination). An example is $BaTiO_3$, with Ba being A, Ti being B and O being X, as shown.

expected to polarize in response to an electric field, thereby causing strain. The polarization is enhanced by molecular alignment, which is in turn enhanced by the presence of the fibers. This method involves widely available fibers and matrix and is thereby attractive economically and practically. The converse piezoelectric effect occurs in the through-thickness direction of a continuous carbon fiber nylon-6 matrix composite. Nylon-6 is a type of nylon.

A subset of piezoelectric materials is ferroelectric. In other words, a ferroelectric material is also piezoelectric. However, it has the extra ability to have the electric dipoles in adjacent unit cells interact with one another in such a way that adjacent dipoles tend to align themselves. This phenomenon is called self-polarization, which results in ferroelectric domains within each of which all the dipoles are in the same direction. As a consequence, even a polycrystalline material can become a single domain.

Quartz is piezoelectric but not ferroelectric. It is a frequently used piezoelectric material due to its availability in large single crystals, its shapeability and its mechanical strength. On the other hand, barium titanate ($BaTiO_3$), lead titanate ($PbTiO_3$) and solid solutions such as lead zirconotitanate ($PbZrO_3$-$PbTiO_3$, abbreviated PZT) and (Pb,La)-$(Ti,Zr)O_3$ (abbreviated PLZT) are both piezoelectric and ferroelectric.

Barium titanate ($BaTiO_3$) exhibits the Perovskite structure (Fig. 3.37) above 120°C and a distorted Perovskite structure below 120°C. Above 120°C, the crystal structure of $BaTiO_3$ is cubic; below 120°C, it is slightly tetragonal ($a = b = 3.98$ Å, $c = 4.03$ Å). In other words, $BaTiO_3$ exhibits a solid-solid phase transformation at 120°C. The distorted structure is inherently associated with polarization, due to the asymmetry and the consequent separation of the positive and negative charge centers in the unit cell. In contrast, the undistorted cubic structure is symmetrical, with the centers of positive and negative charges overlapping at the body center of the unit cell (Fig. 3.37). Thus, the cubic form does not inherently have polarization. The tetragonal form of $BaTiO_3$ is ferroelectric, which refers to the state in which the adjacent electric dipoles automatically align with one another. The cubic form is paraelectric, which refers to the state in which the electric dipoles do not interact with one another, but they individually respond to the applied electric field. The absence of interaction means that there is no domain, in contrast to the presence of domains in a ferroelectric material. Due to the absence of the interaction between dipoles, the polarization is much weaker for a paraelectric material than a ferroelectric material at the same applied electric field. Due to this phase transformation, $BaTiO_3$ is ferroelectric below 120°C, which is called the Curie temperature. (Jacques and Pierre Curie experimentally confirmed the piezoelectric effect over 100 years ago.)

As shown in Fig. 3.38(a), the polarization decreases to a residual value at the Curie temperature, such that the decrease starts to occur at temperatures considerably below the Curie temperature. The decrease is most abrupt at the Curie temperature, where the change in crystal structure occurs. The decrease in polarization with increasing temperature below the Curie temperature is due to the increase in the thermal energy and the consequent decreasing tendency for adjacent electric dipoles to align. Due to the higher value of the relative permittivity for the tetragonal form compared to the cubic form, the relative permittivity abruptly decreases as the crystal structure changes from the tetragonal form to the cubic form upon heating, as shown in Fig. 3.38(b). At the Curie temperature, the relative permittivity peaks, due to the change in crystal structure causing an anomaly in the relative permittivity. The Curie temperature is 494°C for $PbTiO_3$ and 365°C for PZT (i.e., $PbZrO_3$-$PbTiO_3$ solid solution). It limits the operating temperature range of a piezoelectric material.

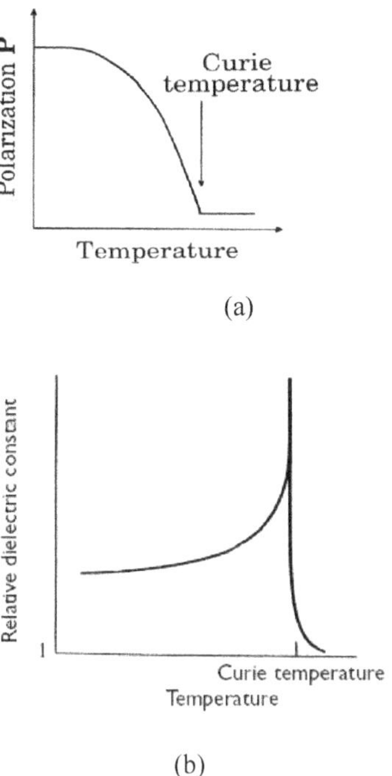

Fig. 3.38 The effect of temperature on a ferroelectric material. (a) The polarization. (b) The relative permittivity (the lowest value of which is 1).

In cubic $BaTiO_3$, the centers of positive and negative charges overlap as the ions are symmetrically arranged in the unit cell, so cubic $BaTiO_3$ is not ferroelectric. In tetragonal $BaTiO_3$, the O^{2-} ions are shifted in the negative c-direction, while the Ti^{4+} ions are shifted in the positive c-direction, thus resulting in an electric dipole along the c-axis (Fig. 3.39). The dipole moment per unit cell can be calculated from the displacement and charge of each ion and summing the contributions from the ions in the unit cell. There are one Ba^{2+}, one Ti^{4+} ion and three O^{2-} ions per unit cell (Fig. 3.37). Consider all displacements with respect to the Ba^{2+} ions. The contribution to the dipole moment of a unit cell by each type of ion is listed in Table 3.3. Hence, the dipole moment per unit cell is 17×10^{-30} C.m. The polarization P is the dipole moment per unit cell, so it is given by the dipole moment per unit cell divided by the volume of a unit cell. Thus,

$$P = \frac{17 \times 10^{-30}\,\text{C.m}}{4.03 \times 3.98^2 \times 10^{-30}\,\text{m}^3} = 0.27\,\text{C.m}^{-2} \qquad (3.140)$$

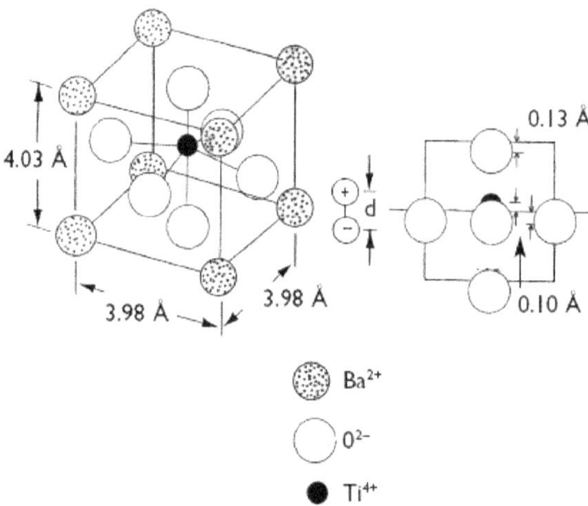

Fig. 3.39 Crystal structure of tetragonal BaTiO₃. The sketch exaggerates the shifts of the atoms from the symmetrical positions.

Table 3.3 Contribution to dipole moment of a BaTiO₃ unit cell by each type of ion.

Ion	Charge (C)	Displacement (m)	Dipole moment (C.m)
Ba^{2+}	$(+2)(1.6 \times 10^{-19})$	0	0
Ti^{4+}	$(+4)(1.6 \times 10^{-19})$	$+0.10(10^{-10})$	6.4×10^{-30}
$2O^{2-}$ (side of cell)	$2(-2)(1.6 \times 10^{-19})$	$-0.10(10^{-10})$	6.4×10^{-30}
O^{2-} (top and bottom of cell)	$(-2)(1.6 \times 10^{-19})$	$-0.13(10^{-10})$	4.2×10^{-30}
			Total $= 17 \times 10^{-30}$

Because tetragonal BaTiO$_3$ is almost cubic in structure, very slight movement of the ions within a grain can change the c-axis of that grain to a parallel or orthogonal direction. When $E > 0$, more grains are lined up with the electric dipole moment in the same direction as E, so $P > 0$. P increases with E much more sharply for a ferroelectric material than a material that has $P = 0$ within each grain at $E = 0$. The latter is called a paraelectric material (Fig. 3.40).

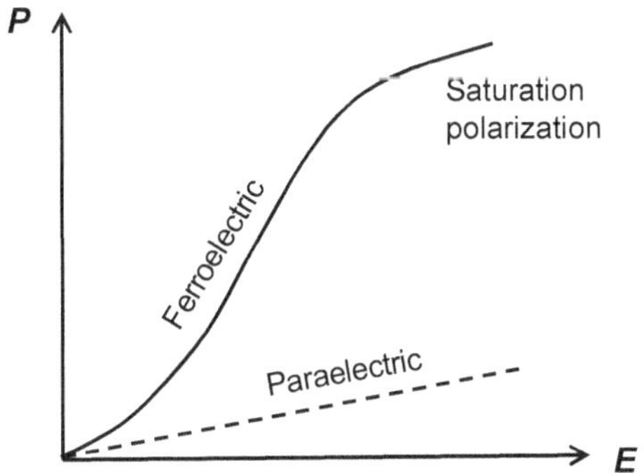

Fig. 3.40 Plots of polarization P versus electric field E for a ferroelectric material (solid curve) and a paraelectric material (dashed curve). The curve is essentially linear for a paraelectric material (as illustrated in Fig. 3.10(a)), but it is nonlinear for a ferroelectric material. The value of κ, which relates to the slope of the curve of P vs. E, is much higher for a ferroelectric material than a paraelectric material. Thus, ferroelectric materials are technologically much more important than paraelectric materials.

The process of having more and more grains with the dipole moment in the same direction as E can be viewed as the movement of the grain boundaries (also called domain boundaries) such that the grains with the dipole moment in the same direction as E grow while the other grains shrink. These grains are also known as domains (or ferroelectric domains).

When all domains have their dipole moment in the same direction as E, the ferroelectric material becomes a single domain (a single crystal) and P has reached its maximum, which is called the saturation polarization. Its value is 0.27 C/m^2 for BaTiO$_3$ and $\sim$ 0.5 C/m^2 for PZT.

Consider a polycrystalline piece of tetragonal BaTiO$_3$, such that the grains are oriented with the c-axis of each grain along one of six orthogonal or parallel directions (i.e., $+x$, $-x$, $+y$, $-y$, $+z$ and $-z$). When the applied electric field E is zero, the polarization P is non-zero within each grain, but the total polarization of all the grains together is zero, since different grains are oriented differently. Therefore, when $E = 0$, $P = 0$.

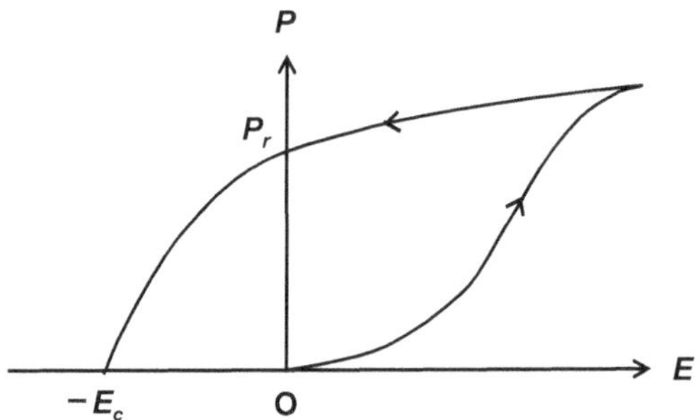

Fig. 3.41 Variation of polarization P with electric field E during increase of E from 0 to a positive value and subsequent decrease of E to a negative value ($-E_c$) at which P returns to 0. Ferroelectric behavior is characterized by a positive value of the remanent polarization P_r and a negative value of $-E_c$.

Upon decreasing E after reaching the saturation polarization, domains with dipole moments not in the same direction as E appear again and they grow as E decreases. At the same time, domains with dipole moments in the same direction as E shrink. This process again involves the movement of domain boundaries. In spite of this tendency, P does not return all the way to zero when E returns to zero. A remanent polarization ($P = P_r > 0$) remains when $E = 0$ (Fig. 3.41 and 3.10(b)).

In order for P to return all the way to zero, an electric field must be applied in the reverse direction. The required electric field is $E = -E_c$, where E_c is called the coercive field (Fig. 3.41). The coercive field should be sufficiently high for a ferroelectric memory. Otherwise, the stored information may be easily lost.

When E is even more negative than $-E_c$, the domains start to align in the opposite direction until the polarization reaches saturation in the reverse direction. This is known as polarization reversal. To bring the negative polarization back to zero, a positive electric field is needed. In this way, the cycling of E results in a hysteresis loop in the plot of P versus E (Fig. 3.42(a)). The area within the loop is the hysteresis energy

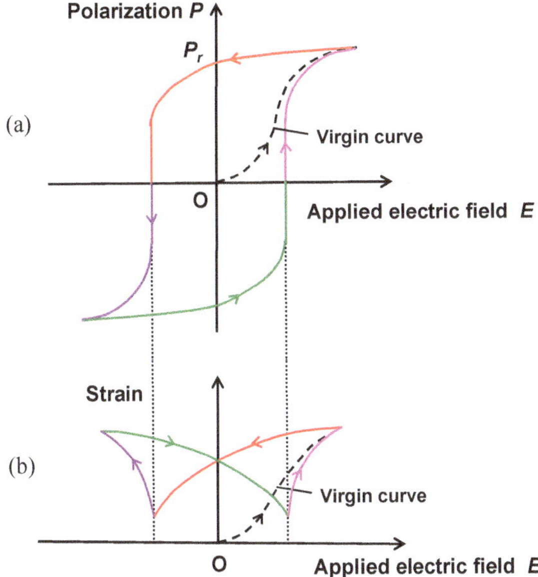

Fig. 3.42 (a) Plot of polarization P versus electric field E during first application of E (dashed curve) and during subsequent cycling of E (solid curves of various colors). (b) Corresponding plot of strain versus electric field E. The colors in (a) and (b) are coordinated. For example, the red part of the curve in (a) corresponds to the red part of the curve in (b).

loss per unit volume due to one cycle of variation in E. The corresponding change in strain $\Delta l/l$ associated with the change in polarization is shown in Fig. 3.42(b), which is called the butterfly curve. The slope of the butterfly curve is d (Eq. (3.105)). The entire plot in Fig. 3.42(b) may be downshifted vertically, reflecting the case of more reversibility in the polarization change. In an ideal case, in the range of low electric field, the strain is linearly related with the electric field, such that the curve goes through the origin, with the slope being equal to the piezoelectric coupling coefficient, as shown in Fig. 3.36(d). This ideal linear curve corresponds to the non-ideal red curve in Fig. 3.42(b).

A ferroelectric material can be used as a computer memory to store digital information, as a positive remanent polarization can represent '0' while a negative remanent polarization can represent '1'. The application of an electric field in the appropriate direction can change the stored information from '0' to '1', or vice versa. A ferroelectric memory is advantageous in its high writing speed, low power consumption and high endurance. However, it suffers from a low memory capacity (e.g., only 256 Kb), due to inadequately reliable material characteristics.

Poling refers to the process in which the domains (i.e., dipoles) are aligned so as to achieve saturation polarization. The process typically involves placing the ferroelectric material in a heated oil bath (e.g., 90°C, below the Curie temperature) and applying an electric field (typically above 2 kV/mm). The heating allows the domains to rotate more easily in the electric field. The electric field is maintained until the oil bath is cooled down to room temperature. An excessive electric field is to be avoided as it can cause dielectric breakdown (Sec. 3.7) in the ferroelectric material.

After poling, depoling occurs spontaneously. The depoling process takes time and the rate of depoling decreases with increasing time. This process is known as aging or piezoelectric aging. The polarization P at time t can be expressed as

$$P = P_o\, e^{-t/\tau}, \tag{3.141}$$

where P_o is the polarization at $t = 0$, t is the time from the start of depoling and τ, known as the time constant for this process, is the time at which P has decreased to $1/e$ of P_o, as shown in Fig. 3.43(a).

Taking logarithm (to the base 10) of both sides of Eq. (3.141) gives

$$\log P = \log P_o - t/[(2.3)\tau]. \tag{3.142}$$

Hence, the plot of $\log P$ vs. t is a straight line of slope equal to $-1/[(2.3)\tau]$, as shown in Fig. 3.43(b). Thus, τ can be determined from theslope of the plot. The greater is τ, the lower is the rate of the depolarization.

During use, a ferroelectric material should not encounter an excessive electric field. Typically, the maximum electric field is 2 kV/mm in the poling direction and 300 V/mm in the direction opposite to the poling direction. An excessive electric field in the reverse direction may disturb the poling. An excessive field in the poling direction may cause dielectric breakdown and irreversible damage.

Through doping (i.e., substituting for some of the barium and titanium ions in $BaTiO_3$ to form a solid solution), the permittivity can

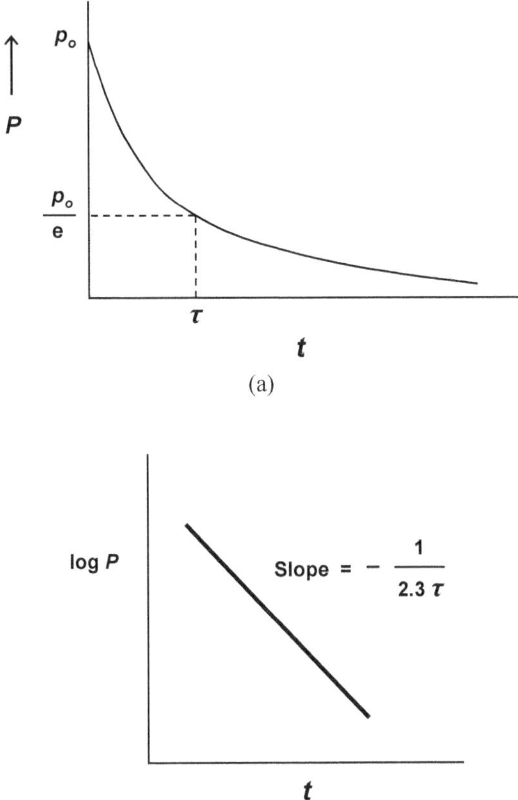

Fig. 3.43 (a) Plot of polarization P vs. time t from the start of depoling. (b) Plot of log P vs. t, with the slope equal to $-1/[(2.3)\tau]$, where τ is the time constant for the depolarization.

be increased, the loss factor (Eq. (3.69)) can be decreased, and the temperature dependence of the loss factor, tan δ, can be flattened. The Ba^{2+} sites in $BaTiO_3$ are known as A sites; the Ti^{4+} sites are known as B sites. Substitutions for one or both of these sites can occur provided that the overall stoichiometry is maintained. Possible substitutions are shown in Fig. 3.44. For example, an A site and a B site can be substituted by an Na^+ ion and an Nb^{5+} ion respectively, since the two sites together should have a charge of 6+. As another example, three B sites can be substituted by an Mg^{2+} ion and two Nb^{5+} ions, since the three sites together should have a charge of 12+. As a further example, three A sites can be substituted by two La^{3+} ions and a cation vacancy, since the three sites together should have a charge of 6+.

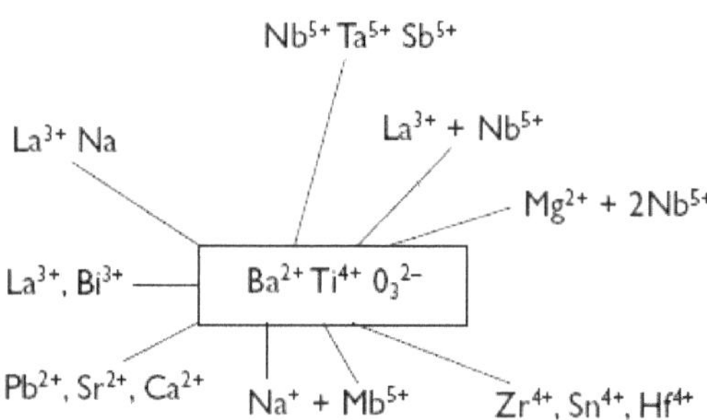

Fig. 3.44 Possible substitutions of the A and B sites in $BaTiO_3$.

A solute can be an electron acceptor or an electron donor. In the substitution of a B site by Fe^{3+}, Fe^{3+} is an acceptor; by accepting an electron, it changes from Fe^{4+} (unstable, but similar to Ti^{4+} in charge) to Fe^{3+}. However, in order to maintain overall charge neutrality, the change of two Fe^{4+} ions to two Fe^{3+} ions must be accompanied by the creation of one O^{2-} vacancy. The O^{2-} vacancies move by diffusion, thereby decreasing the loss factor. Moreover, the Fe^{3+} ion and O^{2-} vacancies form dipoles which align with the polarization of the domain, thus pinning the domain walls. The substitution of two A sites with an La^{3+} ion and an Na^+ ion maintains charge neutrality, such that La^{3+} is the donor and Na^+ is an acceptor. The substitution of an A site by La^{3+} and a B site by Nb^{5+} is also possible; both La^{3+} and Nb^{5+} are donors in this case. The substitution of an A site by La^{3+} without substitution of a B site will require a reduction in oxygen vacancies and/or the creation of cation vacancies in order to maintain charge neutrality. Donors are not as effective as acceptors in pinning the domain walls.

The most common commercial ferroelectric material is PZT (i.e., $PbZrO_3$-$PbTiO_3$ solid solution or lead zirconotitanate). It is also written as $Pb(Ti_{1-z}Zr_z)O_3$ in order to indicate that a fraction of the B sites is substituted by Zr^{4+} ions. Figure 3.45 shows the binary phase diagram of

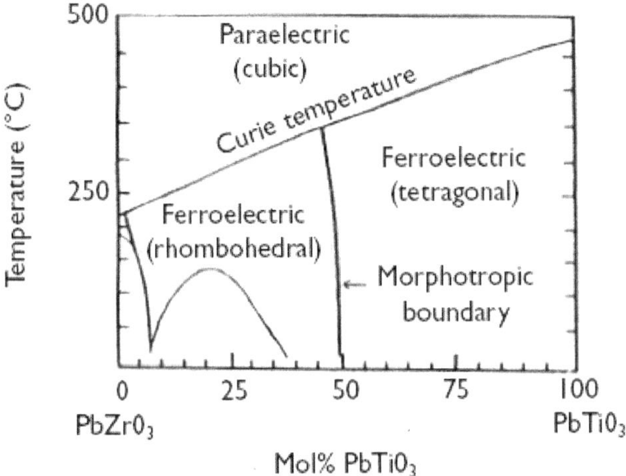

Fig. 3.45 The PbZrO₃-PbTiO₃ binary phase diagram.

the $PbZrO_3$-$PbTiO_3$ system. The cubic phase at high temperatures is paraelectric. The tetragonal and rhombohedral phases at lower temperatures are ferroelectric. The tetragonal phase is titanium rich; the rhombohedral phase is zirconium rich. The tetragonal phase has larger polarization than the rhombohedral phase. A problem with the tetragonal phase is that the tetragonal long axis is 6% greater than the transverse axis, thus causing stress during polarization (so much stress that the ceramic can be shattered). The substitution of Zr for Ti alleviates this problem. The compositions near the morphotropic phase boundary between the rhombohedral and tetragonal phases are those that pole most efficiently. This is because these compositions are associated with a large number of possible poling directions over a wide temperature range. As a result, the piezoelectric coupling coefficient d is highest near the morphotropic boundary (Fig. 3.46).

PZT's are divided into two groups, namely hard PZT's and soft PZT's. This division is according to the difference in piezoelectric properties (not mechanical properties). Hard PZT's generally have low relative permittivity, low loss factor and low piezoelectric coefficients, and are relatively difficult to pole and depole. Soft PZT's generally have high relative permittivity, high loss factor and high piezoelectric coefficients, are relatively easy to pole and depole.

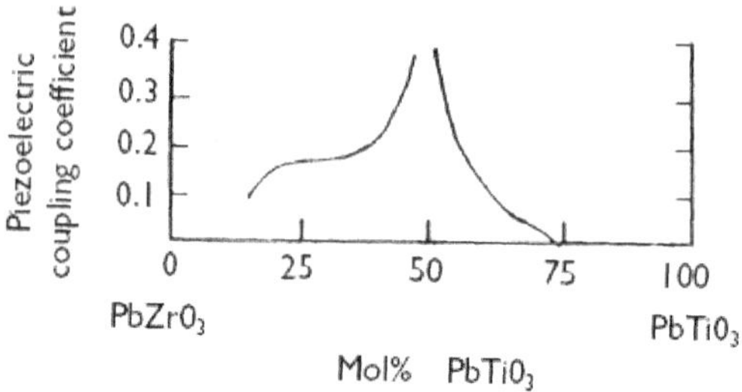

Fig. 3.46 Variation of the piezoelectric coupling coefficient d with the composition ranging from pure $PbZrO_3$ to pure $PbTiO_3$ (i.e., the composition spanning the phase diagram in Fig. 3.45).

Table 3.4 lists the properties of two commercial PZT's, namely PZT-5H (soft PZT) and PZT4 (hard PZT). Hard PZT's are doped with acceptors such as Fe^{3+} for Zr^{4+}, thus resulting in oxygen vacancies. Soft PZT's are doped with donors such as La^{3+} for Pb^{2+} and Nb^{5+} for Zr^{4+}, thus resulting in A-site vacancies. Hard PZT's typically having a small grain size (about 2 μm), whereas soft PZT's typically have a larger grain size (about 5 μm).

Table 3.4 Properties of commercial PZT ceramics

Property	PZT-5H (soft)	PZT4 (hard)
Relative permittivity (κ at 1 kHz)	3400	1300
Dielectric loss (tan δ at 1 kHz)	0.02	0.004
Curie temperature (T_c, °C)	193	328
Piezoelectric coefficients (10^{-12} m/V)		
d_{33}	593	289
d_{31}	−274	−123
d_{15}	741	496
Piezoelectric coupling factors		
k_{33}	0.752	0.70
k_{31}	−0.388	−0.334
k_{15}	0.675	0.71

A relaxor ferroelectric is one having a diffuse ferroelectric phase transition, in addition to having a very strong piezoelectric effect and a very high relative permittivity ($\kappa \le 30{,}000$, compared to $\kappa \le 15{,}000$ for $BaTiO_3$). It has composition $Pb(B_1, B_2)O_3$, where B_1 can be Mg^{2+}, Zn^{2+}, etc., and B_2 can be Nb^{5+}, Ta^{5+}, etc. A particularly well-known relaxor ferroelectric is a solid solution of $PbMg_{1/3}Nb_{2/3}O_3$ (abbreviated PMN), and $PbTiO_3$ (abbreviated PT). The solid solution is abbreviated PMNPT. The strong piezoelectric effect stems from the morphotropic boundary, which occurs at 30 mol% $PbTiO_3$ in the PMNPT system. The average Curie temperature is about 0°C.

Another example of a ferroelectric material is poly(vinylidene fluoride) or $(CH_2CF_2)_n$ (abbreviated PVDF), which is a fluoropolymer (called a piezopolymer, as opposed to a piezoceramic), specifically a semicrystalline (typically 50%–60% crystalline) thermoplastic. Cyrstallinity enhances the piezoelectric behavior. PVDF is polymorphic, exhibiting three solid phases that are known as α, β and γ. The β-phase is the one that is piezoelectric and pyroelectric; it has an orthorhombic crystal structure, with the molecular chains in a zigzag conformation. A conformation refers to a shape of the molecule. Different conformations are due to bond rotation. Crystallinity is needed for poling, so the ferroelectric behavior depends on the degree of crystallinity. In order to be piezoelectric, PVDF needs to be stretched to orient the molecules, followed by poling under tension at an electric field of at least 30 MV/m. The properties of poled PVDF start to degrade above about 80°C, but the ferroelectric behavior remains up to the relatively low melting temperature of 177°C. The glass transition temperature is −35°C. A mer of this polymer molecule is

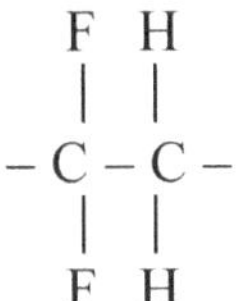

It has two fluorine atoms, which are electronegative and non-symmetrically positioned (i.e., at different sides of the molecular chain) (Fig. 3.47). Due to the electronegativity of fluorine, the dipole moment of the macromolecule is strongly affected by the positions in the fluorine atoms in a chain. In contrast to PZT, the piezoelectric coupling coefficient d (more exactly d_{33}, which is the value for the case of the strain and the electric field being both in the direction of high polarization) of PVDF is negative.

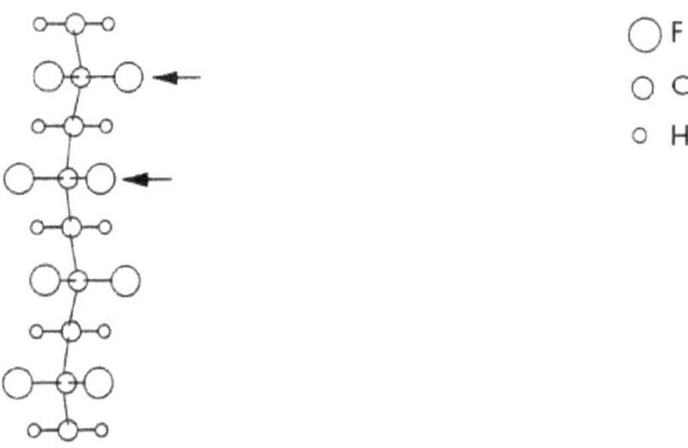

Fig. 3.47 A molecular configuration of polyvinylidene fluoride (PVDF).

PVDF can be processed by polymer processing methods to form large-area lightweight detectors, such as sonar hydrophones, audio transducers, etc. The ferroelectric behavior of PVDF can be enhanced by forming a copolymer, such as polyvinylidene fluoride-trifluoroethylene (abbreviated VF_2-VF_3). In general, piezopolymers suffer from low permittivity, high dielectric loss (particularly at high frequencies), low Curie temperature and low poling efficiency (particularly for specimens with large thickness, > 1 mm).

Example Problems 17 and 18 at the end of the chapter are relevant.

3.8.5 *Piezoelectric composites principles*

From Eq. (3.2), the voltage V across a dielectric of relative permittivity κ is given by

$$V = \kappa Q/C \tag{3.143}$$

From Eq. (3.26),

$$\kappa Q = DA = \kappa \varepsilon_0 EA \tag{3.144}$$

From Eq. (3.8), the capacitance of a parallel-plate capacitor of thickness x is given by

$$C = \kappa \varepsilon_o A/x. \qquad (3.145)$$

Using Eq. (3.144) and (3.145), Eq. (3.143) becomes

$$V = Ex. \qquad (3.146)$$

Using Eq. (3.23), Eq. (3.146) becomes

$$V = Px/[\varepsilon_o (\kappa - 1)] \qquad (3.147)$$

Differentiation with respect to the stress U gives

$$\Delta V/\Delta U = \{P/[\varepsilon_o (\kappa - 1)]\} (\Delta x/\Delta U) + \{x/[\varepsilon_o (\kappa - 1)]\} (\Delta P/\Delta U) \qquad (3.148)$$

Equation (3.148) gives the voltage sensitivity ($\Delta V/\Delta U$), which consists of two terms. The second term involves $\Delta P/\Delta U$, which is the piezoelectric coupling coefficient d. The first term involves $\Delta x/\Delta U$, which is related to the strain per unit stress, i.e., the compliance. Hence, the greater the compliance (i.e., the smaller the modulus), the higher is the voltage sensitivity.

The polymer PVDF is relatively high in compliance. The rhombohedral form of PZT (Fig. 3.45) has greater compliance than the tetragonal form; both forms are ferroelectric. The most common way to increase the compliance is to incorporate a piezoelectric/ferroelectric material in a polymer to form a composite. The mechanical properties in various directions of the composite can be adjusted so as to benefit the piezoelectric behavior. For example, a composite can be in the form of aligned PZT rods in a polymer matrix (Fig. 3.48). Such a composite has the PZT phase connected in only one direction and the polymer phase connected in all three directions. As a result, this composite is referred to as a 1-3 composite (i.e., the active phase having connectivity in 1 dimension, and the passive phase having connectivity in three dimensions). When a hydrostatic stress is applied to a 1-3 composite, the transverse stress (perpendicular to the rods) is absorbed by the polymer while the longitudinal stress (parallel to the rods) is applied mainly to the PZT rods. As a consequence, the piezoelectric response to the hydrostatic stress is enhanced. Furthermore, the polymer decreases the relative permittivity κ, resulting in an increase of g, the voltage coefficient (Eq. (3.124)).

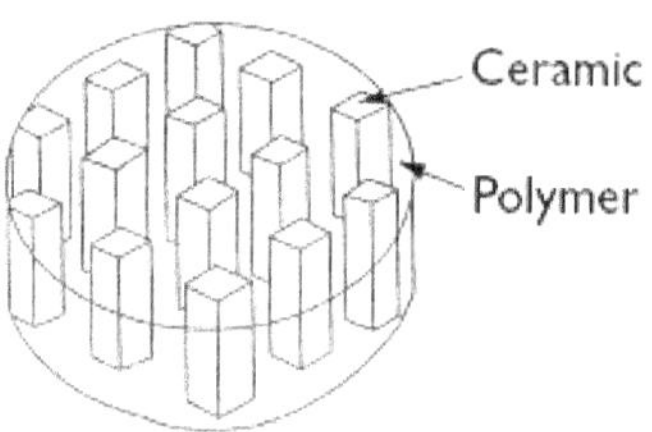

Fig. 3.48 A 1-3 piezoelectric ceramic polymer-matrix composite.

Composites for piezoelectric functions are designed to improve these particular functions (e.g., through tailoring of the modulus in various directions, κ and the Curie temperature), and to improve processability (e.g., the sinterability) and mechanical behavior (e.g., the fracture toughness). These composites can involve two or more components that are all capable of providing the desired function, such that the combination of these components results in enhancement of the function. The composites can also involve a component that is capable of providing the function and another component that is not capable of providing the function, such that the combination results in enhancement of the function.

Although the perovskite ceramics are the main materials for piezoelectric functions, the polymer polyvinylidene fluoride (PVDF) is also capable of these functions. Therefore, both ceramic-ceramic and polymer-ceramic composites are relevant.

Among composites in which a component (e.g., epoxy) is not able to provide the function, polymer-matrix composites dominate, due to their processability. These composites are classified according to the connectivity of each component, as the connectivity affects the functional behavior in various directions and different applications require a different set of directional characteristics.

Among the 32 different connectivities in a two-component material, the 1-3, 2-2 and 0-3 connectivities are particularly common. In this notation, the first number describes the connectivity of the functional (active) component and the second number describes the connectivity of the non-functional (passive) component. For example, in a 1-3 composite, the functional component has connectivity in 1 dimension and the non-functional component has connectivity in 3 dimensions. A composite in the form of aligned lead zirconate titanate (PZT) rods in a polymer matrix has 1-3 connectivity.

Another example of connectivity is 0-3, which refers to particles in a matrix. The particles are not continuous in any direction, while the matrix is continuous in all three directions. To make poling of a 0-3 composite more efficient (i.e., requiring less electric field), a conductive filler such as carbon black can be added to the composite so as to increase the electrical conductivity and have more of the poling field be applied to the piezoelectric/ferroelectric particles.

The piezoelectric materials mentioned above are not cost-effective for reinforcement in a composite. Therefore, a structural composite capable of sensing or actuation is most commonly in the form of a continuous fiber polymer-matrix composite laminate with one or more piezoelectric elements embedded between the plies in the composite during composite fabrication, such that the leads (wires) from the elements come out of the composite. Although a piezoelectric element is typically in a sheet form (of thickness similar to that of a prepreg tape), its presence tends to degrade the mechanical properties of the laminate.

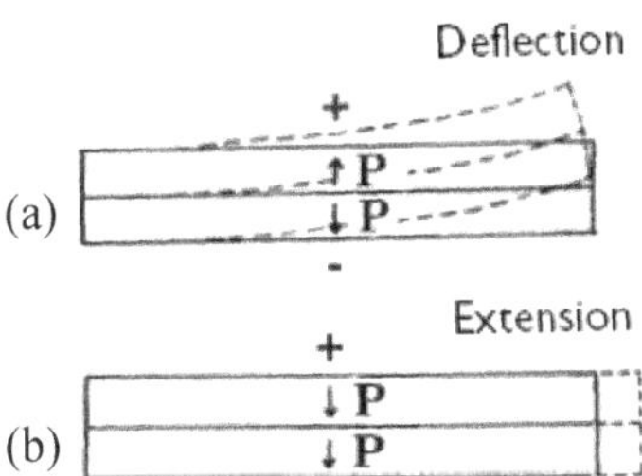

Fig. 3.49 Cantilever beam configuration for actuation using a bi-strip of piezoelectric material. (a) Polarizations P of the two strips of the bi-strip are in opposite directions, thereby causing deflection at the tip of the beam upon electric field application. (b) Polarizations P of the two strips are in the same direction, thereby causing extension of the beam upon electric field application.

A piezoelectric element for actuation is commonly in the form of two strips of piezoelectric material, such that one strip is bonded on top of the other and the bi-strip (called a bimorph) is fixed at one end to form a cantilever beam configuration. Furthermore, the polarization of the two strips is in opposite directions, such that the application of an electric field across the thickness of the bi-strip causes one strip to extend axially (in the plane of the strip) while the other strip contracts axially, thereby causing the bi-strip to deflect at the tip far from the fixed end (Fig. 3.49(a)). In contrast, if the strips are polarized in the same direction, both strips extend axially upon application of the electric field and the resulting movement (Fig. 3.49(b)) is much smaller than that for the case of the strips polarized in opposite directions.

A continuous fiber polymer-matrix composite laminate for actuation in the form of bending can be achieved by embedding two piezoelectric/ferroelectric strips in the two sides relative to the plane of symmetry of the laminate, such that one strip extends while the other strip contracts upon application of an electric field. A laminate for actuation in the form of axial extension can be achieved by embedding two strips, such that both strips extend upon application of an electric field.

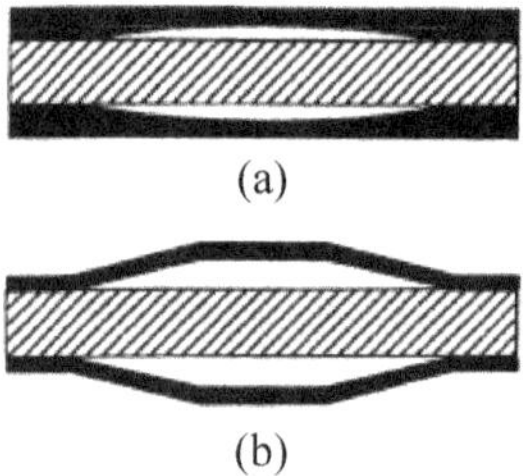

(a)

(b)

Fig. 3.50 A piezoelectric composite material sandwiched by metal faceplates. (a) "Moonie" structure involving air cavities in the shape of a half moon. (b) "Cymbal" structure involving air cavities in the shape of a truncated cone.

Macroscopic composites in the form of a piezoelectric/ferroelectric material sandwiched by metal faceplates (Fig. 3.50) are attractive for enhancing the piezoelectric coupling coefficient. The air gap between the metal faceplate and the piezoelectric/ferroelectric material allows the metal faceplate to serve as a mechanical transformer for transforming and amplifying a part of the applied axial stress into tangential and radial stresses. The metal faceplate transfers the applied stress to the piezoelectric/ferroelectric material, in addition to transferring the displacement to the medium.

3.9 Electrets

The inherent voltage in a material refers to the stable voltage that is present in the material in the absence of an applied electric field. A material that exhibits an inherent voltage is an electret.

Electrets employed in desalination and air filters use their permanent charge polarization to attract and trap the charged particles. The ones used in γ-radiation dosimeters measure the reduction in the charge of an electret upon the neutralization of surface charges due to the air ionized by the γ-radiation. The amount of loss in charge is directly proportional to the radiation present. In addition, electrets are used in microphones, which make use of the effect of mechanical strain on the inherent voltage). Other applications of electrets include dust anchoring, blood platelet adhesion and memories.

Electrets are conventionally in the form of nonconductors, namely polymers and ceramics. Examples are silicon dioxide (SiO_2), $Pb(Ti,Zr)O_3$, $CaTiO_3$, $MgO\text{-}CaO\text{-}SiO_2\text{-}Al_2O_3$ and hydroxyapatite, polyvinylydene fluoride (PVDF), polymethyl methacrylate (PMMA), polyethylene terephthalate (PET), polyethylene (PE) and polypropylene (PP). The electret stems from the asymmetry in the arrangement of the ionic species. For forming the electret, poling (high DC electric field application) is necessary. The poling process involves the input of electrical energy. The poled state (the electret state) is not thermodynamically stable, due to the decreased entropy. Thus, depoling occurs spontaneously. Due to the depoling, repoling is necessary.

The poling (charging) of a polymer electret is conventionally achieved by the use of corona voltage application, electron beam bombardment and triboelectrification. Corona methods are most common. The use of poling during the curing of a thermosetting polymer resin is an alternate method of charging. The poling causes ions to move or polar bonds to rotate, thus resulting in polarization. In order to exploit the movement of ions as the mechanism of polarization, ions must be present. Polymers do not tend to have ions, unless they are doped with ions, which can be provided by appropriate ionic salts.

Cement-based electrets are potentially attractive for attaining multifunctionality in structures, such as an environmentally friendly concrete chimney that can suck particles from the effluent and a concrete wall that can detect high energy electromagnetic radiation. Cement is a silicate material, primarily calcium silicate ($3CaO.SiO_2$ and $2CaO.SiO_2$). Ions (particularly those dissolved in the pore solution) in cement are responsible for the dielectric behavior, including the electret effect. The permittivity is increased by mild heating, due to the increasing ease of movement of the ions. The alignment of the water molecules, which are polar, plays a minor role in the polarization of a cement-based material. If the water molecular alignment plays an important role, the permittivity would decrease with increasing temperature, due to the thermal randomization tendency upon heating. An inherent voltage occurs in cement paste, even in the absence of poling, making the material an electret. The inherent voltage is attributed to the inhomogeneous distribution of the ions in the cement paste.

A relatively new form of electret is in the form of conductors, namely carbons (e.g., graphite [2] and carbon fibers [8]) and metals (e.g., copper [9]). The electret stems from the interaction of a fraction of the carriers with the atoms, the sluggishness of depolarization compared to polarization (Sec. 3.6.4), and the electrical asymmetry resulting from the microstructural asymmetry.

3.10 Effect of the microstructure on the dielectric behavior

The dielectric behavior is affected by the microstructure. In case of the nonconductive materials, the crystallographic order and crystallographic preferred orientation enhance the permittivity. As a consequence, the grain boundaries, which are sites of less crystallographic order, tend to be detrimental to the dielectric behavior of a nonconductive material.

Polymers are trypically nonconductive. Fig. 3.51 shows the effect of the microstructure and compressive stress on the relative permittivity of three-dimensionally (3D) printed polymer, which consists of layers of a controlled thickness. The smaller the layer thickness, the greater is the degree of in-plane molecular alignment (due to the greater shear stress on the resin during printing), and hence the higher is the relative permittivity in the direction perpendicular to the layers. In addition, the application of compressive stress in the direction perpendicular to the layers further enhances the degree of molecular alignment, thereby further increasing the relative permittivity [10]. This is piezopermittivity. The technique can be used for sensing the defects in the printed structure by capacitance measurement, because the capacitance is reduced by the presence of defects [11]. The sensing is amenable to the real-time monitoring of the printing process for the purpose of smart manufacturing, in which the printing conditions are adjusted in real time in response to the sensed defects. The same material exhibits the piezoelectric behavior (Fig. 3.36(d)).

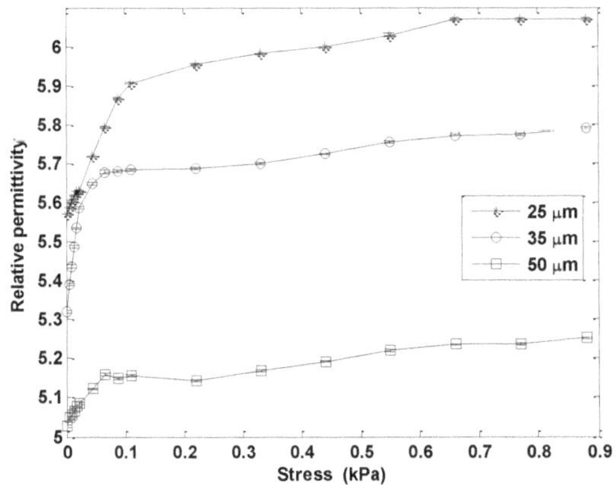

Fig. 3.51 The effect of stress on the relative permittivity during stress increase for each of three nominal layer thicknesses (25, 35 and 50 μm). [10]

In case of the conductive materials, the carrier-atom interaction is the key to the dielectric behavior, with the interaction occurring primarily at the defects, such as the grain boundaries. As a consequence, the grain boundaries promote the permittivity of a conductive material.

Cold work refers to mechanical working at temperatures below about 1/3 of the melting temperature in K. Metals can be subjected to cold work, but ceramics and carbons cannot, due to their brittleness. The cold work is known to affect the microstructure of metals greatly. The microstructural changes due to cold work include grain size reduction, grain elongation along the direction of the deformation, crystallographic preferred orientation enhancement and dislocation density increase.

Cold work increases the permittivity of a conductor, due to the enhanced carrier-atom interaction that is responsible for the polarization. For example, 37% cold work (with the cold work quantitatively defined as the fractional reduction in the cross-sectional area) of copper increases

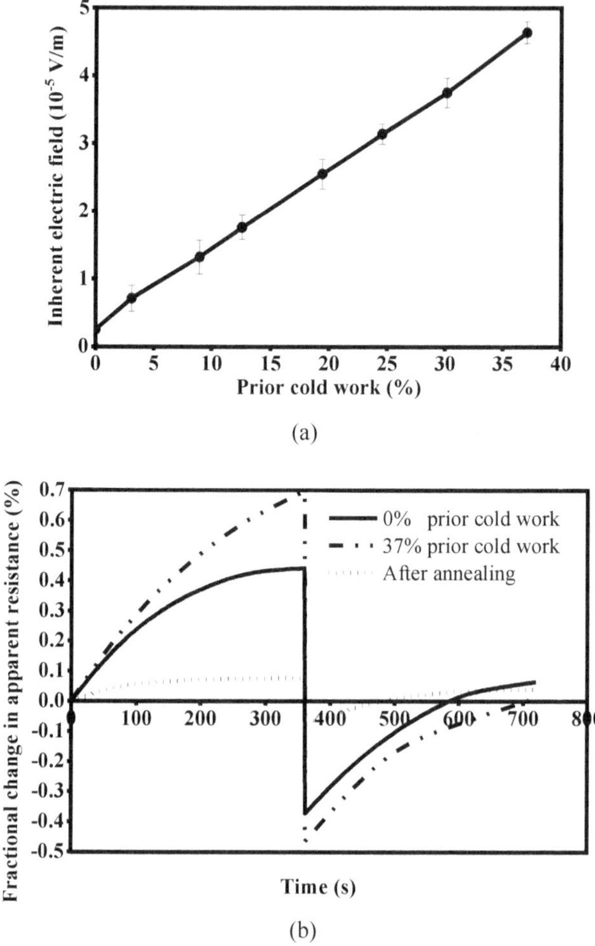

Fig. 3.52 (a) Effect of various amounts of prior cold work by rolling on the electret's electric field of copper for various degrees of cold work. The rolling is conducted prior to the electrical testing. The electric field is measured for inter-electrode distance $l = 50.8$ mm. (b) Effect of 37% prior work by rolling and subsequent annealing on the degree of asymmetry in the polarization/depolarization behavior indicated by the apparent resistivity. [9]

the relative permittivity (2 kHz) from 2.4×10^3 to 2.7×10^3, while the DC resistivity is increased from 1.4×10^{-6} to 1.5×10^{-6} [9]. This means that the permittivity is more sensitive to the microstructure than the resistivity.

The electret's electric field is also increased by cold work, as shown for cold work in the form of rolling. The more is the amount of cold work, the greater is the electret's electric field (Fig. 3.52(a)). Moreover, the positive end of the voltage in the electret is where the rolling starts. This implies that the electrons (a small fraction of the valence electrons of the metal) move along with the atoms during the plastic deformation. The electron movement promotes the electret. The strong effect of cold work means that processing is effective for strengthening an electret.

The strength of the electret correlates with the degree of asymmetry in the polarization/depolarization behavior indicated by the apparent resistivity. The degree of asymmetry is increased by the cold work and is decreased by the subsequent annealing (Fig. 3.52(b)). In addition, the extent of polarization is increased by the cold work and is decreased by the subsequent annealing.

The strong effect of the microstructure on the dielectric behavior of a conductive material is also indicated by the effect of the grain size of polycrystalline graphite. The smaller is the grain size, the higher is the permittivity (Fig. 3.53(a)). The reciprocal of the grain size relates to the amount of grain boundary area per unit volume. The permittivity is essentially linearly related to the reciprocal of the grain size (Fig. 3.53(b)). This implies that the carrier-atom interaction that is responsible for the polarization occurs primarily at the grain boundaries of polycrystalline graphite.

The resistivity also increases with decreasing grain size (Fig. 3.53(d)), but the decrease is roughly linear, in contrast to the strong nonlinear relationship of the permittivity with the grain size (Fig. 3.53(a)). The electret's electric also increases with decreasing grain size (Fig. 3.53(c)), such that the relationship is roughly linear. Hence, the resistivity and electret's electric field are correlated, indicating the importance of Ohm's Law is governing the electret's electric field.

The fraction of carriers that participate in the polarization increases with decreasing grain size (Fig. 3.53(e)), as expected from the dependence of the permittivity on the grain size (Fig. 3.53(a)). This fraction is small, of the order of 10^{-11}.

The combination of Fig. 3.53(b) and 3.53(d) indicates that, among the polycrystalline graphites of various grain sizew, higher permittivity correlates with higher resistivity. However, when carbon materials of various types (all belonging to the graphite allotrope family, as opposed

 Functional Materials

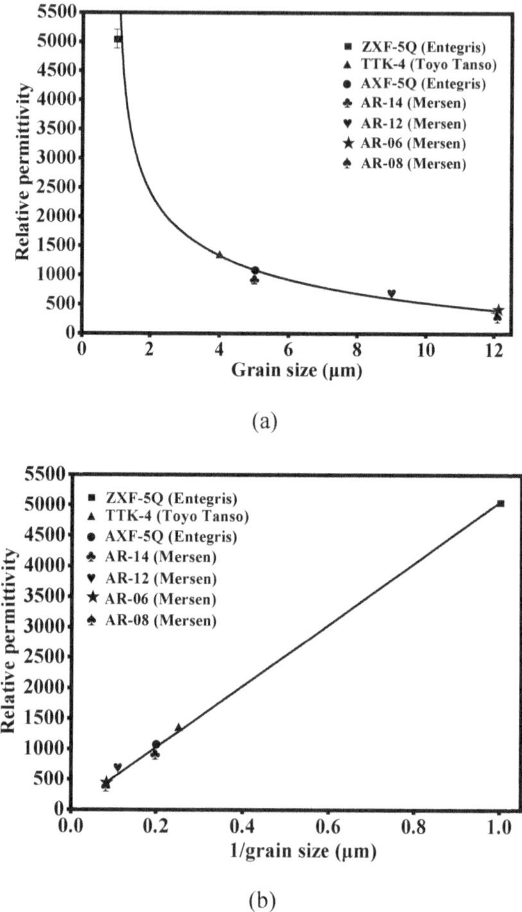

(a)

(b)

Fig. 3.53 Effect of the grain size on the dielectric and conduction behavior of polycrystalline graphite. (a) Relative permittivity (2 kHz) vs. grain size. (b) Relative permittivity (2 kHz) vs. the reciprocal of the grain size. (c) Electret's electric field vs. grain size. (d) DC resistivity vs. grain size. (e) Fraction of carriers that participate in the polarization vs. grain size. [2,12]

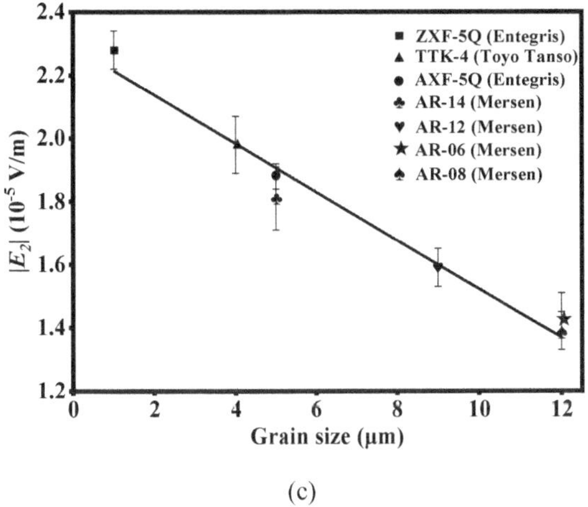

(c)

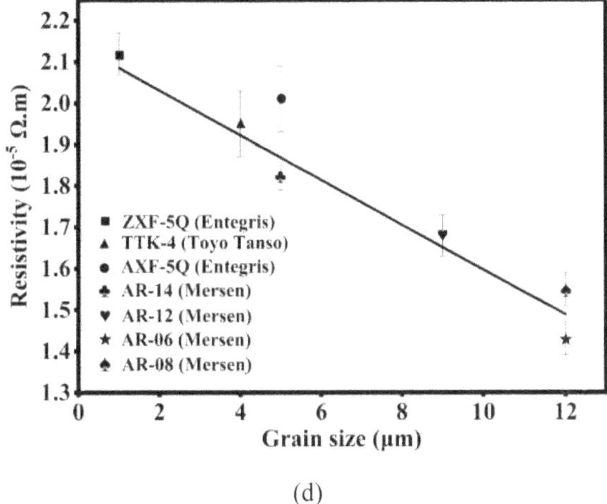

(d)

Fig. 3.53 (*Contined*).

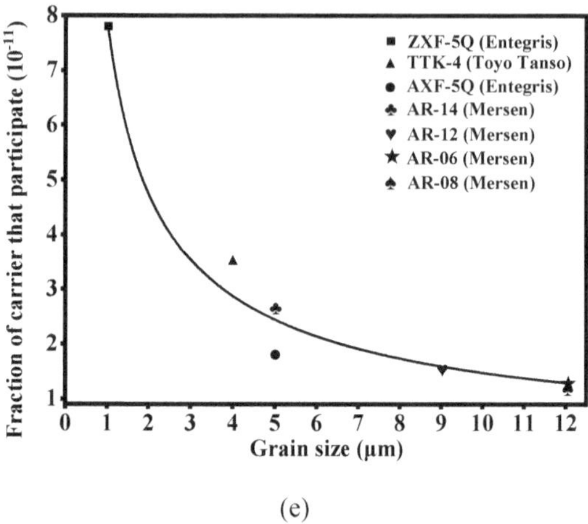

(e)

Fig. 3.53 (*Continued*).

to the diamond allotrope family) are considered together, including poly-crystalline graphites, carbon fibers, pyrolytic graphite, exfoliated-graphite-based flexible graphite and graphite flake thick film, the correlation between the permittivity and resistivity is weak. The permittivity is more sensitive to the microstructural heterogeneities than the resistivity, which only slightly influences the permittivity. In general, smaller crystallite dimensions promote the permittivity. The crystallographic preferred orientation is not influential to the permittivity, though it influences the resistivity. Interfaces (e.g., the grain boundaries in polycrystalline graphite and the interface between the carbon layer ribbons in a carbon fiber) promote the permittivity, being sites for the carrier-atom interaction that enables the polarization. [13]

3.11 Piezoelectret effect

The piezoelectret effect (also known as the piezoelectric electret effect) refers to the effect of stress or strain on the voltage across a material, such that the effect is due to the effect of stress or strain on the electret associated with the material. This effect is to be distinguished from the direct piezoelectric effect, which is due to the effect of stress or strain on the stored charge in the capacitor embodied in the material without the necessity of electret formation. The effect is also to be distinguished from piezopermittivity, which refers to the effect of stress or strain on the permittivity, which is associated with the capacitance. The piezoelectret effect, direct piezoelectric effect and piezopermittivity are all reversible. They convert mechanical energy to electrical energy and allow the sensing of stress or strain. In addition, they are useful for acoustic emission signal detection and accelerometers.

Electrets in the form of polymer films are commercially used in microphones. This application is based on the notion that the electret voltage is affected by the acoustic vibrations. This notion basically involves the concept of the piezoelectret effect.

Analogous to Eq. (3.106), the piezoelectret coupling coefficient is defined as

$$d = It/\Delta F = \Delta Q/\Delta F = \Delta Q/(A\ \Delta U) = C\ \Delta V/(A\ \Delta U)$$

$$= (\kappa\varepsilon_o A/l)\ \Delta V/(A\ \Delta U) = (\kappa\varepsilon_o\ \Delta V)/(l\ \Delta U) \tag{3.149}$$

where κ is the relative permittivity, d is piezoelectret coupling coefficient, l is the length of the specimen in the direction of polarization, ΔV is the change in voltage and ΔU is the change in stress.

3.12 Pyroelectric effect and pyropermittivity

The pyroelectric effect refers to the change in polarization in a material due to a change in temperature. The change in polarization gives rise to a change in voltage across the material in the direction of the polarization. The situation is similar to that in Fig. 3.30, except that heat rather than stress is the input. In this way, thermal energy is converted to electrical energy.

As in the discussion on the effect of stress on the polarization, the effect of temperature on the polarization is given by

$$\Delta P = (\kappa - 1)\ (\Delta Q/A) + (\Delta\kappa)\ (Q/A) + \Delta\kappa\ \Delta Q/A, \tag{3.111}$$

where ΔP is the change in polarization due to temperature change ΔT, and $\Delta\kappa$ is the change in κ due to ΔT.

In case that $\Delta\kappa$ is negligible, while ΔQ is not negligible, Eq. (3.111) becomes

$$\Delta P = (\kappa - 1)\, (\Delta Q/A). \qquad (3.112)$$

The ΔQ gives rise to an imbalance between the charge on the dielectric material surface and the charge in the electrode, thus resulting in a voltage or current output, i.e., the pyroelectric effect.

In case that ΔQ is negligible, while $\Delta\kappa$ is not negligible, Eq. (3.111) becomes

$$\Delta P = (\Delta\kappa)\, (Q/A). \qquad (3.113)$$

The change of κ with the temperature is a phenomenon known as pyropermittivity [3]. The $\Delta\kappa$ gives rise to a change in capacitance ΔC (Eq. (3.2)), which in turn causes a change in the amount of charge on the dielectric material surface (Fig. 3.4). The resulting imbalance between the charge on the dielectric material surface and the charge in the electrode gives rise to a voltage or current output. This effect, which stems from the fact that the dielectric material provides a capacitance that is temperature dependent, is to be distinguished from the pyroelectric effect. Both pyropermittivity and the pyroelectric effect convert thermal energy to electrical energy.

In spite of the difference in mechanism, the ΔQ and $\Delta\kappa$ effects cannot be distinguished by measurement of the electrical output, since both effects convert thermal energy to electrical energy. Distinction requires separate measurement of κ at various temperatures.

If $\Delta\kappa$ is small compared to κ and ΔQ is small compared to Q (a case which is practically not very important), the third term on the right side of Eq. (3.111) may be neglected, so that

$$\Delta P = (\kappa - 1)\, (\Delta Q/A) + (\Delta\kappa)\, (Q/A). \qquad (3.114)$$

However, if both $\Delta\kappa$ and ΔQ are large (a case which is rare), the third term dominates in Eq, (3.111), so that

$$\Delta P = \Delta\kappa\, \Delta Q/A. \qquad (3.115)$$

Equation (3.114) and (3.115) describe effects that involve both ΔQ and $\Delta\kappa$.

At the same temperature, ΔQ and $\Delta\kappa$ may have the same or opposite signs. In case that they have the same sign, with $\Delta\kappa$ being small compared to κ and ΔQ being small compared to Q, the ΔQ and $\Delta\kappa$ effects combine to result in a larger ΔP (Eq. (3.114)); in case that they have opposite signs, the ΔQ and $\Delta\kappa$ effects combine to result in a smaller ΔP (Eq. (3.114)). In case that both $\Delta\kappa$ and ΔQ are large, ΔP is negative when either $\Delta\kappa$ or ΔQ is negative, and is positive when both $\Delta\kappa$ and ΔQ are either positive or negative (Eq. (3.115)).

A pyroelectric material is also piezoelectric, as suggested by the similarity in the equations for the effect of stress and that of temperature on the polarization.

The pyroelectric coefficient p (lower case) is defined as

$$p = \Delta P/\Delta T. \tag{3.150}$$

Table 3.5 gives the values of p for various materials. From Eq. (3.112) and (3.143),

$$p = \Delta P/\Delta T = [(\kappa -1)/A]\,(\Delta Q/\Delta T) = [(\kappa -1)/A]\,C\,(\Delta V/\Delta T), \tag{3.151}$$

where C is the capacitance and A is the area. Rearrangement of Eq. (3.151) gives

$$\Delta V/\Delta T = pA/([\kappa -1)C]. \tag{3.152}$$

Table 3.5 Pyroelectric coefficient (10^{-6} C/m^2.K)

BaTiO$_3$	20
PZT	380
PVDF	27

Equation (3.151) means that a low value of κ is preferred for a large voltage output. For PZT having $p = 380 \times 10^{-6}$ C/(m^2.K), $\kappa = 1{,}300$, $A = 1$ cm$^2 = 10^{-4}$ m^2 and a thickness corresponding to $C = 100$ pF $= 10^{-10}$ F, Eq. (3.142) gives

$$\Delta V/\Delta T = [380 \times 10^{-6}\ \text{C/(m}^2.\text{K)}]\,(10^{-4}\ \text{m}^2) / [(1300 - 1)\,(10^{-10}\ \text{F})]$$

$$= 0.29\ \text{V/K}.$$

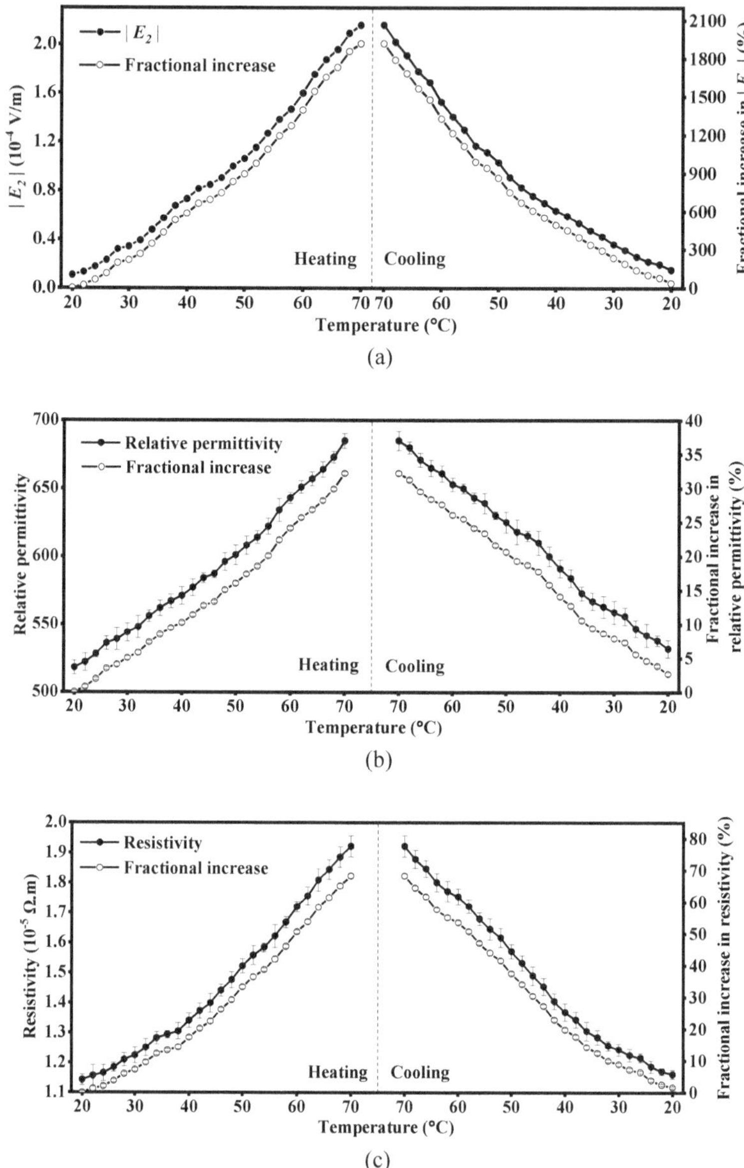

(a)

(b)

(c)

Fig. 3.54 Effects of mild heating and subsequent cooling of the dielectric and conduction behavior of polycrystalline graphite. (a) Electret's electric field (i.e., pyroelectret effect). (b) Relative permittivity (i.e., pyropermittivity). (c) DC resistivity (i.e., pyroresistivity). [3]

If $\Delta T = 0.001$ K, then $\Delta V = (0.29$ V/K$)$ $(0.001$ K$) = 0.29$ mV, which is a large enough voltage to be easily measured. This example calculation shows that a temperature change of only 0.001 K can be detected. Thus, the pyroelectric effect allows very sensitive temperature sensing. Even a person passing by a pyroelectric sensor can be detected, due to the body heat associated with the person. Thus, pyroelectric sensors are used for intruder detection, infrared detection, etc.

The pyroelectric effect, pyropermittivity and the Seebeck (thermoelectric) effect (Sec. 2.11) all allow the conversion of thermal energy to electrical energy. However, the Seebeck effect is different in the scientific origin from the other two phenoemena. Pyropermittivity and the pyroelectric effect relate to the dielectric behavior, whereas the Seebeck effect relates to the electrical conduction behavior. Furthermore, pyropermittivity and the pyroelectric effect require a change in temperature, whereas the Seebeck effect requires a temperature gradient.

The pyroelectret and pyropermittivity are illustrated in Fig. 3.54 for polycrystalline graphite. The electret's electric field, relative permittivity and DC resistivity all increase reversibly with increasing temperature. This means that the dielectric behavior is strengthened by the mild heating, while the conduction behavior is hampered. The hampering of the conduction behavior is due to the decrease in the moblity of the carriers as the temperature increases. The strengthening of the dielectric behavior while the resistivity is increased is in line with the known fact that the dielectric behavior and conduction behavior are not independent of one another. In the case of a conductor, both dielectric behavior and conduction behavior hinge on the carriers.

Example Problems 29 and 30 at the end of the chapter are relevant.

3.13 Electrostrictive behavior

Electrostrictive behavior (or electrostriction) is a phenomenon in which the dimension changes in response to an applied electric field, due to the energy increase associated with the polarization induced by the electric field in the material. The polarization may be due to atoms changing their shape (e.g., becoming egg-shaped rather than spheres), bonds between ions changing in length, or orientation of the permanent electric dipoles in the material. An electrostrictive material is usually centro-symmetrical in crystal structure, in contrast to classical piezoelectric materials (such as the Perovskite ceramics), which are noncentrosym-metrical. The dimensional changes can be in all directions, in contrast to the directional dimensional change in the piezoelectric effect. Also in contrast to piezoelectric behavior, no voltage or current results from electrostriction, i.e., an applied stress does not cause an electric field. Thus, electrostriction can be used for actuation, but not sensing. Electrostriction is a second-order effect, i.e., the strain $\Delta l/l$ is proportional to the square of the electric field E, or

$$\Delta l/l = W E^2, \tag{3.153}$$

where W is the electrostrictive coefficient, as illustrated in the low-field regime in Fig. 3.55. The second-order effect is due to the anharmonicity of the "springs" that connect the adjacent ions. The anharmonicity is associated with an asymmetric potential well, so that a spring tends to extend more easily than contracting. Thus, upon increase in energy, the average bond distance increases. In contrast, piezoelectricity is a first order effect, with the strain being linearly related to the electric field (Eq. (3.105)). The strain due to electrostriction is small compared to that due to piezoelectricity. For example, a field of 10^4 V/m produces 23 nm per meter in quartz (piezoelectric), but electrostrictive glass produces only 1 nm per meter. However, electrostrictive materials exhibit essentially no hysteresis upon cycling the electric field (Fig. 3.56(b)), whereas piezoelectric materials exhibit hysteresis due to non-linearity between strain and electric field at electric fields above 1 kV/cm (Fig. 3.56(a)). The non-linearity of the converse piezoelectric behavior is also shown in Fig. 3.42(b).

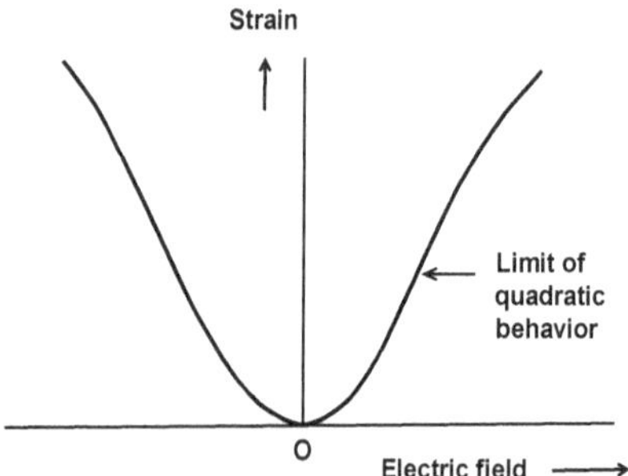

Fig. 3.55 Strain induced by an electric field in electrostrictive behavior.

The most important electrostrictive materials are Perovskite ceramics based on $Pb(Mg_{1/3}Nb_{2/3})O_3$, i.e., PMN or lead magnesium niobate, which exhibits strains as high as 0.1% (i.e., 10 μm per cm) at moderate fields ($\leq 10^6$ V/m). The phenomenon is called giant electrostriction. These materials are also relaxor ferroelectrics. They are ferroelectric below the Curie temperature, but electrostrictive above the Curie temperature. No poling treatment is needed for electrostrictive materials and these materials do not age (depole).

Example Problem 28 at the end of the chapter is relevant.

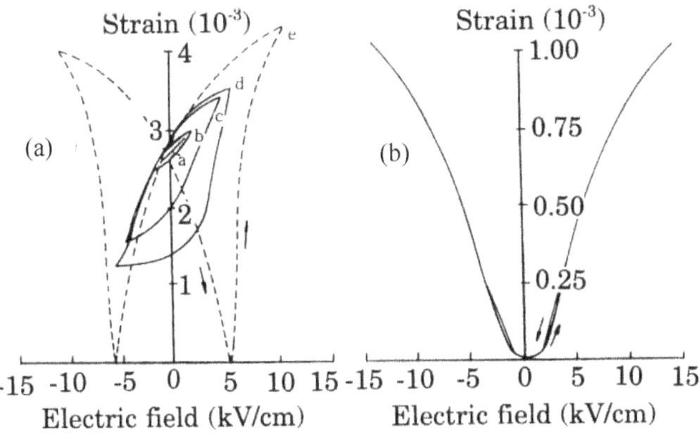

Fig. 3.56 Comparison of (a) the converse piezoelectric behavior and (b) the electrostrictive behavior.

3.14 Electrorheology

Electrorheology refers to the phenomenon in which the rheological behavior changes reversibly upon application of an electric field. The phenomenon is exhibited by an electrorheological fluid (abbreviated ER fluid), which is a suspension of dielectric particles (size typically around 10 μm) in a liquid. Upon application of an electric field, the fluid can change from the consistency of a liquid to that of a gel, with the apparent viscosity increasing greatly. The phenomenon, which is reversible upon removal of the electric field, is known as the Winslow effect. The response time is of the order of milliseconds (very impressive!). The higher is the electric field, the greater is the yield stress, the slightly lower is the plastic viscosity (after yielding), the higher is the apparent viscosity (Fig. 3.57) and the greater is the shear modulus (before yielding) (Fig. 3.58). Typically, an electric field of 1 kV/mm is required. The electric field should be below the breakdown electric field of air, i.e., 3 kV/mm. The yield stress is typically more than 10 kPa. The increase in apparent modulus is large, e.g., from 10 to 300 Pa.s. At a sufficiently high shear rate, shear thinning occurs. Thus, a structural material with an embedded pocket of electrorheological fluid has mechanical properties that can be varied by an electric field. This characterisitic is particularly attractive for vibration damping or structures.

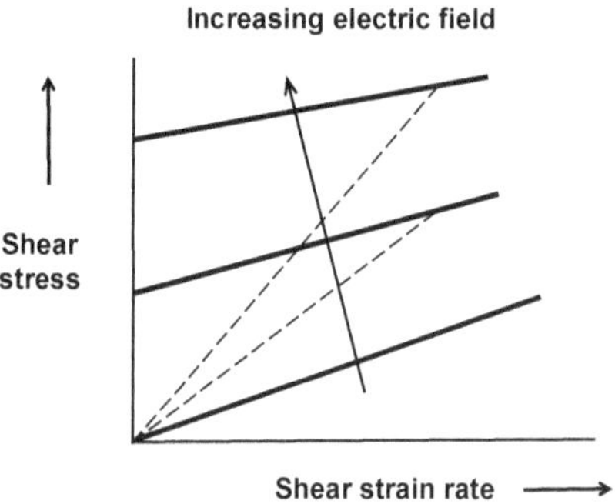

Fig. 3.57 Variation of the shear stress with the shear strain rate (or the shear rate) for an electrorheological fluid at various electric fields. The slope of a dashed line is the apparent viscosity at a particular shear rate.

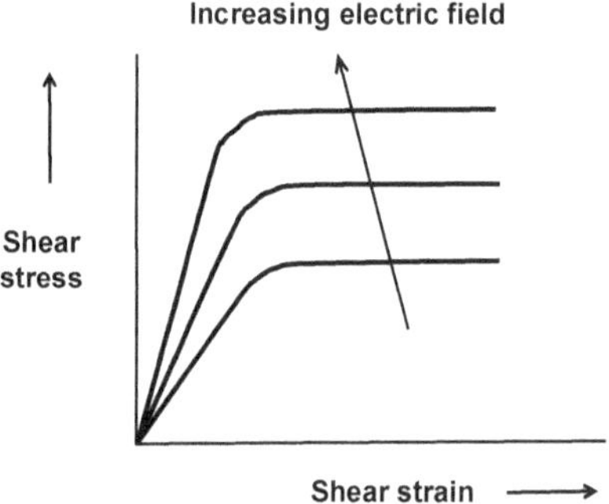

Fig. 3.58 Shear stress-strain curves for an electrorheological fluid at various electric fields.

Examples of dielectric particles are corn starch, silica and titania. The particle surface is frequently coated with an organic activator compound (water or other polar fluids such as ethanol) to enhance the polarization. The liquid is commonly an oil, such as silicone oil. The volume fraction of the solid ranges from 0.1 to 0.4, typically around 0.3. Both the yield stress and the apparent viscosity increase with increasing volume fraction of the solid.

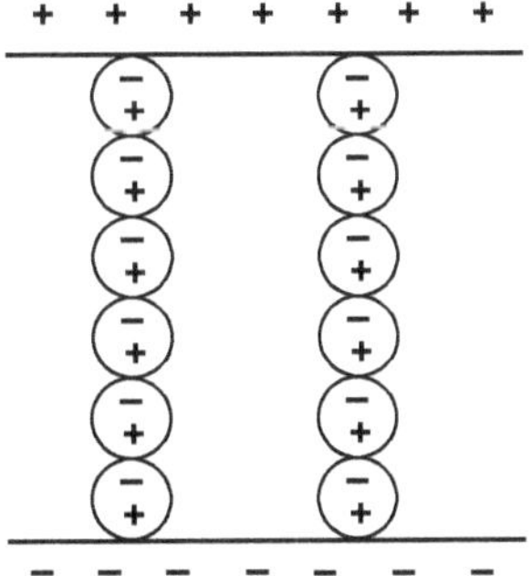

Fig. 3.59 Columns of polarized solid particles in an electrorheological fluid in the presence of an electric field in the direction of the columns.

Under an electric field, each solid particle may be polarized due to charge migration within the particle. As a result, each particle is an electric dipole and particles are attracted to one another by dipole-dipole interaction to form columns in the direction of the electric field, as illustrated in Fig. 3.59. The column formation is also called particle fibrillation. The columns resist shear, but it cannot explain the rapid response time observed. In another mechanism, the solid particles are not polarized, but each particle is associated with an electric double layer (positive surface charges on a particle surrounded by fixed negative charges from the liquid), as illustrated in Fig. 3.60. Mobile charges (with net positive charge) associated with the liquid between adjacent particles allow attraction between the particles. In the presence of an electric field, the particles align in the field direction, forming columns, as shown in Fig. 3.60. The mechanism in Fig. 3.60 is more common than that in Fig. 3.59. The formation of columns is responsible for the change in rheological behavior upon application of an electric field, since the columns resist shear in the direction perpendicular to the columns. The zero-field apparent viscosity of an electrorheological fluid is typically around 4 kPa at an electric field of 3.5 kV/mm. The current density is typically 0.1 A/m^2 at 3.5 kV/mm. The response time is typically 2 ms.

Applications of ER fluids include vibration isolation devices (e.g., shock absorbers), active feedback control damper systems, electrically triggered clutches, pumps, hydraulic valves, miniature robotic joints and robotic control systems. ER fluids suffer from their temperature dependence and the insufficient long-term stability of the ER response.

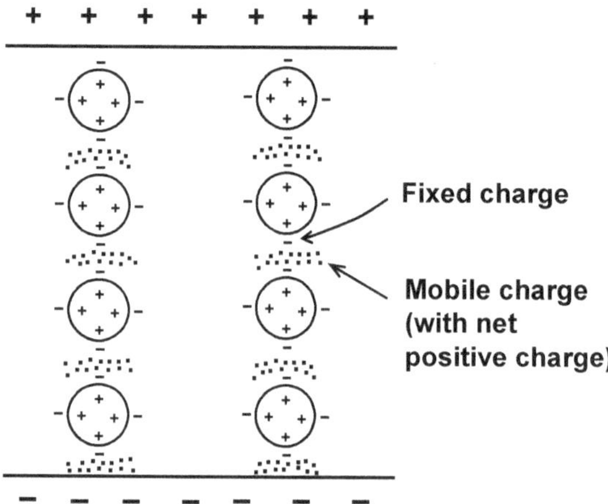

Fig. 3.60 Columns of solid particles with an electric double layer in an electrorheological fluid in the presence of an electric field in the direction of the columns.

3.15 Solid electrolytes

Solid electrolytes that allow ionic conduction but essentially no electronic conduction are needed for batteries and fuel cells. Although polymers are typically incapable of ionic or electronic conduction, a special class of polymers (known as ionomers) that are ionic conductors are available. An example is Nafion ($C_7HF_{13}O_5S$. C_2F_4), which has a tetrafluoroethylene (Teflon) molecular backbone incorporating SO_3H (sulfonic acid) groups. Protons (H$^+$ ions) hop from one sulfonic acid site to another, thereby enabling proton conduction, which is needed for the electrolyte in an important type of fuel cell that is known as the proton exchange membrane (PEM) fuel cell. Some ceramics are good ionic conductors due to the presence of mobile ions; examples are sodium chloride, β-alumina, zirconium dioxide and silver iodide. and ceramics may be used as Ceramic-ceramic and polymer-ceramic composites have been designed to enhance the ionic conductivity. Due to the higher mobility of ions in a liquid than a solid, solid electrolytes tend to be much less conductive than liquid electrolytes.

3.16 Composite materials for dielectric applications

As metals are electrically conducting, whereas polymers and ceramics tend to be not conducting, composite dielectric materials are mainly polymer-matrix and ceramic-matrix composites. The chemical bonding in these materials involves covalent and ionic bonding. The polymer-matrix composites are mainly polymer-ceramic composites. The ceramic-matrix composites are mainly ceramic-ceramic composites.

The composites can be in bulk, thick-film (typically 1–50 μm thick) and thin-film (typically less than 2000 Å thick) forms. The bulk form is needed for cable jackets and printed circuit boards. Both bulk and thick-film forms are needed for substrates and piezoelectric devices. Both thick-film and thin-film forms are needed for wire insulation, interlayer dielectrics and capacitors. Thick films are typically made by the casting of pastes. Thin films are typically made by vapor deposition.

3.16.1 *Composites for electrical insulation*

Composites for electrical insulation include those for cable jackets, wire insulation, substrates, interlayer dielectrics, encapsulations and printed circuit boards. Dielectric composites are tailored to attain low κ, high dielectric strength (high value of the electric field at breakdown), low AC loss (low loss tangent, tan δ, which relates to the energy loss), a low value of the coefficient of thermal expansion (CTE, a low value is needed to match those of semiconductors and other components in an electronic package), and preferably high thermal conductivity (for heat dissipation from microelectronics) as well.

3.16.1.1 *Polymer-matrix composites*

Compared to ceramics, polymers tend to have lower κ, higher CTE, lower thermal conductivity and lower stiffness. Therefore, ceramics are used as fillers in polymer-matrix composites to decrease CTE, increase the thermal conductivity and increase the stiffness.

The most widely used composites for electrical insulation are those containing continuous fibers (glass, quartz, aramid, etc.), which serve as a reinforcement in the composite laminate. These composites are mainly used for printed circuit boards. The fibers are usually woven, though non-woven ones are used also. Short fibers are even less commonly used than continuous fibers for printed circuit boards, but they are used for other forms of insulation. In order to lower κ, fibers of low permittivity, such as hollow glass fibers, are used. Epoxy and unsaturated polyester resin are commonly used for the matrix of printed wiring boards, but polyimide is used for high-temperature applications.

Particulate fillers are often used in dielectric composites other than laminates. In particular, hollow microspheres made of glass, ceramics or polymer are used as fillers to reduce κ. Silica particles, as generated from tetraethoxy silane (TEOS) via a sol-gel process, are used as a filler to lower CTE.

Composites that are thermally conducting but electrically insulating are critically needed for heat dissipation from microelectronic packages. Aluminum nitride (AlN) and boron nitride (BN) particles are used as fillers to increase the thermal conductivity and lower CTE. A high filler volume fraction is needed to attain a high thermal conductivity, but it reduces the workability of the paste, which is the form most commonly used in electronic packaging. Therefore, the development of such composites is challenging.

Polymer concrete, which is a low-cost polymer-matrix composite containing sand and gravel as fillers, is a dielectric structural material. It is an alternative to porcelain for outdoor and indoor electrical insulation applications.

Dielectric composites also include network composites, such as those involving gel silica, foams and interpenetrating polymer networks. These composites typically involve polymers.

3.16.1.2 *Ceramic-matrix composites*

Different ceramic materials differ in their processability, κ, CTE and thermal conductivity. Thus, ceramic-ceramic composites are designed to improve the processability, decrease κ, decrease CTE or increase the thermal conductivity.

Dielectric ceramic-matrix composites are ceramic-ceramic composites. They include glass-matrix composites, composites involving a low-softening-point borosilicate glass and a high-softening-point high silica glass, composites with boron nitride (BN) and aluminum nitride (AlN) matrices which are attractive for their high thermal conductivity, AlN-cordierite composites and alumina-matrix composites. Cordierite (Table 3.1) is attractive due to its low κ and relatively low processing temperature.

3.16.2 *Composites for capacitors*

The capacitors of this section are dielectric capacitors rather than double-layer capacitors. The former can operate at high frequencies, whereas the latter cannot.

Conventional capacitors involve thick film dielectrics, whereas integrated capacitors involve thin film dielectrics. Polymer-matrix composites are more common than ceramic-matrix composites, due to the ease of processing.

3.16.2.1 *Polymer-matrix composites*

Polymers tend to have better processability, lower κ, higher CTE and lower thermal conductivity than ceramics, so polymer-ceramic composites are designed to increase κ, lower CTE or increase the thermal conductivity. These composites are polymer-matrix composites containing ceramic particles that exhibit high κ. The most common ceramic particles are the perovskite ceramics, such as $BaTiO_3$, lead magnesium niobate (PMN) and lead magnesium niobate-lead titanate (PMN-PT). Kraft paper, which is not a ceramic and not high in κ, is also commonly used, due to its low cost and amenability to composite fabrication by simply winding the paper. For the particulate composites, a small particle size is preferred.

3.16.2.2 *Ceramic-matrix composites*

Ceramic-ceramic composites for capacitors involve a ceramic with high κ (e.g., $BaTiO_3$ and Ta_2O_5) and another ceramic (e.g., SiC and Al_2O_3) chosen for processability, surface passivation or other attributes. Ceramic-matrix metal particle (e.g., nickel) composites are also available for capacitor use, due to the increase of κ by the metal particle addition.

3.16.3 *Composites for piezoelectric functions*

Piezoelectricity pertains to the conversion between mechanical energy and electrical energy. The applications are in sensors and actuators. The sensing relates to strain/stress sensing.

3.16.3.1 *Polymer-matrix composites*

Polymer-matrix composites for piezoelectric functions mainly involve a polymer matrix, which itself cannot provide the function, as the majority component. The polymer serves to increase the compliance, which relates to the voltage sensitivity. Such polymers include epoxy, polytetra-fluoroethylene (PTFE), polyethylene, polypropylene, polyvinyl chloride and nylon. However, the composites can involve PVDF or other functional polymers (being able to provide the function) as the matrix or as a component within the matrix.

The functional behavior of PVDF is poorer than that of the ceramic single crystals, but PVDF can be processed by polymer processing methods to form large-area lightweight detectors, such as sonar hydro-phones, audio transducers, etc. The piezoelectric behavior of PVDF can be enhanced by forming a copolymer, such as polyvinylidene fluoride-trifluoroethylene (abbreviated VF_2-VF_3). In general, piezopolymers suffer from high AC loss (particularly at high frequencies), low Curie temperature and low poling efficiency (particularly for specimens with large thickness, > 1 mm).

The filler in the composites is typically a perovskite ceramic (such as PZT), which can be in the form of wires, fibers or particles, as needed for the connectivity. The unit size (e.g., particle size) of the filler affects the functional behavior, partly due to the surface layer on a filler unit.

Although bulk composites have received most attention, composite coatings are increasingly important, due to the material cost saving and the need of coating particles, fibers, wires and other non-functional surfaces.

3.16.3.2 *Ceramic-matrix composites*

Ceramic-matrix composites in this category are ceramic-ceramic composites in which either one or more components is (are) piezoelectric. Composites in which both components are functional include $PbZrO_3$/

PbTiO$_3$ and related oxide-oxide systems, although these materials may be considered alloys (as governed by the phase diagram) rather than composites.

Composites in which only one component is functional include barium strontium titanate matrix composites containing metal oxide fillers such as Al$_2$O$_3$ and MgO. They also include PZT-matrix composites containing glass, which serves as a sintering aid, as a binder in the case of thick-film composites, as a barrier layer to prevent reaction between certain constituents, and as a component to decrease κ and to increase the Curie temperature. In addition, they include PZT-matrix composites containing silver particles, which serve as a sintering aid.

3.16.4 *Composites for microwave switching and electric field grading*

Composites that exhibit permittivity that can be tuned by voltage variation are of use to switching devices which reversibly change their microwave properties under DC bias. These composites are mainly polymer blends.

The electric field encountered by an electrically insulating part may be non-uniform, so that dielectric breakdown occurs at the locations of high electric field. The field distribution can be homogenized by using a dielectric material which exhibits permittivity that varies nonlinearly with the electric field. This is known as electric field grading.

3.16.5 *Composites for electromagnetic windows*

Electromagnetic windows require transparency in certain frequency ranges, particularly the microwave range, due to the relevance to radomes. These materials are dielectrics, as conductors tend to be strong reflectors and are thus not transparent. To enhance the mechanical properties, which are needed for impact/ballistic protection, these materials are commonly polymer-matrix composites containing dielectric reinforcing fibers, such as ultra-high-strength polyethylene fibers.

Review questions

1. Why is it difficult to achieve a capacitor that operates at a high frequency and has a high capacitance?

Answer: This is because the permittivity of the dielectric material in the capacitor decreases with increasing frequency.

2. Why is a low value of the relative permittivity desirable for electronic packaging?
   Answer: A low value of the permittivity alleviates the problem of signal propagation delay that is associated with the RC time constant.

3. Describe an application of a piezoelectric material.
   Answer: Applications include stress/strain sensing, mechanical energy harvesting and actuation.

4. What is the direct piezoelectric effect?
   Answer: The direct piezoelectric effect provides an electrical output in response to a mechanical input.

5. Why does the relative permittivity of a material decrease with increasing frequency?
   Answer: This trend is because the electric dipoles in the material need to flip (change direction) in response to the AC electric field and the flipping is not fast enough when the frequency is too high.

6. Define polarization in relation to the dielectric behavior of a material.
   Answer: Polarization refers to the separation of the positive charge center and negative charge center in the material. Mathematically, it refers to the electric dipole moment per unit volume.

7. Define dipole friction.
   Answer: Dipole friction refers to the difficulty for the electric dipoles in a material to flip (change direction) fast enough when the frequency of the AC electric field is high.

8. Define the dielectric strength.
   Answer: The dielectric strength is the maximum electric field above which an electrical insulator becomes conductive.

9. What is the main application of a dielectric material that exhibits a high value of the permittivity?
   Answer: Such a material is valuable for capacitors.

10. What is the difference between the electrostrictive behavior and the converse piezoelectric behavior?
    Answer: In the electrostrictive behavior, the strain has the same sign, regardless of the sign of the input electric field. However, in the converse piezoelectric behavior, the strain has opposite signs corresponding to the two opposite signs of the input electric field.

11. What is the main advantage of the electrostrictive behavior compared to the converse piezoelectric behavior?
    Answer: The electrostrictive behavior has less hysteresis.

12. What is meant by an electret?
    Answer: An electret is a permanent electric dipole.

13. What is the difference between the piezoelectret behavior and the direct piezoelectric effect?
    Answer: The piezoelectret behavior involves the electret's inherent electric field changing in response to the applied stress/strain. In contrast, the direct piezoelectric effect involves the stored charge changing in response to the applied stress/strain.

14. What is the difference between the direct piezoelectric behavior and the pyroelectric behavior?
    Answer: The direct piezoelectric behavior has the input being stress/strain and the output being electrical. The pyroelectric behavior has the input being a temperature change and the output being electrical.

15. What is the difference between the pyroelectric behavior and the thermoelectric behavior (Sec. 2.11)?
    Answer: The pyroelectric behavior is a manifestation of the dielectric behavior. The thermoelectric behavior is a manifestation of the electrical conduction behavior.

16. How does the elastic modulus affect the piezoelectric behavior?
    Answer: A low value of the elastic modulus promotes the piezo-electric behavior.

17. How does the relative permittivity affect the pyroelectric behavior?
    Answer: A high value of the relative permittivity promotes the pyroelectric behavior (Eq. (3.111)).

18. How does the area affect the output of a pyroelectric device?
    Answer: A small area promotes the output (Eq. (3.111)).

19. Why should the particles in an electrorheological fluid be not conductive electrically?
    Answer: The conductivity would hamper the polarization.

Example problems

Problem 1 (Capacitance)

A parallel-plate capacitor consists of two parallel metal plates that are separated by a gap of 0.21 mm. The gap is filled with air. What area of each of the plates is needed in order for the capacitance to be 1.50 nF. Note: 1 nF = 10^{-9} F; $\kappa = 1$ for air; $C = \varepsilon_0\,\kappa A/l$, where ε_0 is the permittivity of free space (8.85×10^{-12} F/m), κ is the relative permittivity, A is the area of the capacitor, and l is the thickness of the gap.

Solution:

$$A = Cl/(\varepsilon_0\,\kappa) = (1.5 \times 10^{-9}\ \text{F})\,(2.1 \times 10^{-4}\ \text{m})/(8.85 \times 10^{-12}\ \text{F/m})$$

$$= 0.036\ \text{m}^2$$

$$= 360\ \text{cm}^2.$$

Problem 2 (Capacitance)

A parallel-plate capacitor has a dielectric material of relative permittivity $\kappa = 85.7$, area $A = 475$ mm^2 and thickness $l = 1.25$ mm sandwiched by two electrodes (i.e., two conductive sheets). The measured capacitance C_m of this capacitor is 0.124 nF. It is the capacitance of three capacitances that are electrically in series. These three capacitances are C_v (the volumetric capacitance of the dielectric material) and the capacitance C_i of each of the two electrode-dielectric interfaces. Assume that the two interfaces are identical. Based on the equation for capacitances in series,

$$1/C_m = 1/C_v + 2/C_i, \tag{1}$$

where C_v relates to the relative permittivity κ of the dielectric material by the equation

$$C_v = \varepsilon_o \kappa A/l, \tag{2}$$

where ε_o is the permittivity of free space (8.85×10^{-12} F/m). Calculate C_i. Note: 1 nF = 10^{-9} F.

Solution:

$$C_v = \varepsilon_o \kappa A/l$$

$$= (8.85 \times 10^{-12} \text{ F/m}) \, (85.7) \, (475 \text{ mm}^2)/(1.25 \text{ mm})$$

$$= (288209 \times 10^{-12} \text{ F})/(10^3)$$

$$= (2.88 \times 10^{-7} \text{ F})/(10^3)$$

$$= 2.88 \times 10^{-10} \text{ F}$$

$$= 0.288 \text{ nF}$$

$$2/C_i = 1/C_m - 1/C_v$$

$$= [(1/0.124) - (1/0.288)]\,(\text{nF})^{-1}$$

$$= (8.065 - 3.472)\,(\text{nF})^{-1}$$

$$= 4.593\,(\text{nF})^{-1}$$

$$1/C_i = (4.593/2)\,(\text{nF})^{-1}$$

$$= 2.297\,(\text{nF})^{-1}$$

$$C_i = 0.435\,\text{nF}$$

Problem 3 (Capacitance)

Figure 3.61 shows the plot of $1/C_m$ vs. l, where C_m is the measured capacitance and l is the inter-electrode distance. The area is 4.76 mm × 0.0025 mm. Calculate the relative permittivity and interfacial capacitance C_i for low carbon steel (the upper curve in Fig. 3.61). The permittivity of free space = 8.85×10^{-12} F/m.

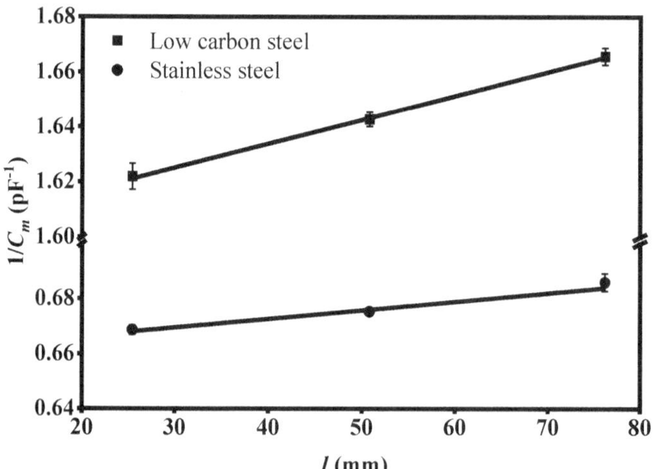

Fig. 3.61 The plot of $1/C_m$ vs. l for steels.

Solution:

Slope $= (1.66 - 1.62)\,(\text{pF})^{-1}/(76 - 25)\,\text{mm}$

$\qquad = 7.8 \times 10^{11}\,(\text{F.m})^{-1}$

$\qquad = 1/(\varepsilon_0 \kappa A)$

$\kappa = 1/[\varepsilon_0 A\,(7.8 \times 10^{11}\,(\text{F.m})^{-1})]$

$\qquad = 1/[(8.85 \times 10^{-12}\,\text{F/m})\,(4.76\,\text{mm} \times 0.0025\,\text{mm})\,(7.8 \times 10^{11}\,(\text{F.m})^{-1})]$

$\qquad = 1.22 \times 10^{7}$

Intercept (at $l = 0$) $= 2/C_i = 1.60\,(\text{pF})^{-1}$

$\qquad\qquad C_i = 1.3\,\text{pF}$

Problem 4 (Polarization)

The polarization is defined as the electric dipole moment per unit volume. The dipole moment is defined as the magnitude of charge at each of the two charge center multiplied by the distance between the two charge centers. The polarization is 0.21 C/m^2 in a material of thickness 36 μm and diameter 500 μm. This polarization is along the thickness direction. What is the electric dipole moment?

Solution:

$\qquad$ Dipole moment $=$ polarization $\times$ volume

$\qquad\qquad = (0.21\,\text{C/m}^2)\,(3.6 \times 10^{-5}\,\text{m})\,\pi(5 \times 10^{-4}\,\text{m/2})^2$

$\qquad\qquad = 1.5 \times 10^{-12}\,\text{C.m}$

Problem 5 (Polarization)

An electric field of 4.3×10^6 V/m is applied to $BaTiO_3$ with relative permittivity 2,300. How much polarization results? The polarization P is given by the equation

$$P = (\kappa - 1)\, \varepsilon_o E,$$

where E is the electric field and κ is the relative permittivity.

Solution:

$$P = (\kappa - 1)\, \varepsilon_o E$$

$$= (2300 - 1)\,(8.85 \times 10^{-12}\ C/V.m)\,(4.3 \times 10^6\ V/m)$$

$$= 0.087\ C/m^2$$

Problem 6 (Polarization)

The electric polarization P of a dielectric material increases linearly with the applied electric field E, such that

$$P = \varepsilon_0\,(\kappa - 1)\, E,$$

where κ (2655) is the relative permittivity and ε_0 (8.85×10^{-12} F/m) is the permittivity of free space. In order for P to be 0.013 C/m^2, what value of E is needed? The correct unit should be used.

Solution:

$$E = P/[\varepsilon_0\,(\kappa - 1)]$$

$$= (0.013\ C/m^2)/[(8.85 \times 10^{-12}\ F/m)\,(2654)]$$

$$= 5.5 \times 10^5\ V/m = 0.55\ MV/m$$

Problem 7 (Capacitors in series)

(a) Derive the equation for capacitors (C_1, C_2 and C_3) in series.

(b) Consider three identical capacitors (each of capacitance C) in series. Derive the total capacitance C_t in terms of C.

Solution:

(a)

$$C_1 \quad C_2 \quad C_3$$

Capacitors in series
C = total capacitance

$$C_1 = \frac{\varepsilon_0 \kappa A}{l_1} \quad \rightarrow \quad \frac{1}{C_1} = \frac{l_1}{\varepsilon_0 \kappa A}$$

$$C_2 = \frac{\varepsilon_0 \kappa A}{l_2} \quad \rightarrow \quad \frac{1}{C_2} = \frac{l_2}{\varepsilon_0 \kappa A}$$

$$C_3 = \frac{\varepsilon_0 \kappa A}{l_3} \quad \rightarrow \quad \frac{1}{C_3} = \frac{l_3}{\varepsilon_0 \kappa A}$$

$$C = \frac{\varepsilon_0 \kappa A}{l_1 + l_2 + l_3} \quad \rightarrow \quad \frac{1}{C} = \frac{l_1 + l_2 + l_3}{\varepsilon_0 \kappa A}$$

$$= \frac{l_1}{\varepsilon_0 \kappa A} + \frac{l_2}{\varepsilon_0 \kappa A} + \frac{l_3}{\varepsilon_0 \kappa A}$$

$$= \frac{1}{C_1} + \frac{1}{C_2} + \frac{1}{C_3}$$

(b)

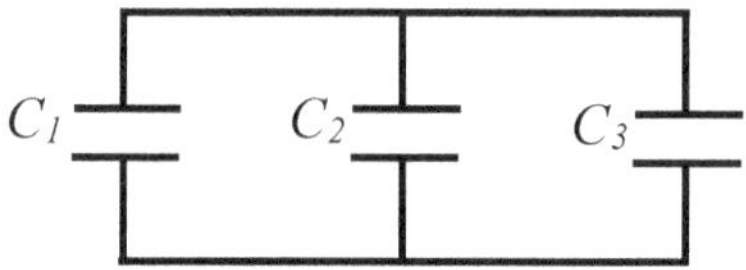

$$\frac{1}{C_t} = \frac{1}{C} + \frac{1}{C} + \frac{1}{C}$$

$$C_t = \frac{C}{3}$$

Problem 8 (Capacitors in parallel)

(a) Derive the equation for capacitors (C_1, C_2 and C_3) in parallel.

(b) Consider three identical capacitors (each of capacitance C) in parallel. Derive the total capacitance C_t in terms of C.

Solution:

(a)

Capacitors in parallel
C = total capacitance

$$C_1 = \frac{\varepsilon_0 \kappa A_1}{l}$$

$$C_2 = \frac{\varepsilon_0 \kappa A_2}{l}$$

$$C_3 = \frac{\varepsilon_0 \kappa A_3}{l}$$

$$C = \frac{\varepsilon_0 \kappa (A_1 + A_2 + A_3)}{l}$$

$$= \frac{\varepsilon_0 \kappa A_1}{l} + \frac{\varepsilon_0 \kappa A_2}{l} + \frac{\varepsilon_0 \kappa A_3}{l}$$

$$= C_1 + C_2 + C_3$$

(b)

$$C_t = 3C.$$

Problem 9 (Permittivity for capacitors in series)

(a) Consider three pieces of dielectric material disks that are stacked up. The three disks are of the same area A, but have thicknesses l_1, l_2 and l_3 in the capacitance direction, and relative permittivity κ_1, κ_2 and κ_3, respectively. Derive the expression (Rule of Mixtures) for the relative permittivity of the combination of the three stacked disks.

(b) In case that the three dielectric material disks in the stack are the same in area, thickness and relative permittivity κ, give the expression for the relative permittivity of the combination of three disks.

Solution:

(a) The three disks constitute three capacitors in series.

$$\dashv\vdash\dashv\vdash\dashv\vdash$$

$$C_1 \quad C_2 \quad C_3$$

$$C_t = \text{total capacitance}$$

$$\frac{1}{C_t} = \frac{1}{C_1} + \frac{1}{C_2} + \frac{1}{C_3}$$

$$\frac{l_1 + l_2 + l_3}{\varepsilon_0 \kappa_t A} = \frac{l_1}{\varepsilon_0 \kappa_1 A} + \frac{l_2}{\varepsilon_0 \kappa_2 A} + \frac{l_3}{\varepsilon_0 \kappa_3 A}$$

$$\frac{1}{\kappa_t} = \frac{l_1}{l_1 + l_2 + l_3}\frac{1}{\kappa_1} + \frac{l_2}{l_1 + l_2 + l_3}\frac{1}{\kappa_2} + \frac{l_3}{l_1 + l_2 + l_3}\frac{1}{\kappa_3}$$

$$= \frac{v_1}{\kappa_1} + \frac{v_2}{\kappa_2} + \frac{v_3}{\kappa_3}$$

where $v_1 = \dfrac{l_1}{l_1 + l_2 + l_3}$ = vol. fraction of 1

$v_2 = \dfrac{l_2}{l_1 + l_2 + l_3}$ = vol. fraction of 2

$v_3 = \dfrac{l_3}{l_1 + l_2 + l_3}$ = vol. fraction of 3

(b)

$$1/\kappa_t = (1/3)(1/\kappa) + (1/3)(1/\kappa) + (1/3)(1/\kappa) = 1/\kappa$$

$$\kappa_t = \kappa$$

Problem 10 (Capacitors in series and in parallel)

(a) Two capacitors with capacitances $C_1 = 19$ μF and $C_2 = 36$ μF are in series. Calculate the capacitance C of the combination of the two capacitors.

(b) Two capacitors with capacitances $C_1 = 19$ μF and $C_2 = 36$ μF are in parallel. Calculate the capacitance C of the combination of the two capacitors.

Solution:

(a) $1/C = 1/C_1 + 1/C_2 = (1/19 + 1/36)$ μF$^{-1} = 0.0804$ μF^{-1}

$C = 12$ μF

(b) $C = C_1 + C_2 = (19 + 36)$ μF $= 55$ μF

Problem 11 (Composite material with three components in series)

Calculate the relative permittivity of a composite material consisting of three components that are in series. The relative permittivity is in the series direction. Component 1 has relative permittivity 3.1 and volume fraction 0.65. Component 2 has relative permittivity 4.3 and volume fraction 0.24. Component 3 has relative permittivity 2.3 and volume fraction 0.11.

Hint: $1/\kappa = v_1/\kappa_1 + v_2/\kappa_2 + v_3/\kappa_3$

Solution:

$$1/\kappa = v_1/\kappa_1 + v_2/\kappa_2 + v_3/\kappa_3$$

$$= (0.65/3.1) + (0.24/4.3) + (0.11/2.3) = 0.31$$

$$\kappa = 3.2$$

Problem 12 (Composite material with components in parallel)

A composite material consists of two components that are arranged in parallel electrically. The relative permittivity is 3.67 and 56.8 for components 1 and 2, respectively. In order for the composite to exhibit relative permittivity 48.8, what is required for the volume fraction of component 1?

Hint:

Let v_1 be the volume fraction of component 1.

Let v_2 be the volume fraction of component 1.

$$v_1 + v_2 = 1$$

Relative permittivity of composite
$= v_1$ (relative permittivity of component 1)
$\quad + v_2$ (relative permittivity of component 2)

Solution:

$$48.8 = v_1 (3.67) + (1 - v_1)(56.8)$$

$$= v_1 (-53.13) + 56.8$$

$$v_1 = 0.15$$

Problem 13 (Composite material)

A composite material consists of component 1 and component 2.
Hint:
Relative permittivity:

$$\kappa = v_1 \kappa_1 + v_2 \kappa_2 \,(\text{parallel})$$

$$1/\kappa = v_1/\kappa_1 + v_2/\kappa_2 \,(\text{series})$$

(a) Component 1 has relative permittivity 5.289; component 2 has relative permittivity 29.668. The volume fraction is 0.689 for component 1. Calculate the relative permittivity of the composite if the two components are in parallel.

(b) Component 1 has relative permittivity 5.289; component 2 has relative permittivity 29.668. The volume fraction is 0.689 for component 1. Calculate the relative permittivity of the composite if the two components are in series.

(c) Component 1 has electrical resistivity 4.687×10^{-3} $\Omega.\text{cm}$; component 2 has electrical resistivity 1.265×10^{6} $\Omega.\text{cm}$. The volume fraction is 0.689 for component 1. Calculate the electrical resistivity of the composite if the two components are in parallel.

(d) Component 1 has electrical resistivity 4.687×10^{-3} $\Omega.\text{cm}$; component 2 has electrical resistivity 1.265×10^{6} $\Omega.\text{cm}$. The volume fraction is 0.689 for component 1. Calculate the electrical resistivity of the composite if the two components are in series.

Solution:

(a) Relative permittivity of the composite if the two components are in parallel

$$= v_1 \kappa_1 + v_2 \kappa_2$$

$$= v_1 \kappa_1 + (1 - v_1) \kappa_2$$

$$= (0.689)(5.289) + (1 - 0.689)(29.668)$$

$$= 3.644 + 9.227$$

$$= 12.87$$

(b) Relative permittivity of the composite if the two components are in series $= \kappa$

$$1/\kappa = v_1/\kappa_1 + v_2/\kappa_2$$

$$= v_1/\kappa_1 + (1 - v_1)/\kappa_2$$

$$= 0.689/5.289 + (1 - 0.689)/29.668$$

$$= 0.1303 + 0.0105$$

$$= 0.1408$$

$$\kappa = 1/0.1408$$

$$= 7.10$$

(c) Resistivity of the composite if the two components are in parallel $= \rho$

$$1/\rho = v_1/\rho_1 + v_2/\rho_2$$

$$= v_1/\rho_1 + (1 - v_1)/\rho_2$$

$$= [0.689/(4.687 \times 10^{-3}\ \Omega.cm)] + [(1 - 0.689)/(1.265 \times 10^{6}\ \Omega.cm)]$$

$$= 1.470 \times 10^2 \ \Omega^{-1}.\text{cm}^{-1} + 2.458 \times 10^{-7} \ \Omega^{-1}.\text{cm}^{-1}$$

$$= 1.470 \times 10^2 \ \Omega^{-1}.\text{cm}^{-1}$$

$$\rho = 1/(1.470 \times 10^2 \ \Omega^{-1}.\text{cm}^{-1})$$

$$= 6.80 \times 10^{-3} \ \Omega.\text{cm}$$

(d) Resistivity of the composite if the two components are in series

$$= v_1 \rho_1 + v_2 \rho_2$$

$$= v_1 \rho_1 + (1 - v_1) \rho_2$$

$$= (0.689)\,(4.687 \times 10^{-3} \ \Omega.\text{cm}) + (1 - 0.689)\,(1.265 \times 10^6 \ \Omega.\text{cm})$$

$$= 3.229 \times 10^{-3} \ \Omega.\text{cm} + 3.934 \times 10^5 \ \Omega.\text{cm}$$

$$= 3.93 \times 10^5 \ \Omega.\text{cm}$$

Problem 14 (Current through a capacitor)

The current I passing through a capacitor of capacitance C is given by

$$I = C \, \Delta V / \Delta t,$$

where V is the voltage and t is the time. Derive this equation.
Hint: $\kappa Q = CV$, where κ is the relative permittivity and Q is the magnitude of charge in case of a vacuum.

Solution:

$$I = \Delta(\kappa Q)/\Delta t = C \, \Delta V / \Delta t.$$

Problem 15 (Dielectric breakdown)

A dielectric material in the form of a disc of thickness 0.55 mm exhibits dielectric strength 275 V/mil (1 mil = 0.001 inch, 1 inch = 25.4 mm). What is the highest voltage that can be applied across the thickness of this disc without dielectric breakdown (i.e., without the material becoming electrically conductive)?

Solution:

Highest voltage = Highest electric field × thickness

$\quad$ = (275 V/mil) (0.55 mm)

$\quad$ = (275 V/mil) (1 mil/0.001 in) (1 inch/25.4 mm) (0.55 mm)

$\quad$ = 5950 V

Problem 16 (Loss tangent, i.e., tan δ)

A non-ideal dielectric material of relative permittivity $\kappa = 4500$ at 100 Hz can be considered a capacitor C (0.005 F) and a resistor R (10^4 Ω) in parallel. The frequency is 1000 Hz. The amplitude of the applied voltage is 2.3 V. Calculate the phase angle δ between the resultant current and the current through the capacitor C.
Hint: $\tan \delta = 1/(\omega CR)$.

Solution:

$$\tan \delta = \frac{1}{\omega CR} = \frac{1}{(2\pi 10^3)(0.005)(10^4)} = 3.2 \times 10^{-6}$$

$$\delta = 1.8 \times 10^{-4}$$

Problem 17 (Perovskite)

(a) For the HCP structure, how many octahedral interstitial sites are there per conventional unit cell that is based on a regular hexagon?

(b) For the diamond structure, how many tetrahedral interstitial sites are there per conventional cubic unit cell?

(c) For the cubic perovskite structure, how many anions are there per conventional cubic unit cell?

Solution:

(a) 6

(b) 8

(c) 3

Problem 18 (Perovskite)

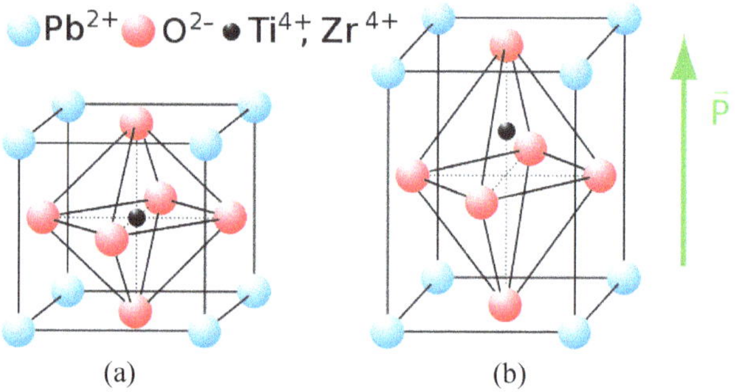

Fig. 3.62 (a) Cubic (Perovskeit Structure). (b) Tetragonal (Distorted Perovskite Structure). (https://en.wikipedia.org/wiki/Lead_titanate, public domain)

The distorted Perovskite structure has a tetragonal (almost cubic) unit cell, such that there is no center of symmetry in the unit cell (Fig. 3.62(b)). PZT (lead zirconotitanate, also known as lead zirconium titanate) is a substitutional solid solution containing 50 mol% lead titanate ($PbTiO_3$) and 50 mol% lead zirconate ($PbZrO_3$). Both lead titanate and lead zirconate exhibit the distorted Perovskite structure.

(a) Describe the chemical bonding in PZT.

(b) How many Pb atoms are there per unit cell?

(c) What is the fraction of the unit cells in PZT that contain Ti atom(s)?

(d) PZT is piezoelectric at room temperature. However, it loses its piezoelectric character when the temperature is too high. Explain from the viewpoint of the scientific origin.

(e) What is the direction of the electric dipole moment in a unit cell of PZT?

(f) On the average, a unit cell of PZT has electric dipole moment μ and lattice constants a (dimension of each edge of the base of the tetragonal prism) and c (height of the tetragonal prism). Express the saturation polarization of PZT in terms of μ, a and c. Hint: An electric dipole refers to a pair of positive and negative charges that are separate from one another. The electric dipole moment is the product of the magnitude of charge at each end of the dipole and the distance between the positive and negative ends of the dipole. The polarization refers to the dipole moment per unit volume. The saturation polarization refers to the maximum polarization, which occurs when all the unit cells are perfectly aligned.

Solution:

(a) Ionic bonding

(b) 1

(c) 0.5

(d) Above the Curie temperature, PZT is cubic in structure and has a center of symmetry in the unit cell.

(e) c-axis

(f) Saturation polarization $= \mu/(a^2 c)$

Problem 19 (Effect of stress on the permittivity)

A dielectric material changes its relative permittivity κ from 1.35 to 1.56 due to stress. The material is a cylindrical disc. The electric field is fixed and is along the cylindrical axis of the disc. Assume that the dimensional change of the disc due to the change in κ is negligible.

(a) What is the fractional change in polarization that results from this change in κ?

(b) What is the fractional change in capacitance that results from this change in κ? Note:

$$P = \varepsilon_0 \, (\kappa - 1) \, E,$$

where E is the electric field, P is the polarization and ε_0 is the permittivity of free space.

$$C = \varepsilon_0 \kappa A / \ell,$$

where C is the capacitance, A is the cross-sectional area of the disc, and ℓ is the height of the disc.

Solution:

(a) $P = \varepsilon_0 \, (\kappa - 1) \, E$

$\Delta P/P = \Delta \, \kappa/(\kappa - 1)$

$\qquad = (1.56 - 1.35)/(1.35 - 1)$

$\qquad = 0.6$

(b) $\qquad C = \varepsilon_0 \kappa A / \ell$

$\Delta C / C = \Delta \kappa / \kappa$

$\qquad = (1.56 - 1.35)/1.35 = 0.16$

Problem 20 (Direct piezoelectric effect)

A piezoelectric material (with longitudinal $d = 240 \times 10^{-12}$ C/Pa.m^2 (m/V)) is in the form of a wafer. It is subjected to a stress of 4.5×10^8 Pa in the direction perpendicular to the wafer. How much polarization is produced across the wafer?

Hint: $\Delta P = d\, U$, where U is the stress.

Solution:

$$\Delta P = d\, U$$

$$= (2.4 \times 10^{-10} \text{ C/Pa.m}^2)\,(4.5 \times 10^8 \text{ Pa})$$

$$= 1.1 \times 10^{-1} \text{ C/m}^2$$

Problem 21 (Converse piezoelectric effect)

In the converse piezoelectric effect,

$$\text{Strain} = d \text{ (electric field)},$$

where d is the piezoelectric coupling coefficient. A piezoelectric material exhibits $d = 2400$ pC/N. The material disc of thickness 0.55 mm is subjected to a voltage difference of 15 V across its thickness. What is the strain that results in the thickness direction?

Note: 1 pC $= 10^{-12}$ C; C/N $=$ m/V.

Solution:

$$\text{Electric field} = 15 \text{ V}/0.55 \text{ mm}$$

$$= 27 \text{ V/mm} = 2.7 \times 10^4 \text{ V/m}$$

$$\text{Strain} = d \text{ (electric field)}$$

$$= (2.4 \times 10^{-9} \text{ m/V}) \, (2.7 \times 10^4 \text{ V/m})$$

$$= 6.5 \times 10^{-5}$$

Problem 22 (Piezoelectric effect equation)

The equation for the converse piezoelectric effect is

$$\text{Strain} = d \times \text{electric field},$$

where d is the piezoelectric coupling coefficient. Derive the equation for the direct piezoelectric effect.
Hint: Mechanical energy = Electrical energy.

Solution:

Mechanical energy = Electrical energy

Stress × strain = Polarization × electric field

Stress = Polarization × electric field/strain

$$= \text{Polarization} \, (1/d)$$

Polarization = $d \times$ stress

Problem 23 (Piezoelectric effect – voltage coefficient)

A piezoelectric material has elastic modulus $M = 760$ GPa and piezoelectric coupling coefficient $d = 230 \times 10^{-12}$ m/V. What is its voltage coefficient g?
Hint: $g = 1/(Md)$.

Solution:

$g = 1/(Md)$

$\quad = 1/[(760 \text{ GPa}) (230 \times 10^{-12} \text{ m/V})]$

$\quad = 5.7 \times 10^{-3} \text{ m}^2/\text{C}$

The unit of g is $(\text{m}^2/\text{N}) (\text{V/m}) = \text{m}^2 \text{ V/J} = \text{m}^2/\text{C}$

Problem 24 (Piezoelectric effect – voltage coefficient)

A piezoelectric material converts mechanical energy to electrical energy, such that a stress of 1.2 MPa causes an electric field of 2.6×10^3 V/m in the material in the stress direction. Calculate the piezoelectric voltage coefficient g.
Hint: $g = E/\Delta U$, where E is the electric field and ΔU is the change in stress.

Solution:

$g = E/\Delta U,$

where E is the electric field and ΔU is the change in stress (i.e., 1.2 MPa). Hence,

$g = (2.6 \times 10^3 \text{ V/m})/(1.2 \text{ MPa})$

$\quad = 2.2 \times 10^{-3} \text{ m}^2/\text{C}$

Problem 25 (Permittivity, dielectric strength and piezoelectricity)

A polymer has relative permittivity (κ) 3.79. It is in the form of a circular disc of diameter 28.94 mm and thickness 0.785 mm. The dielectric strength of this polymer is 2.58×10^3 V/m. This polymer is piezoelectric, with piezoelectric coupling coefficient (d) 128 pC/N. The permittivity of free space (ε_0) is 8.85×10^{-12} F/m.

Hints:

$$C = \varepsilon_0\, \kappa\, A/l$$

$$P = \varepsilon_0\, (\kappa - 1)E,$$

where E is the electric field.

$$\text{Energy in capacitor} = (1/2)CV^2,$$

where C is the capacitance and V is the voltage.

$$\text{Strain} = dE$$

The dielectric strength is the highest electric field that the material can sustain without losing its insulation ability.

$$C/N = m/V$$

(a) Calculate the permittivity of this material.

(b) Calculate the capacitance of this disk in the direction perpendicular to the plane of the disc.

(c) What is the maximum electrical energy that can be stored in this disk?

(d) This disc is subjected to an electric field of 2.365×10^2 V/m along the direction perpendicular to the plane of the disk. Calculate the polarization in this disc in the presence of this electric field.

(e) What is the thickness of the disk in the presence of the electric field described in (d).

Solution:

(a) $\varepsilon = \kappa\varepsilon_o = (3.79)\,(8.85 \times 10^{-12}\ \text{F/m}) = 3.35 \times 10^{-11}\ \text{F/m}$

(b) $C = \varepsilon_o\,\kappa\,A/l$

$= (8.85 \times 10^{-12}\ \text{F/m})\,(3.79)\,[\pi\,(2.894 \times 10^{-2}\ \text{m}\,/2)^2]\,/\,(7.85 \times 10^{-4}\ \text{m})$

$= 2.81 \times 10^{-11}\ \text{F}$

(c) Highest voltage $=$ (highest electric field) (thickness)

$$= (2.58 \times 10^3\ \text{V/m})/(7.85 \times 10^{-4}\ \text{m})$$

$$= 3.29 \times 10^6\ \text{V}$$

Energy in capacitor $= (1/2)CV^2$

$$= (1/2)\,(2.81 \times 10^{-11}\ \text{F})\,(3.29 \times 10^6\ \text{V})^2$$

$$= 152\ \text{J}$$

(d) $P = \varepsilon_o\,(\kappa - 1)E$

$= (8.85 \times 10^{-12}\ \text{F/m})\,(3.79 - 1)\,(2.365 \times 10^2\ \text{V/m})$

$= 5.84 \times 10^{-9}\ \text{C/m}^2$

(e) Strain $= dE$

$$= (128\ \text{pC/N})\,(2.365 \times 10^2\ \text{V/m})$$

$$= 3.03 \times 10^{-8}$$

New thickness $= (0.785\ \text{mm})\,(1 + 3.03 \times 10^{-8})$

$$= 0.785\ \text{mm}$$

Problem 26 (Piezoelectric effect - polarization)

A PZT material has the piezoelectric coupling coefficient d equal to 4×10^{-10} C/N.

(a) What is the change in polarization that results from a change in the applied tensile stress from 0.88 to 1.46 MPa along the c-axis? The c-axis is the direction in the PZT tetragonal unit cell that is perpendicular to the square base of the unit cell. The polarization is in the direction of the stress.

(b) The PZT material is a disk of diameter 10 mm and height 1.1 mm. It is compressed along the cylindrical axis of the disk, which is the c-axis. The compressive stress is changed from 1.57 MPa to 2.45 MPa. How much is the corresponding change in the electric dipole moment of the material? Is the change in the electric dipole moment an increase or a decrease?

Note:

$$P = d\sigma,$$

where P is the polarization, d is the piezoelectric coupling coefficient and σ is the stress.

$$1 \text{ MPa} = 10^6 \text{ Pa}$$

$$1 \text{ Pa} = 1 \text{ N/m}^2$$

Solution:

$$(a) \; \delta P = d\delta\sigma$$

$$= (4\times 10^{-10} \; \tfrac{C}{N}) \, (1.46\text{-}0.88) \, (10^6 \; \tfrac{N}{m^2})$$

$$= 2.32 \times 10^{-4} \; \tfrac{C}{m^2}$$

(b) $\delta P = d\delta\sigma$

$$= \left(4 \times 10^{-10} \; \frac{C}{N}\right)[(-2.45) - (-1.57)]\left(10^6 \; \frac{N}{m^2}\right)$$

$$= -3.5 \times 10^{-4} \; \frac{C}{m^2}$$

P is dipole moment per unit volume

Change in dipole moment $= (\delta P)(\text{volume})$

$$= (-3.5 \times 10^{-4} \; \frac{C}{m^2}) \, \pi \, (5 \times 10^{-3} \text{ m})^2 \, (1.1 \times 10^{-3} \text{ m})$$

$$= -3.0 \times 10^{-11} \text{ C.m}$$

Decrease

Note: The unit for polarization is C.m/m^3. The unit for stress is N/m^2. Thus, the unit for d is (C.m/m^3)/(N/m^2) = C/N.

Problem 27 (Piezoelectric coupling coefficient)

A piezoelectric material provides a strain of 1.56% under an applied electric field of 2,650 V/m. What is the piezoelectric coupling coefficient?

Solution:

Strain $= d \, E$, where d is the piezoelectric coupling coefficient and E is the electric field.

$d = \text{strain}/E$

$= (1.56\%)/(2650 \text{ V/m})$

$= 5.9 \times 10^{-6} \text{ m/V}$

Problem 28 (Electrostriction)

Electrostriction produces a strain of 0.1% at an electric field of 2 MV/m. Calculate the electrostrictive coefficient in the unit m^2/V^2. Assume that the relationship between the strain and electric field is quadratic.

Solution:

Electrostrictive coefficient $W = (0.1\%)/(2\ MV/m)^2$

$$= 2.5 \times 10^{-4}\ m^2/(MV)^2$$

$$= 3 \times 10^{-16}\ m^2/V^2$$

Problem 29 (Pyroelectric effect)

A pyroelectric material with pyroelectric coefficient (p) $330 \times 10^{-6}\ C/m^2.K$ and relative permittivity (κ) 260 is subjected to a temperature change of 0.01 K. What is the resulting change in electric field?
Hint: $\Delta P = p\ \Delta T$

$$\Delta P = (\kappa - 1)\ \varepsilon_0\ \Delta E$$

Solution:

$$\Delta P = p\ \Delta T$$

$$= (330 \times 10^{-6}\ C/m^2.K)\ (10^{-2}\ K) = 330 \times 10^{-8}\ C/m^2$$

$$\Delta P = (\kappa - 1)\ \varepsilon_0\ \Delta E$$

$$\Delta E = \frac{\Delta P}{(\kappa - 1)\varepsilon_0}$$

$$= \frac{330 \times 10^{-8}\ C/m^2}{(260 - 1)(8.85 \times 10^{-12}\ C/V.m)}$$

$$= 1440\ V/m$$

Problem 30 (Pyroelectric effect)

For PZT having pyroelectric coefficient $p = 380 \times 10^{-6}$ C/(m^2.K), relative permittivity $\kappa = 1{,}300$, area $A = 2$ cm^2 and a thickness corresponding to capacitance $C = 100$ pF, what voltage results from a temperature change of 0.002K?

Hint: $\Delta V/\Delta T = pA/[(\kappa - 1)C]$

Solution:

$$\Delta V/\Delta T = pA/[(\kappa - 1)C]$$

$$\Delta V/\Delta T = [380 \times 10^{-6}\,\text{C/(m}^2.\text{K)}]\,(2 \times 10^{-4}\,\text{m}^2)/[(1300 - 1)\,(10^{-10}\,\text{F})]$$

$$= 0.58\ \text{V/K}$$

If $\Delta T = 0.002$K, then $\Delta V = (0.58\text{V/K})\,(0.002\ \text{K})$

$$= 1.2\ \text{mV}$$

Problem 31 (Impedance)

Derive the expressions for the real and imaginary parts of the impedance for the circuit below.

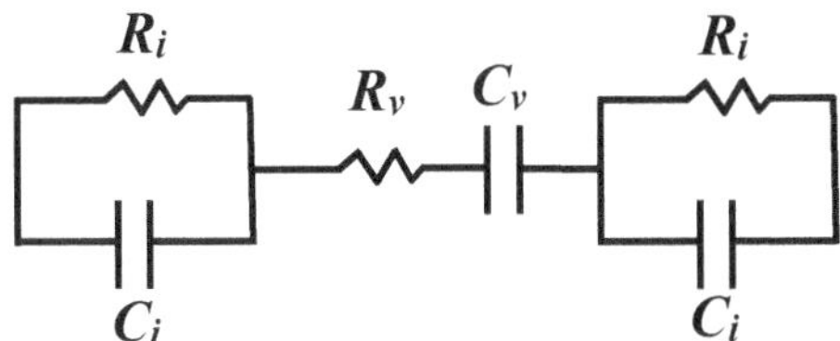

Solution:

Consider the following two parts of the overall circuit.

Part A

$$\text{--}\mathsf{W}\text{--}|\vdash\text{--}$$
$$\overset{R \quad C}{}$$

Part B

$$Z_A = R_V - \frac{i}{\omega C_V}$$

$$\frac{1}{Z_B} = \frac{1}{R_i} + i\omega C_i = \frac{1 + i\omega R_i C_i}{R_i}$$

$$\Rightarrow Z_B = \frac{R_i}{1 + i\omega R_i C_i}$$

Total:

$$Z = Z_A + 2Z_B$$

$$Z = \left(R_V - \frac{i}{\omega C_V} \right) + \frac{2R_i}{1 + i\omega R_i C_i}$$

$$= R_V - \frac{i}{\omega C_V} + \frac{2R_i(1 - i\omega R_i C_i)}{(1 + i\omega R_i C_i)(1 - i\omega R_i C_i)}$$

$$= \left(R_V + \frac{2R_i}{1 + \omega^2 R_i^2 C_i^2} \right) - i \left(\frac{1}{\omega C_V} + \frac{2\omega R_i^2 C_i}{1 + \omega^2 R_i^2 C_i^2} \right)$$

Real part of Z

$$= R_V + \frac{2R_i}{1 + \omega^2 R_i^2 C_i^2}$$

Imaginary part of Z

$$= -\left(\frac{1}{\omega C_V} + \frac{2\omega R_i^2 C_i}{1 + \omega^2 R_i^2 C_i^2}\right)$$

References

1. Chung DDL. Development, design and applications of structural capacitors. Applied Energy 2018;231:89-101.

2. Xi X, Chung DDL. Dielectric behavior of graphite, with assimilation of the AC permittivity, DC polarization and DC electret. Carbon 2021;181:246-259.

3. Xi X, Chung DDL. Pyropermittivity and pyroelectret behavior of graphite. Carbon 2021;174:357-367.

4. Xi X, Chung DDL. Dynamics of the electric polarization and depolarization of graphite. Carbon 2021;172:83-95.

5. Wen S, Chung DDL. Electric polarization in carbon fiber reinforced cement. Cem Concr Res 2001;31(2):141-147.

6. Chung DDL, Xi X. Piezopermittivity for capacitance-based strain/stress sensing. Sensor Actuators A 2021;332:113028.

7. Chakraborty P, Zhou C, Chung DDL. Converse piezoelectric behavior of three-dimensionally printed polymer and comparison of the in-plane and out-of-plane behavior. Mater Sci Eng 2020; B252:114447.

8. Xi X, Chung DDL. Electret behavior of unpoled carbon fiber with and without nickel coating. Carbon 2020;159:122-132.

9. Xi X, Chung DDL. Effects of cold work, stress and temperature on the dielectric behavior of copper. Mater Chem Phys 2021; 270:124793.

10. Chakraborty P, Gundrati NB, Zhou C, Chung DDL. Effect of stress on the capacitance and electric permittivity of three-dimensionally printed polymer, with relevance to capacitance-based stress monitoring. Sensors and Actuators A 2017;263C:380-385.

11. Chakraborty P, Zhao G, Zhou C, Chung DDL. Unprecedented sensing of interlayer defects in three-dimensionally printed polymer by capacitance measurement. Smart Mater Struct 2018;27(11):115012, 7 pp.

12. Xi X, Chung DDL. Role of grain boundaries in the dielectric behavior of graphite. Carbon 2021;173:1003-1019.

13. Chung DDL, Xi X. Factors that govern the electric permittivity of carbon materials in the graphite allotrope family. Carbon, 2021; 184:245-252.

Chapter 4

Electromagnetic Behavior

This chapter covers the science and applications that pertain to the electromagnetic behavior of materials. The radio wave and microwave regime of electromagnetic radiation is specifically addressed, due to their relevance to wireless communication and electromagnetic interference shielding. The optical regime (ultraviolet, visible and infrared) is addressed in Chapter 5. The science and applications in this chapter relate to Faraday's Law and the interaction of electromagnetic radiation with materials.

4.1 Electromagnetic applications

Electromagnetic applications include the following.

(i)   *Electromagnetic interference (EMI) shielding*, i.e., electromagnetic disturbance that interrupts, obstructs or otherwise degrades the effective performance of electronic or electrical equipment, including ventilators, defibrillators and pacemakers. Shielding is needed to keep electronic devices from interfering with others and to keep other devices from interfering with yours. Sources of EMI include power supplies, fluorescent lights, electric motors, electromechanical switches, relays, integrated circuits, pagers, cell phones, power lines and transformers. EMI shielding in the radio wave regime is increasingly needed as the abundance of radiation sources and the sensitivity of electronics increase. Relevant to the radio wave regime are AM/FM radio, TV antenna, cellphones, satellite GPS (Global Positioning System), Bluetooth (a wireless standard for exchanging data between mobile and fixed devices over short distances) devices, Wi-Fi (IEEE-802.11-based wireless networking technologies), and and microwave devices. Shielding is needed for both radiation sources and electronics.

(ii) *Antennas* that operate in the radio wave and microwave regime. They are important for wireless communication, with applications including autonomous cars, smart wearables (wearable or stretchable electronics), radio-frequency identification device (RFID) tags, virtual/augmented reality, and the Internet of Things. The drive for the 5th generation mobile networks (5G) adds to the importance of the antenna technology. In particular, flexible antennas are needed for smart wearables and other portable electronics.

(iii) *Cell-phone proof buildings and buildings that deter electromagnetic spying*, i.e., buildings that are shielded by the use of appropriate construction materials so that cell phones cannot be used in the building (as desired in a concert hall) and electromagnetic radiation emitted by telephones, fax machines and computers in the building cannot be analyzed by spies outside the building.

(iv) *Low electromagnetic observability or Stealth*, i.e., near invisibility to radar, which uses reflected electromagnetic waves to identify the position, speed and direction of both moving and stationary objects, such as aircraft, ships, motor vehicles, weather formations and terrain, with the Doppler shift of the frequency indicating the speed). To a radar, a Stealth aircraft may look like a bird.

(v) *Radio frequency identification (RFID)*, i.e., the use of a tag that contains digital information (e.g., product code) in a chip, such that the information can be transmitted by using an antenna on the tag upon interrogation by a transceiver).

(vi) *Radomes*, i.e., a structural enclosure that protects a radar antenna and allows radio wave radiation to pass through with little or no attenuation. Hence, electromagnetic transparency is needed for radomes.

(vii) *Lateral guidance*, i.e., the use of radio wave reflecting concrete in either the center part or the edge part of a traffic lane of a road to guide vehicles, each of which is installed with an emitter and a receiver of radio wave (Fig. 4.1). This is useful for catching drunken drivers, enhancing traffic safety at intersections, and automatic steering (which is needed for automatic highways). Conventional concrete is not able to reflect radio wave. However, concrete can be rendered this ability by the addition of a conductive component (such as carbon fiber) to the concrete mix. For example, by paving the central part of a traffic lane with radio-wave-reflecting concrete and installing radio wave transmitter and receiver in each vehicle, the vehicle would know if it is not travelling along the center of the lane. By using feedback control, the vehicle is maintained along the center of the traffic lane, thereby achieving automatic steering. This approach is less expensive than the alternate approach of embedding magnets along a traffic lane and guiding the vehicle through sensing the magnetic field.

(viii) *Electromagnetic pulse (EMP).* EMP refers to a short burst of electromagnetic energy, as in lightning strikes and nuclear explosions. Materials for EMI shielding are closely related to materials for EMP shielding.

(ix) *Conversion of mechanical energy to electrical energy,* i.e., the use of the phenomenon described by Faraday's Law, so that an AC electric current is generated by a conductor loop rotating in the presence of a magnetic field. This principle is used in classical electrical generators and in windmills.

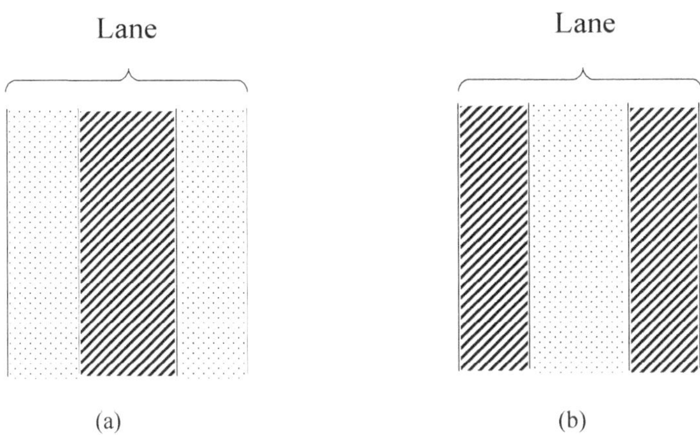

Fig. 4.1 Lateral guidance of a motor vehicle moving in a traffic lane of a road. The shaded region is radio wave reflecting concrete. The dotted region is conventional concrete, which is a poor reflector of radio wave. (a) Configuration with the radio wave reflecting concrete in the center of the lane. (b) Configuration with the radio wave reflecting concrete at the edges of the lane.

(x) *Metal detection*, i.e., a device for detecting metals and based on Faraday's Law.

(xi) *Eddy current inspection*, i.e., a nondestructive testing (abbreviated NDT) method based on Faraday's Law for detecting flaws in an electrically conductive material.

(xii) *Conversion of AC voltage of one voltage amplitude to one of another voltage amplitude*, i.e., a transformer, which is based on Faraday's Law.

4.2 Electromagnetic radiation

4.2.1 *Electromagnetic waves*

Electromagnetic radiation is in the form of waves. An electromagnetic wave is a transverse wave, which refers to the wave with the disturbance being perpendicular to the direction of propagation of the wave. For example, a string hanging down vertically from one's hand and being disturbed by horizontal back-and-forth movement by the hand is associated with a transverse wave that propagates vertically down the length of the string, with the disturbance being horizontal (transverse). For an electromagnetic wave, the relevant quantities that describe the disturbance are the electric field and the magnetic field. This means that the electric field and magnetic field are in directions that are perpendicular to the direction of propagation, as illustrated in Fig. 4.2, which shows a snapshot at a particular time during the propagation. The electric field and the magnetic field have the same frequency. In Fig. 4.2, the electric field is in the y direction, the magnetic field is in the z direction and both electric field wave and magnetic field wave propagate in the x direction. As required by Maxwell's Equations (which describe the nature of electromagnetism), the electric and magnetic fields are in a plane perpendicular to the direction of propagation. Both electric field and magnetic field waves are transverse waves (not longitudinal waves).

The electric field in an electromagnetic wave may be in any direction in the yz plane, although it is drawn in the y direction in Fig. 4.2. If the electric field is not restricted to any particular direction in the yz plane, it is said to be unpolarized. If it is restricted to a particular direction the yz plane, it is said to be linearly polarized. The electric field wave in

Fig. 4.2 is linearly polarized, with the electric field vector in the y direction.

The waves in Fig. 4.2 are said to be plane waves, because the wavefront is a plane, which is perpendicular to the direction of propagation, as shown in Fig. 4.3. A plane wave is in contrast to a spherical wave, the wavefront of which is a sphere.

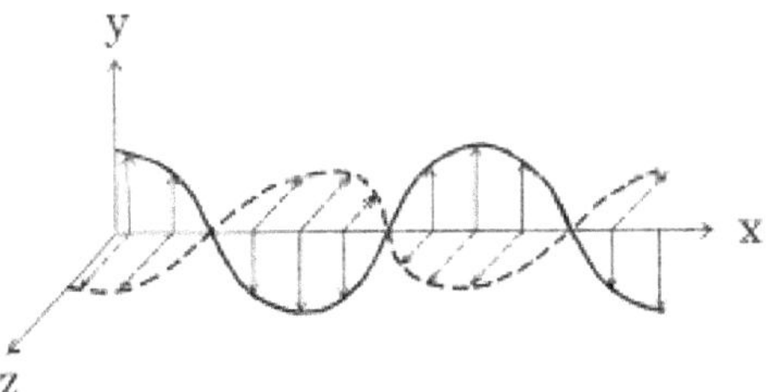

Fig. 4.2 An electromagnetic wave (a transverse wave) propagating in the x-direction. The solid curve depicts the electric field component of the wave. The dashed curve depicts the magnetic field component of the wave.

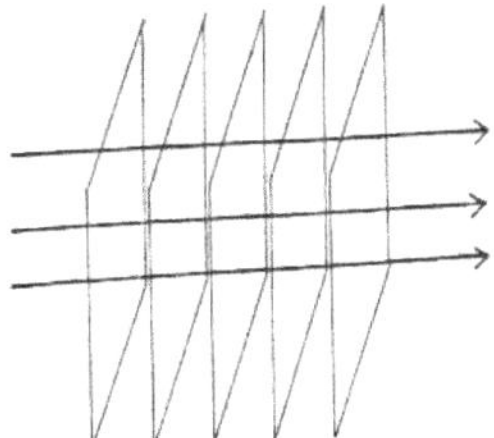

Fig. 4.3 A plane wave. The planes represent wavefronts spaced a wavelength apart. The arrows represent rays.

Figure 4.4(a) shows, for the sake of simplicity, the variation with position x along the path of propagation at a particular time for the electric field component only. The wavelength λ, as shown in Fig. 4.4(a), is the distance of one cycle of the wave. Figure 4.4(b) shows the variation with time at a particular point along the path of propagation. The time for one cycle of the wave to pass by this point is called the period, which is equal to $1/f$, where f is the frequency in the unit cycles per second (Hertz or, in short, Hz). The angular frequency ω is defined as the number of radians per second and is equal to $2\pi f$. Hence, the period can be expressed as $2\pi/\omega$, as shown in Fig. 4.4(b).

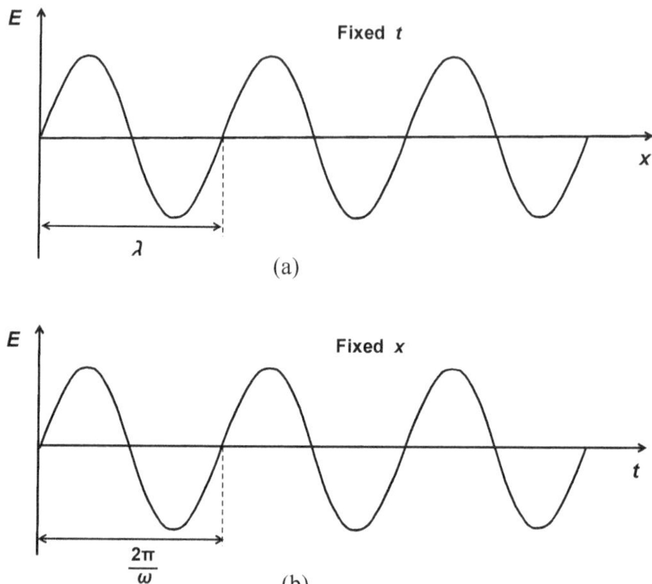

Fig. 4.4 The electric field (E) component of an electromagnetic wave. (a) The wave at a particular time t, showing the variation with position x. (b) The wave at a particular position x, showing the variation with time t.

4.2.2 *Photon energy, frequency and wavelength*

A photon is a quantum of electromagnetic radiation and is characterized by its energy (i.e., the photon energy, E), frequency (f) or wavelength (λ), which are related by the equations below:

$$E = hf, \tag{4.1}$$

where h is the Planck's constant (a universal constant equal to 6.63×10^{-4} J.s = 4.14×10^{-15} eV.s);

$$f = s/\lambda, \tag{4.2}$$

where s is the speed of propagation of the wave. This speed depends on the medium in which the wave propagates. It can also vary with the direction within the medium. For free space (vacuum), $s = c$ is

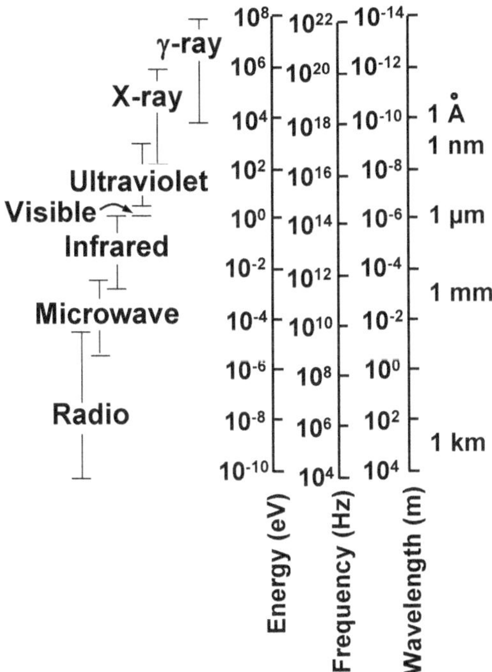

Fig. 4.5 The electromagnetic spectrum, which covers a large range of photon energy/frequency/wavelength.

3×10^8 m/s. If the medium is not a vacuum, $s < c$. According to Eq. (4.1), E is proportional to v. Equation (4.2) means that, for a given value of s, the quantities f and λ are inversely related to one another. Figure 4.1 shows the correspondence among the three quantities E, f and λ. In addition, Fig. 4.5 shows the names given to electromagnetic radiation in various regimes of E, v or λ. The visible region only constitutes a small part of the electromagnetic spectrum, which includes γ-rays, x-rays, ultraviolet, visible, infrared, microwave and radio wave radiation, in order of increasing wavelength from 10^{-14} to 10^4 m. In particular, radio waves are in the regime with low E, low v and large λ. This regime is important for radio and television transmission, because the atmosphere is transparent in this regime.

The radio wave regime includes frequencies ranging from 10^4 to 10^{10} Hz (i.e., wavelengths ranging from 10^{-1} to 10^4 m). The microwave regime includes frequencies ranging from 10^9 to 10^{12} Hz (i.e., wavelengths ranging from 10^{-3} to 10^0 m). The two regimes overlap. The most important frequencies are in the GHz range. For example, the frequency of 1 GHz corresponds to the wavelength of 0.30 m. Such wavelengths are long. This is why shields can be in the form of a metal wire mesh or a metal sheet with holes that are so large that one can see through the sheet, as for the door of a microwave oven. Due to the long wavelengths, materials for EMI shielding should be designed with consideration of the macroscopic structure, although most research in EMI shielding materials has focused on the microstructure or nanostructure.

Example problems 1–3 at the end of the chapter are relevant.

4.2.3 *Propagation of electromagnetic radiation*

Electromagnetic radiation (particularly that above 30 MHz in frequency) may propagate by radiation, using a transmitter (i.e., a transmitting antenna, which converts an electric current to electromagnetic waves) and a receiver (i.e., a receiving antenna, which converts an electromagnetic wave to a current), as in broadcasting. The function of a transmitting antenna is based on the principle that electrons that decelerate causes the emission of electromagnetic radiation and deceleration occurs when the electrons that are associated with an AC electric current in the antenna, which is an electrical conductor, change directions. The function of a receiving antenna is based on the principle that the electric field component of the electromagnetic radiation incident on a conductor interacts with the electrons (which are electrically charged) in the conductor, thereby causing the electrons to move (i.e., drift, Sec. 2.1) and hence produce a current.

Electromagnetic radiation (particularly that below 30 MHz) may also propagate by conduction, if there is a physical connection between the points involved. A physical connection may be a waveguide, which is an electrical conductor in the form of a hollow tube, as illustrated in Fig. 4.6(a). The waveguide is equipotential (the same in the electric potential) on its surface, so it causes the radiation to propagate along the axis of the tube. A common type of connection is a coaxial cable (Fig. 4.6(b)), which is a cylindrical waveguide with a central conductor line. This geometry causes the electric field to be radial, being directed between the central conductor and the outer conductor, as the radiation propagates along the axis of the cable.

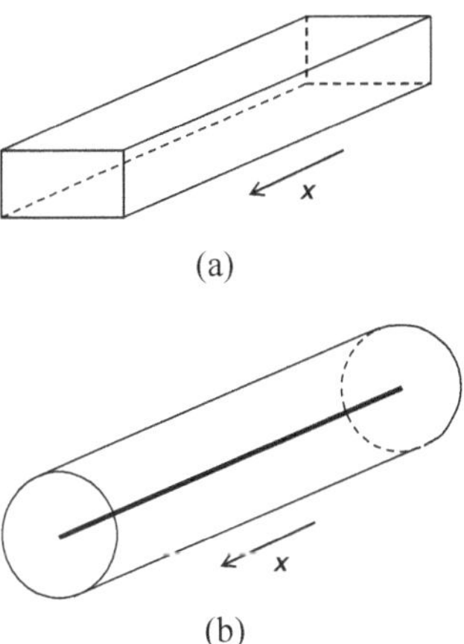

Fig. 4.6 Waveguide configurations. (a) A hollow conductor tube without a central conductor line. (b) A hollow conductor tube with a central conductor line (thick line) and known as a coaxial cable.

An electromagnetic wave may go through a capacitor (Sec. 3.1). This stems from Eq. (3.2), the rearrangement of which gives

$$\kappa Q = CV. \tag{4.3}$$

Differentiation with respect to time t gives the current I as

$$I = \Delta(\kappa Q)/\Delta t = C\,\Delta V/\Delta t. \tag{4.4}$$

Equation (4.4) means that the current through a capacitor is proportional to the rate of change of V. For a DC voltage, the rate of change of V is zero and the current is thus zero. In other words, an AC wave can go through a capacitor, which is an open circuit to a DC current.

4.2.4 *Skin effect*

Electromagnetic radiation at high frequencies penetrates only the near surface region of an electrical conductor. This is known as the Skin Effect. The effect is because of the eddy current (arising from a changing magnetic field) canceling the current flow in the center region of the conductor, while reinforcing the current flow in the skin region. The electric field of a plane wave penetrating a conductor drops exponentially with increasing depth into the conductor. The depth at which the field drops to 1/e of the incident value is called the skin depth (δ). In other words, the electric field E at a depth of x is given by

$$E = E_i\, e^{-x/\delta}, \tag{4.5}$$

where E_i is the incident electric field. According to electromagnetic theory, the skin depth δ is given by

$$\delta = 1/\sqrt{(\pi f \mu \sigma)} \tag{4.6}$$

where f = frequency,

μ = magnetic permeability = $\mu_0 \mu_r$,

μ_r = relative magnetic permeability,

μ_o = $4\pi \times 10^{-7}$ H/m, and

σ = electrical conductivity in $\Omega^{-1}.m^{-1}$.

Hence, the skin depth decreases with increasing frequency and with increasing conductivity or permeability. For copper, $\mu_r = 1$, $\sigma = 5.8 \times 10^7\ \Omega^{-1}m^{-1}$, so δ is 2.09 μm at a frequency of 1 GHz. For nickel of $\mu_r = 100$, $\sigma = 1.15 \times 10^7\ \Omega^{-1}.m^{-1}$, so δ is 0.47 μm at 1 GHz. The small value of δ for nickel compared to copper is mainly due to the ferromagnetic nature of nickel. Example Problem 10 at the end of the chapter concerns the calculation of δ based on Eq. (4.6).
   Example problems 9 and 10 at the end of the chapter are relevant.

4.2.5 *Faraday's Law and Lenz's Law*

Electromagnetic radiation may also propagate by induction, which is based on Faraday's Law. This law is one of the basic laws of electro-magnetism. It states that a changing magnetic flux through a conductor loop causes a voltage, and hence a current, in the loop. For example, due to a change in the current through the large conductor loop in Fig. 4.7, the magnetic field resulting from the current changes, thus causing a change in the magnetic flux through the loop (i.e., the product of the magnetic field and the area of the loop). A nearby conductor loop (the small loop in Fig. 4.7) intercepts some of the magnetic flux lines emanating from the large loop, so the magnetic flux through the small loop also changes. Thus, a current appears in the small loop. The current in the small loop is in the opposite direction from that of the large loop. This is due to the negative sign in the Faraday's Law, which is expressed as

$$V = -\Delta\Phi/\Delta t, \tag{4.7}$$

where V is the voltage generated, Φ is the magnetic flux (with unit Weber, or V.s, and defined as the product of the magnetic flux density B and the area through which the magnetic flux lines pass through) and t is the time. The unit of B is weber/m^2, or Tesla. The quantity $\Delta\Phi/\Delta t$ is the rate of change of the magnetic flux. In other words, the induced voltage is equal to the negative of the rate of change of the magnetic flux. The use of a coil with N turns results in N times the voltage from a single loop with the same area, i.e.,

$$V = -N\Delta\Phi/\Delta t. \tag{4.8}$$

The current in the coil is equal to the voltage divided by the resistance of the coil. The product of the voltage and the current is the power. In general, the change of flux can be due to a change in magnetic field or a change in area.

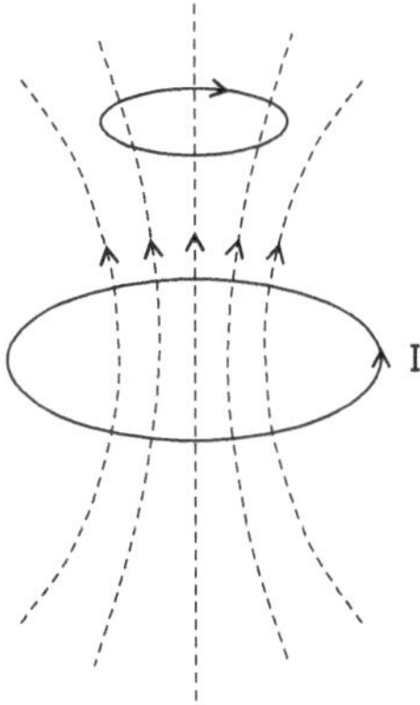

Fig. 4.7 The change in magnetic flux inducing a current. The dashed lines are the magnetic flux lines. The current is denoted *I*.

The negative sign in Eq. (4.7) is a description of Lenz's law, which is a consequence of the principle of the conservation of energy. It means that the current induced in a coil will create a magnetic field that opposes the change of magnetic flux in the coil. For example, in case of a decreasing magnetic flux, the induced current will be in a direction to oppose further decrease in the flux; in case of an increasing magnetic flux, the induced current will oppose further increase in the flux. In other words, the north pole of a magnet approaching a loop induces a north pole in the near face of a loop. If the north pole of the magnet induces instead a south pole in the near face of the loop, the magnet and loop will attract one another, thereby accelerating the magnet's approach and causing the magnetic field to increase more quickly. As a consequence, the current in the loop increases. Thus, both the kinetic energy of the magnet and the Joule heating at the loop increase, so that a small energy input causes a large energy output, violating the law of the conservation of energy.

Example problems 4–7 at the end of the chapter are relevant.

4.3 Applications of Faraday's Law and Lenz's Law

4.3.1 *Metal detection*

In case that the second conductor loop in Fig. 4.7 is a piece of an electrical conductor, current is induced in the surface region of the conductor by a changing magnetic flux, such as one provided by the AC current in a conductor coil. This current is known as the eddy current. By using a higher frequency for the AC current, the rate of change of the magnetic flux is greater and the eddy current is higher. The direction of the eddy current is such that the magnetic field that it generates opposes the applied magnetic field. The magnetic field from the eddy current can be picked up by a detector, due to the current produced at the detector. This concept is utilized in metal detectors, which detect metals due to the electrical conductivity of metals and the consequent eddy current in the metals. In contrast, polymers and ceramics cannot be detected, due to their poor electrical conductivity.

4.3.2 *Eddy current inspection*

In case that the material is an electrical conductor, the presence of defects or variation in the composition in the conductor affects the eddy current, thus allowing the detection of defects or composition variation in the material. However, the information obtained about the defects or the composition variation is qualitative (not quantitative) and distinction between defect effects and composition variation effects cannot be made. For example, cracks may be detrimental for a certain material while composition variation may not be detrimental. The inability for discerning the two effects makes the condition assessment unclear.

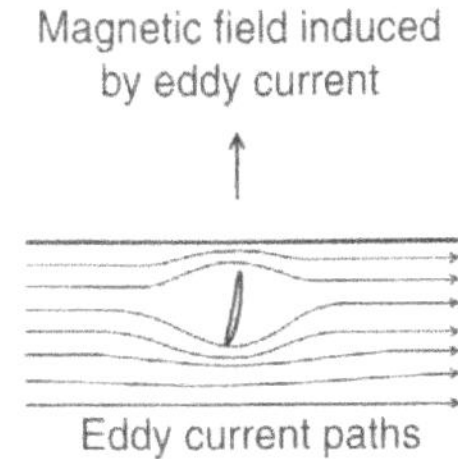

Fig. 4.8 Effect of a defect on the eddy current paths.

Figure 4.8 illustrates the distortion of the eddy current path around a defect, which is a region of high local electrical resistivity. This constitutes a nondestructive testing (abbreviated NDT) (or nondestructive evaluation, abbreviated NDE) method known as eddy current inspection. This inspection method involves placing a coil (called the excitation coil or the magnetizing coil) carrying an AC current in close proximity to the surface of the part under inspection. Variation in the eddy current can be detected by either using a second coil (called the search coil) or by measuring the change in current in the excitation coil. The typical frequency used in eddy current inspection ranges from 1 kHz to 3 MHz. The penetration depth of the eddy current (the skin depth described by Eq. (4.6)) is typically from 5 μm to 1 mm. Thus, the inspection is limited to the surface region. The lower the frequency, the greater is the penetration depth. Eddy current inspection is limited to the inspection of electrical conductive materials. Figure 4.9 shows two common configurations of eddy current inspection. The configuration in Fig. 4.9(a) is suitable for cylindrical parts, whereas that in Fig. 4.9(b) is suitable for parts with flat surfaces. Eddy current inspection is also used for measuring the thickness of a nonconductive coating on a conductive substrate. This thickness measurement is based on the notion that the signal provided by the eddy current in the conductive substrate is diminished by the presence of a nonconductive coating between the substrate and the detector. In general, eddy current inspection is advantageous is the portability of the equipment, the absence of consumables and the suitability for automation.

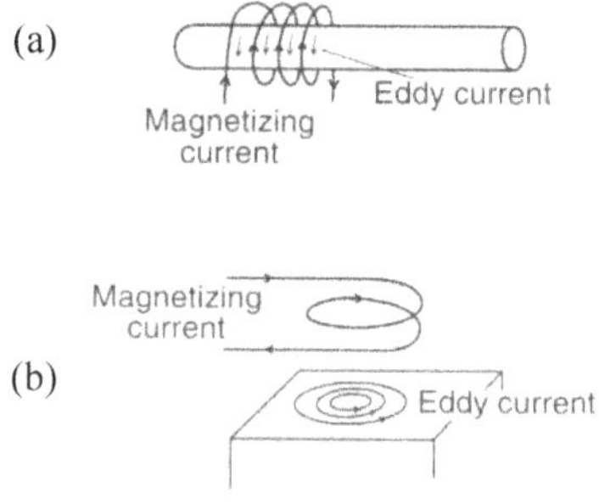

Fig. 4.9 Eddy current inspection. (a) The eddy current is circumferential in a direction opposite to the magnetizing current in the coil. (b) The eddy current is circular, in a direction opposite to the magnetizing current in the coil.

4.3.3 *Transformer*

A transformer is a device that changes the amplitude of an AC voltage wave from the value of the input to that of the output. A step-down transformer changes from a higher amplitude to a lower amplitude. A step-up transformer changes from a lower amplitude to a higher amplitude. In general, a transformer involves two conductor coils that are wound around the same piece of magnetic material (known as the transformer core), which serves to direct the magnetic flux lines from the primary coil (the input) to the secondary coil (the output), as shown in Fig. 4.10. Due to the difference in the number of loops between the two coils, the voltage at the secondary coil differs from that of the primary coil, even if the magnetic flux through the two coils may be the same (due to the two coils having the same area). The ratio of the number of loops in the primary coil to that in the secondary coil equals the ratio of the input voltage to the output voltage. The input must be AC (not DC), in order for the flux to change with time. That the flux changes with time is necessary for the occurrence of the effect described by Eq. (4.8).

Example problem 8 at the end of the chapter is relevant.

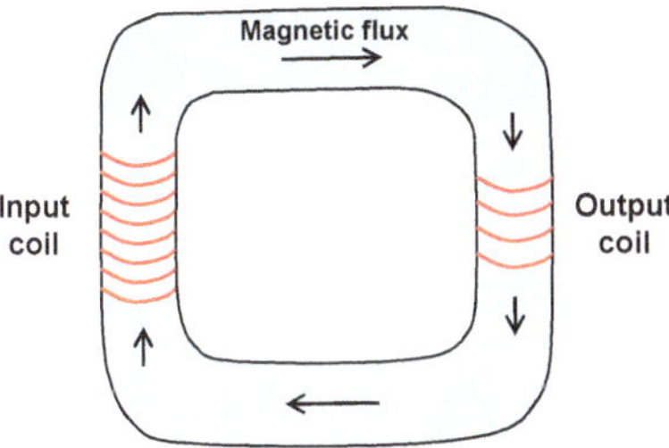

Fig. 4.10 A transformer with 8 turns in the input coil and 4 turns in the output coil. Thus, this is an example of a step-down transformer, with the output voltage amplitude being half of that of the input.

4.3.4 *Conversion of mechanical energy to electrical energy*

The classical electric generator involves a conductor loop rotating in the presence of a magnetic field, as illustrated in Fig. 4.11. As the loop rotates while the magnetic field is in a fixed direction, the magnetic flux through the loop changes. The flux is zero when the loop is parallel to the magnetic field (Fig. 4.11(a)) and is maximum when the loop is perpendicular to the magnetic field (Fig. 4.11(b)). Let the angular frequency of the rotation be ω (in unit of radians/s). As the loop rotates, the projection of the loop on the plane perpendicular to the magnetic field has an area A that is given by where A_o is the area of the loop and t is the time. Differentiation with respect to t gives

$$A = A_o \sin \omega t, \tag{4.9}$$

$$\Delta A / \Delta t = \omega A_o \cos \omega t. \tag{4.10}$$

Hence, based on Faraday's Law,

$$\Delta \Phi / \Delta t = B \, \Delta A / \Delta t = B\omega A_o \cos \omega t, \tag{4.11}$$

where B is the magnetic flux density (magnetic inductance). Based on Eq. (4.8),

$$V = -N\Delta\Phi/\Delta t = -N\omega BA_o \cos \omega t. \tag{4.12}$$

Therefore, an AC voltage wave results. This principle is also used in windmills, which convert mechanical energy to electricity. The large size of the blades of a windmill allows greater effect of the wind on the rotation of the blades. The speed of rotation of the blades is typically only 10–22 revolutions/min, so a gearbox is usually used to step up the rotational speed in the device (an electric generator) for conversion of the mechanical energy to electricity.

Another configuration for this energy conversion is shown in Fig. 4.12. As the magnet vibrates vertically, the flux through the windings changes. Due to Faraday's Law, this results in a voltage at the winding. Due to the linear motion, this device is known as a linear generator. This configuration is used for harvesting energy from structural vibration, which is particularly significant at the natural vibration frequency of the structure. For example, with the magnet hung from a part of a bridge by using a spring and the windings attached to the part of the bridge below the magnet, the mechanical energy associated with the vibration of the bridge is converted to electricity. This device is also known as an electromagnetic energy harvester.

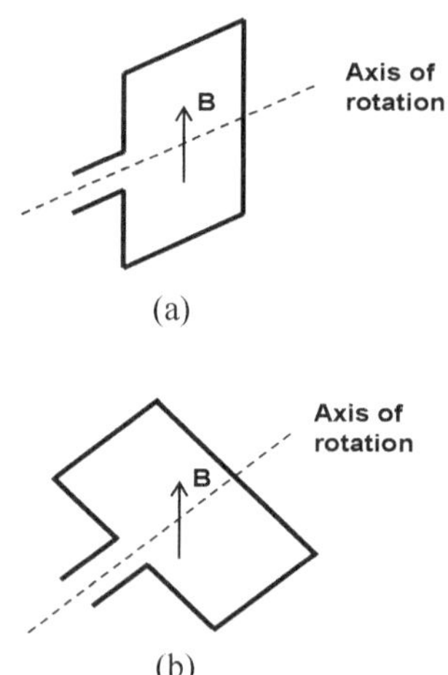

Fig. 4.11 Conversion of mechanical energy to electrical energy by using Faraday's Law.

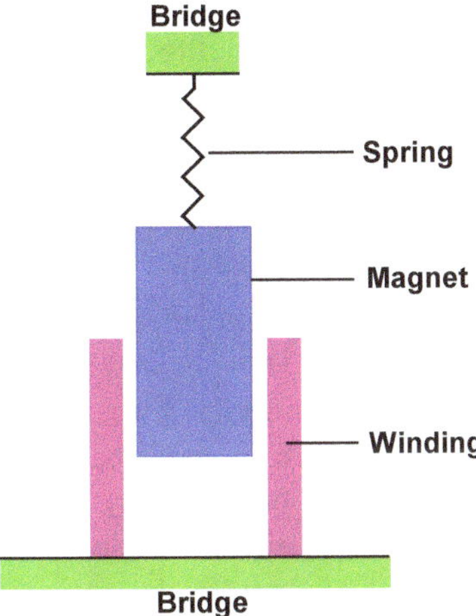

Fig. 4.12 Linear generator used as an electromagnetic energy harvester. The axis of the winding is in the plane of the page.

4.4 Electromagnetic shielding

Electromagnetic interference (EMI) shielding refers to the reflection and/or absorption of electromagnetic radiation by a material, which thereby acts as a shield against the penetration of the radiation through the shield. EMI shielding is to be distinguished from magnetic shielding, which refers to the shielding of magnetic fields at low frequencies (e.g., 60 Hz). Materials for EMI shielding are different from those for magnetic fielding.

Absorption and reflection are the two main mechanisms for the interaction of radio wave or microwave radiation with materials. Both absorption and reflection contribute to shielding. However, absorption rather than reflection is needed for antennas. An antenna functions in both the receiving and transmitting of radiation. Its receiving and transmitting characteristics are similar, due to the reciprocity theorem in electromagnetism.

The reflection mechanism results in radiation bouncing off from the shielding material. The reflected radiation may be undesirable to the environment, particularly to the people present. Therefore, from the safety viewpoint, the absorption mechanism is preferred for shielding. For low observability (Stealth), reflection is also not desired, if the reflected radiation reaches the radar and hence becomes detected.

Decreasing the impedance mismatch between the material and the surrounding medium is an effective method for decreasing the reflection. However, shielding materials are commonly conductive materials (e.g., metals and carbons), for which the impedance mismatch tends to be substantial. Material nanostructuring through layering in the nanoscale provides an avenue for tailoring a material for reducing the reflection.

The magnetic character promotes the absorption of the radiation, due to the interaction of the magnetic dipoles in the material with the magnetic field in the radiation. The pertinent aspects of the magnetic character include the magnetic dipole friction (which increases with increasing frequency and depends on the size of the magnetic domains and the mobility of the domain boundaries), dipole-dipole interaction, magnetic susceptibility and magnetic continuity.

The electric polarizability of a material also promotes the absorption, due to the interaction of the electric dipoles in the material with the electric field in the radiation. The pertinent aspects of the dielectric character include the electric dipole friction (which increases with increasing frequency), dipole-dipole interaction, dielectric connectivity and electric susceptibility.

The dielectric behavior relates to the conduction behavior through the Kramers-Kronig relationship, which is based on the notion that the permittivity is the real part of the complex permittivity, and that the conduction behavior contributes to the dielectric loss, which is indicated by the imaginary part of the complex permittivity. For example, among carbon materials obtained by heat treatment at different temperatures, the conductivity, and the real and imaginary parts of the permittivity all increase with increasing heat-treatment temperature (i.e, with increasing degree of crystallographic order).

Among materials that are not magnetic, high conductivity is commonly considered in the industry as the main key to obtain high shielding. This emphasis on the conductivity is partly because of the ease of measuring the conductivity. However, high conductivity is not necessarily the primary criterion for shielding among non-magnetic materials. There are other factors, as discussed below.

A factor relates to the surface area of the shielding material or the interface area in the shielding material. Due to the Skin Effect, a high specific surface/interface area (i.e., area per unit volume) promotes the volume of the part of the specimen that interacts with the radiation, thereby enhancing the shielding effectiveness. The interface refers to that between the shielding component (e.g., a filler) and non-shielding component (e.g., the matrix) in the material. The non-shielding component is relatively transparent to the radiation, due to the low degree of interaction with the radiation. For a given porosity in a porous material, a smaller feature size of the porous microstructure gives a higher specific surface area, and hence greater shielding effectiveness. For a given filler volume fraction in a composite, a smaller unit size of the filler gives a higher specific interface area, thereby promoting shielding. This is partly why nanofillers are attractive for composites for shielding.

Another factor relates to the electrical connectivity, which promotes the shielding, due to the continuity of the electric field lines in the material. This continuity also allows the flow of eddy current induced by the magnetic field, thereby promoting magnetic power loss. Therefore, electrical percolation (electrical connectivity) in a composite with a conductive filler and a nonconductive matrix is advantageous for shielding. The percolation threshold is lower when the aspect ratio of the filler is higher. A nanofiber tends to have a higher aspect ratio than a microfiber. This is another reason for the attractiveness of nanofillers.

For a shielding material that is magnetic, the magnetic continuity of the magnetic component is needed, because the magnetic field lines are necessarily continuous. Thus, magnetic percolation in a composite with a magnetic filler and a non-magnetic matrix is important for shielding.

The magnetic, dielectric and conduction behavior and the specific surface/interface area all contribute to governing the shielding effectiveness. Furthermore, each of behavior depends on the frequency and temperature. The effects of the frequency and temperature are not only relevant to applications, but they shed light on the mechanisms of the shielding. It is possible that, for a given material, the dominant shielding mechanism depends on the combination of frequency and temperature.

Materials for EMI shielding include functional materials and multifunctional structural materials. Functional shielding materials for small-scale applications have received most attention. However, structural shielding materials allow multifunctional structures in large-scale applications. Continuous carbon fiber composites and cement-based materials are dominant among structural shielding materials.

In spite of the large number of publications on the EMI shielding behavior of numerous materials, a systematic study of the effects of the magnetic, dielectric and conduction behavior and the specific surface/ interface area on the shielding effectiveness at various combinations of

Table 4.1 Correlation of EMI shielding effectiveness with magnetic, electrical and microstructural attributes. [1]

Attribute	Nickel nanofiber	Nickel fiber A	Nickel fiber B
Diameter (μm)	0.4	2	20
EMI shielding effectiveness (1–2 GHz), 19 vol.% filler in composite	92	72	5
Magnetic hysteresis energy loss, for the nickel part of the composite (J/m^3)	7370	4930	1020
Magnetic coercive field (kA/m)	16.9	4.7	0.5
Resistivity (DC), 19 vol.% filler in composite (Ω.cm)	1.5×10^{-3}	2.9×10^{-4}	5.3×10^{-2}
Grain size (μm)	0.016	0.018	2
No. of grains across diameter	9	110	10

frequency and temperature is severely inadequate. Table 4.1 shows an example of a limited form of such a systematic study, as discussed below.

The shielding effectiveness, magnetic properties (hysteresis magnetic energy loss and magnetic coercive force), electrical resistivity, and microstructure (grain size and number of nickel grains across the diameter of the filamentous nickel) are shown in Table 4.1 for filamentous nickel of three different diameters (0.4, 2 and 20 μm). The 0.4-μm diameter filamentous nickel is nickel-coated carbon nanofiber with the carbon core being only 0.1 μm in diameter. This nanofiber is also known as nickel nanofiber or nickel filament, since nickel amounts to 94 vol.% of the nanofiber. The 2-μm diameter filamentous nickel is a nickel fiber. The 20-μm diameter filamentous nickel is also a nickel fiber. At the same filler content of 19 vol.% and for the same nonconductive polymer matrix (polyethersulfone), the EMI shielding effectiveness (1–2 GHz), hysteresis energy loss and coercive force all decrease monotonically with increasing diameter of the filamentous nickel. However, the DC resistivity is lowest at the intermediate diameter of 2 μm, because of two competing factors, namely (i) the resistivity is reduced by better touching of the filamentous nickel units, as enabled by a smaller diameter (higher aspect ratio), and (ii) the resistivity is increased by the presence of a larger number of contact points, as resulting from a smaller diameter. For the same volume fraction, a smaller diameter, which corresponds to a higher aspect ratio, enhances the touching, but also increases the number of contact points, each of which being associated with electrical resistance. Thus, a compromise between these two factors, as provided by the intermediate diameter, gives the lowest resistivity. The grain size is similar for the diameters of 0.4 and 2 μm (grain size: 0.016 and 0.018 μm), but it is much larger (grain size: 2 μm) for the diameter of 20 μm. The number of grains across the diameter is similar for the diameters of 0.4 and 20 μm (number of grains: 9 and 10), but it is much larger for the diameter of 2 μm (number of grains: 110). The resistivity does not correlate with the diameter, grain size or number of grains across the diameter. The comparative results in Table 4.1 reveal that high EMI shielding effectiveness correlates with high values of the hysteresis energy loss and coercive force, and with low diameter, but the shielding effectiveness does not correlate with the resistivity, grain size or the number of grains across the diameter. At the same filler volume fraction, a smaller diameter corresponds to a larger specific interface area. Due to the Skin Effect, the shielding effectiveness is promoted by a larger specific interface area. The influence of the Skin Effect dominates over the influence of the conductivity in governing the shielding effectiveness. Similarly, the hysteresis energy loss and coercive force are promoted by a larger specific interface area.

The effect of the frequency has received much more attention than the effect of the temperature. Prior work has emphasized the development of material formulations for improved shielding, without adequate attention to the scientific basis for the material development.

The frequency dependence tends to be particularly significant at frequencies around which certain absorption mechanism changes (e.g., a change from dipolar polarization at low frequencies to electronic polarization at high frequencies). Ceramics tend to have stronger frequency dependence than carbons and metals, due to the partially ionic nature of ceramics and the sluggishness of ionic movement compared to electronic movement. For typical practical applications, a strong variation of the shielding with frequency is not desired. The attaining of high shielding effectiveness over a wide frequency range is a common goal in much of the research on EMI shielding materials.

Low reflection is attractive from the safety viewpoint. Magnetic materials that are conductive (e.g., metallic magnetic materials) tend to reflect substantially, due to the high impedance mismatch. Thus, magnetic materials that are low in the conductivity (e.g., ceramic magnetic materials) are preferred for reducing the reflection. Similarly, dielectric (polarizable) materials that are conductive tend to reflect substantially. However, the reflection can be substantial while the absorption is even more substantial. Therefore, it is important to measure both the absorption loss and reflection loss in order to understand the mechanism of the shielding.

The absorption of electromagnetic radiation results in heating, i.e., the conversion of the electromagnetic energy to thermal energy. The temperature change can affect the extent of absorption and reflection. By enhancing the absorption through an appropriate temperature rise, the shielding is promoted, as shown for graphene.

A Faraday cage is a volume of space that is totally enclosed by an electronic conductor. In the presence of an external electric field, the charges in the conductor move, thus reflecting the radiation and cancelling the electric field inside the cage. As a result, the cage acts as a shield against the penetration of electromagnetic radiation. Most commonly, Faraday cages involve metal meshes rather than solid metal foils, though the shielding is less perfect. Plaster with wire mesh and concrete with steel rebars are architectural materials that act as Faraday shields. The holes in the mesh should be small compared to the wavelength of the radiation, as in the case of the mesh in the door of a microwave oven. Faraday cages are encountered in our daily life. For example, a lightning that hits a car does not affect the people in the car; cellular phones and radio have worse reception in a car or tunnel.

Carbon materials are important among shielding and antenna materials, due to their ability to absorb radiation over a large range of frequencies [1,2]. Metals are widely used for these applications, due to their high electrical conductivity. Metals are even more conductive than carbons. In relation to carbon materials, the low density, chemical inertness, low thermal expansion, high electrical/thermal conductivity, and amenability to nanostructuring are among their attractions. Figure 4.13 illustrates the interaction of the electromagnetic radiation with the electrons in continuous carbon fibers, which are important for lightweight structural composite materials.

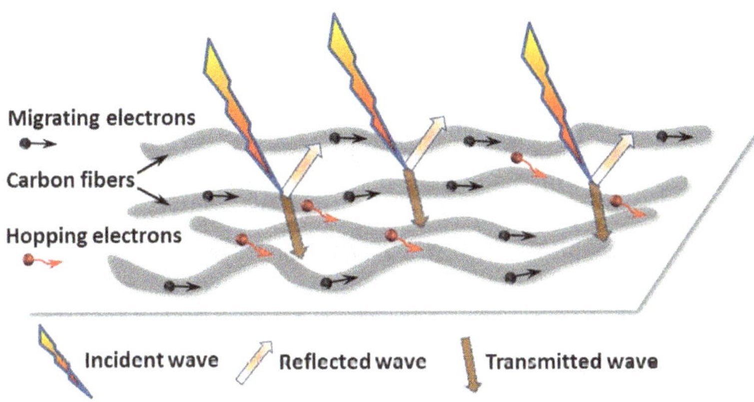

Fig. 4.13 Schematic illustration of the electromagnetic interaction with the carbon fibers. The electrons (mobile charge carriers) are indicated by circles (with the migrating electrons and hopping electrons distinguished). The electromagnetic wave (incident, reflected and transmitted) is indicated by the arrows. [3]

Ceramics, cement and polymers are typically not very effective, unless they are combined with a functional material. Due to the availability of numerous types of microcarbons and nanocarbons, shielding materials in the form of metal-carbon, ceramic-carbon, cement-carbon and polymer-carbon combinations have received much attention. Continuous carbon fiber composites and cement-based materials are dominant among structural shielding materials. [1]

Among the carbon materials, exfoliated-graphite-based flexible graphite (FG, i.e., flexible sheet made by the compaction of exfoliated graphite in the absence of a binder) is particularly effective. Exfoliated graphite is obtained by intercalating graphite flakes, followed by rapid heating that causes exfoliation. Each piece of exfoliated graphite is known as a worm, due to the extensive expansion along the c-axis and the resulting long length along this direction. Upon compaction of the worms, mechanical interlocking occurs among the adjacent worms, thereby resulting in a sheet in the absence of a binder. Due to its flexibility, the sheet is known as "flexible graphite". Because of the high degree of compaction during the fabrication of flexible graphite, the sheet exhibits strong crystallographic preferred orientation of the carbon layers in the plane of the sheet. As a result of the preferred orientation, the sheet is highly anisotropic, with the electrical conductivity being much higher in the plane of the sheet than the perpendicular direction. In addition, the process of exfoliation results in a degree of open porosity (in addition to the closed porosity), so that the surface area is increased. A high surface area is attractive for shielding, because of the Skin Effect. For these reasons, flexible graphite exhibits shielding effectiveness as high as 130 dB, which is essentially the upper limit of the power loss measurement. Furthermore, due to the cellular structure and preferred orientation, the sheet is resilient in the direction perpendicular to the plane of the sheet. The resiliency enables flexible graphite to serve as an EMI gasket material, as needed for the electromagnetic sealing of a shielded enclosure, which suffers from electromagnetic shielding deficiency at the joints (seams) within the enclosure. The flexibility of FG is also advantageous, particularly for flexible antennas. Furthermore, the low thermal expansion coefficient, high thermal conductivity and chemical inertness of FG add to the attraction of this material. In contrast, polymer-matrix composites rendered the shielding ability due to the presence of an adequate volume fraction of a conductive filler tend to be limited in resilience, thermal conductivity and chemical stability, and tend to be high in the thermal expansion coefficient. Inherently conductive polymers also have similar shortcomings.

A volume resistivity of the order of 1 Ω.cm is typically considered sufficient for shielding from a industrial viewpoint. However, electrical conductivity is not the scientific criterion for shielding, as conduction requires connectivity in the conduction path (percolation in case of a composite material containing a conductive filler), whereas shielding does not. Although shielding does not require connectivity, it is enhanced by connectivity. Metals are by far the most common materials for EMI shielding. They function mainly by reflection due to the free electrons in them. Metal sheets are bulky, so metal coatings made by electroplating, electroless plating or vacuum deposition are commonly used for shielding. The coating may be on bulk materials, fibers or particles. Coatings tend to suffer from their poor wear or scratch resistance. However, scratches tend to be acceptable, due to the long wavelength of the radiation.

An example of a magnetic material is nickel, which is a ferromagnetic metal (Sec. 6.3). The coating of carbon fiber with nickel enhances both the conductivity and magnetic character. The nickel coating greatly enhances the shielding effectiveness. Another example of a magnetic materials is iron oxide (Fe_3O_4), which is a ferrimagnetic ceramic (Sec. 6.5). Both ferromagnetic and ferrimagnetic materials are strongly magnetic. Although ceramics are advantageous to metals in terms of their corrosion resistance, they are less amenable to processing than metals.

The power loss refers to the loss in power due to the interaction of the incident radiation with the shielding material. It consists of the electric power loss and the magnetic power loss. The relative importance of these two types of power loss depends on the electrical and magnetic character of the material. For example, the magnetic power loss is significant for a magnetic material, but insignificant for a nonmagnetic material. In addition, the relative importance of the two types of power loss depends on the relative amplitudes of the electric field and magnetic field in the incident radiation.

The power loss due to shielding is expressed in decibels (dB) and is referred to as the shielding effectiveness, or the total shielding effectiveness (SE_T). The part of this loss due to absorption is known as the absorption loss (SE_A). The part of this loss due to reflection is known as the reflection loss (SE_R). The loss, regardless of the mechanism, as commonly expressed in decibels (abbreviated dB), is defined as

$$\text{Loss (dB)} = -10 \log (P/P_i), \tag{4.13}$$

where P is the power output (i.e., the power that remains after the interaction of the radiation with the material, whether the interaction involves absorption, reflection or other mechanisms) and P_i is the power input. The logarithm is to the base 10. With $P < P_i$, the loss given by Eq. (4.13) is positive. The smaller is P, the greater is the loss. The use of a logarithmic scale facilitates the description of large losses. As shown in Example Problem 12 at the end of this chapter, $P/P_i = 0.0001$ (a large loss) corresponds to 40 dB.

In case of reflection alone, P_r is the power that remains (i.e., the power output) after reflection from the front surface, so $P_i - P_r$ is the power that is reflected. Hence, according to Eq. (4.13), the loss in dB due to reflection (i.e., the reflection loss, abbreviated SE_R) is given by

$$SE_R = -10 \log (P_r/P_i). \tag{4.14}$$

Let the reflectivity (also known as the reflection coefficient, i.e., the fraction of the incident power that is reflected) be R. Hence,

$$R = (P_i - P_r)/P_i = 1 - P_r/P_i \tag{4.15}$$

The combination of Eq. (4.14) and (4.15) gives

$$SE_R = -10 \log (1 - R). \tag{4.16}$$

Equation (4.16) may be considered an alternative way to define SE_R.

In case of absorption alone (without reflection included in the consideration),

$$P_a = P_i \, e^{-\alpha x}, \tag{4.17}$$

where P_a is the power after absorption by the path of distance x (i.e., the output after absorption) and α is the linear absorption coefficient (also called the absorption coefficient, with unit m^{-1}), which describes the tendency of a material to absorb electromagnetic radiation and is a material property that depends on the frequency. Equation (4.17) is known as the Beer-Lambert Law. Its derivation is provided in Sec. 5.2. This quantity α is to be distinguished from the temperature coefficient of electrical resistivity, which is also commonly abbreviated as α (Eq. 2.29). Equation (4.17) means that the power decreases exponentially as the radiation travels through the medium. Thus, with the combination of Eq. (4.13) and (4.17), the loss in dB due to absorption (i.e., the absorption loss, abbreviated SE_A) is given by

$$SE_A = -10 \log (P_a/P_i)$$

$$= -10 \log e^{-\alpha x}$$

$$= (10/2.303) \, \alpha x$$

$$= 4.34 \, \alpha x. \tag{4.18}$$

Thus, SE_A is proportional to x. Thus, the absorption loss per unit distance is given by

$$SE_A/x = 4.34 \, \alpha. \tag{4.19}$$

This means that the absorption loss per unit distance is simply proportional to α, which is a material property with the unit m^{-1}.

A common pitfall relates to the **incorrect** definition of α as

$$P_a = P_i\, e^{-2\alpha x}, \tag{4.20}$$

which differs from the correct definition in Eq. (4.17) by a factor of 2 in the exponent. This incorrect definition stems from the incorrect notion that the electric field, which is proportional to the square root of the power, is proportional to $e^{-\alpha x}$.

The α as given by Eq. (4.19) decreases with increasing thickness of the specimen in the direction of propagation of the radiation, as shown in Fig. 4.14 for aluminum. This trend is because of the skin effect, so that the power of the radiation in the specimen decreases as the radiation traverses in the specimen.

In case of reflection at the surface followed by absorption by a distance x into the specimen, as illustrated in Fig. 4.15, the output power is $P_r\, e^{-\alpha x}$, where P_r is the power that remains after reflection. Using Eq. (4.13), the total loss due to the reflection together with the subsequent absorption is thus given by

$$\text{Total loss} = -10 \log (P_r e^{-\alpha x}/P_i)$$

$$= -10 \log [(P_r/P_i)\, e^{-\alpha x}]$$

$$= -10 [\log (P_r/P_i) + \log e^{-\alpha x}] \tag{4.21}$$

Based on Eq. (4.14) and (4.18), Eq. (4.21) becomes

$$\text{Total loss} = SE_R + SE_A. \tag{4.22}$$

That the loss contributions in dB are additive is a reason for the use of the dB unit. The greater is the reflectivity, the smaller is P_r and the greater is the reflection loss. The higher is α, the smaller is P_a and the greater is the absorption loss.

Let T be the transmittivity, also known as the transmission coefficient, i.e., the fraction of the incident power that is transmitted through the specimen. Let A be the absorptivity (also known as the absorption coefficient, i.e., the fraction of the incident power that is absorbed by the specimen, to be distinguished from the linear absorption coefficient α, which is defined in Eq. (4.17)). Let R be the reflectivity (also known as the reflection coefficient, i.e., the fraction of the incident power that is reflected by the specimen at its surface).

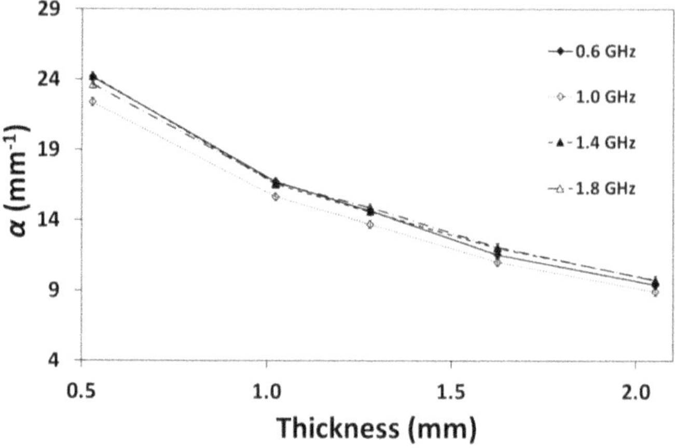

Fig. 4.14 Effect of the absorption distance (thickness) on α as given by Eq. (4.19). The results are for aluminum [4]. Due to the pitfall described in Eq. (4.20) and involved in Ref. [4], this figure is the corrected form of the corresponding figure in Ref. [4]. The correction simply involves doubling the values of α.

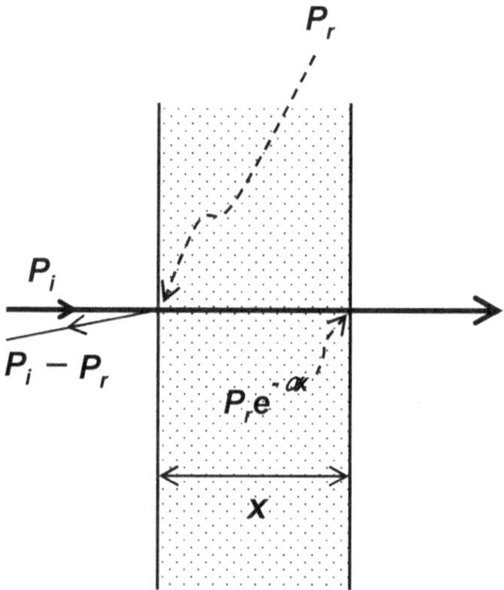

Fig. 4.15 The total loss due to reflection followed by absorption. P_r is the power that remains after reflection. $P_r\,e^{-\alpha x}$ is the power that remains after both reflection and absorption.

Other than reflection and absorption, a mechanism of shielding is multiple reflections, which refer to the reflections at various surfaces or interfaces in the shield. This mechanism requires the presence of a large surface area or interface area in the shield. An example of a shield with a large surface area is a porous or foam material. An example of a shield with a large interface area is a composite material containing a filler which has a large surface area. The loss due to multiple reflections can be neglected when the distance between the reflecting surfaces or interfaces is large compared to the skin depth. Typically, it can be neglected when the absorption loss exceeds 10 dB.

In case that the multiple reflections contribution is negligible,

$$A = 1 - R - T. \tag{4.23}$$

Equation (4.18) for SE_A may be rewritten as

$$SE_A = -10 \log (1 - A_{eff}), \tag{4.24}$$

where A_{eff} refers to the absorbance relative to input being the power after the reflection, which precedes the absorption. In contrast, A in Eq. (4.23) refers to the absorptivity, i.e., the fraction of the incident power that is absorbed

In case that the multiple reflections contribution is negligible, A_{eff} is given by

$$A_{eff} = A/(1 - R). \tag{4.25}$$

The combination of Eq. (4.23) and (4.25) gives

$$A_{eff} = (1 - R - T)/(1 - R). \tag{4.26}$$

Hence,

$$1 - A_{eff} = 1 - [(1 - R - T)(1 - R)]$$

$$= T/(1 - R). \tag{4.27}$$

The combination of Eq. (4.24) and Eq. (4.27) gives

$$SE_A = -10 \log [T/(1 - R)]. \tag{4.28}$$

Equation (4.28) may be considered an alternative way to define SE_A. Equation (4.28) for SE_A is often used along with Eq. (4.16) for SE_R.

The macroscopic structure of a material also affects the shielding performance. The fiber planar arrangement and stacking arrangement in the macroscale can greatly affect the shielding effectiveness of a conductive fiber composite. Figure 4.16 shows the shielding effectiveness for three planar arrangements of a continuous carbon fiber tow, with a tow consisting of 12,000 fibers. In Arrangement 1, the tow is in the form of a planar coil, such that the tow is continuous throughout the coil. In Arrangement 2 (including variations 2A and 2B), the tow is in the form of parallel straight lines, such that the tow is continuous between one line and its adjacent one. Arrangements 2A and 2B correspond to the direction of the tow being, respectively, 0° and 90° apart on the two opposite sides of the printer paper on which the tow is mounted. Thus, Arrangement 2A is described as being unidirectional, and Arrangement 2B is described as being crossply.

If the continuous fiber is magnetic, as for nickel-coated carbon fiber (with the nickel coating being ferromagnetic, Sec. 6.3), the planar coil

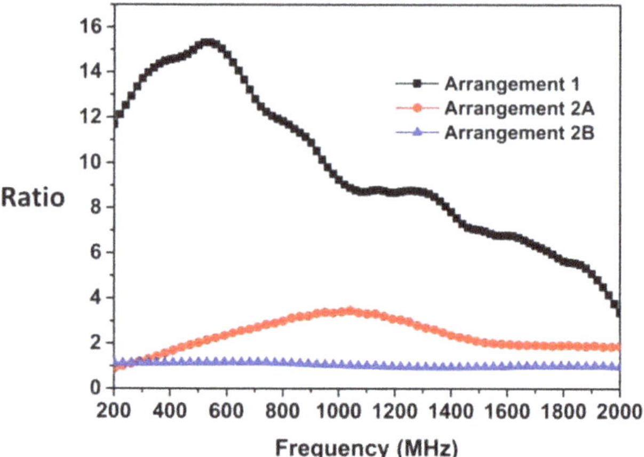

Fig. 4.16 Ratio of absorption losses of nickel-coated carbon fiber to pristine carbon fiber (without coating) for Arrangement 1 (planar coil), Arrangement 2A (unidirectional) and Arrangement 2B (crossply). [3]

configuration is attractive, giving a higher ratio of the absorption loss of the nickel-coated carbon fiber to that of the pristine (uncoated) carbon fiber than the linear configurations (whether unidirectional or crossply) (Fig. 4.16) [3]. The advantage of a magnetic constituent for shielding also applies when the magnetic constituent is discontinuous in the form of short fibers [5] or particles [6].

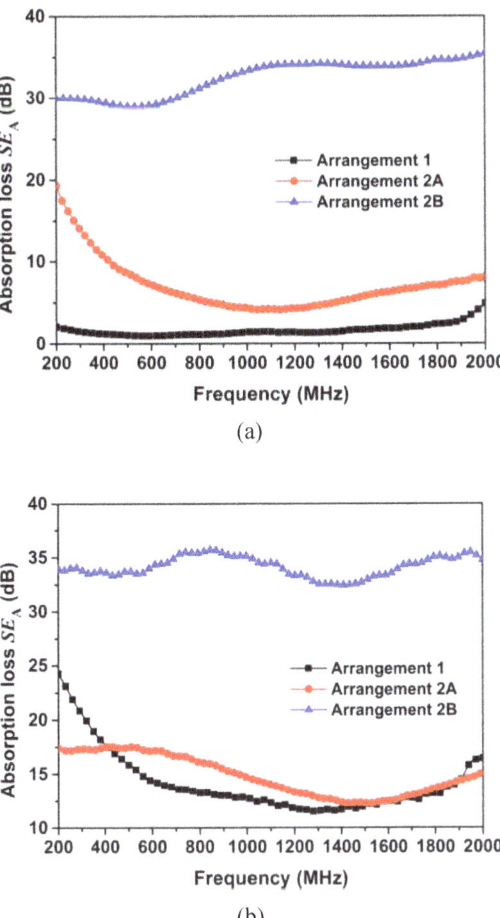

Fig. 4.17 Absorption loss vs. frequency for Arrangement 1 (planar coil), Arrangement 2A (unidirectional) and Arrangement 2B (crossply). (a) Pristine carbon fiber (without coating), (b) Nickel-coated carbon fiber. [3]

When the radiation is unpolarized, as is usually the case, the crossply configuration (with the fibers in two perpendicular directions) gives greater absorption loss than the unidirectional configuration, which in turn gives greater absorption loss than the planar coil configuration. This is shown in Fig. 4.17 for both pristing and nickel-coated carbon fiber. The inferiority of Arrangement 1 to Arrangement 2 is clearer when the coating is absent. As expected from electromagnetic theory, the planar coil configuration (Arrangement 1) is more effective than the linear configuration (Arrangement 2) for magnetic interaction. Thus, the magnetic interaction contributes less to shielding than the electrical interaction, though the magnetic interaction is increased when the fiber is coated with nickel. Thus, Arrangement 2A gives superior shielding performance than Arrangement 1 for the uncoated fiber. Arrangment 2B (crossply) gives more interaction with the electric field than Arrangement 2A, since the incident radiation is unpolarized. As a consequence, Arrangment 2B gives much higher SE_A than Arrangement 2A for both uncoated and nickel-coated fibers.

It is commonly assumed that the interaction of electromagnetic radiation in the radiowave or microwave regime with a metal is dominated by reflection, due to the high conductivity of the metal and the consequent impedance mismatch between the metal and the surrounding air. This mismatch causes reflection of the radiation. However, the absorption loss dominates over the reflection loss, even for metals. The absorption contribution is defined as the absorption loss divided by the total loss, which is the sum of the absorption loss and reflection loss. For example, for aluminum, the absorption contribution is up to 91% (Fig. 4.17). The absorption loss dominance is attributed to the dielectric behavior of metals, as discussed in Sec. 3.1 in relation to conductive dielectrics.

According to the classical electromagnetic theory, the absorption loss SE_A of a conductive material can be calculated by using the equations

$$SE_A = 8.68x\sqrt{\pi f \mu \sigma} = 131.4x\sqrt{\mu_r \sigma_r f} \qquad (4.29)$$

where f is the frequency in Hz, x is the thickness of the specimen in m, σ is the electrical conductivity, σ_r is the electrical conductivity relative to copper, and μ is magnetic permeability (with $\mu = \mu_0 \mu_r$, where $\mu_0 = 4\pi \times 10^{-7}$ H/m and μ_r is the relative magnetic permeability). This equation indicates that SE_A increases with increasing frequency, if μ_r and

σ_r do not decrease with increasing frequency. It also indicates that SE_A at a given frequency increases with both the permeability and conductivity.

The combination of Eq. (4.29) and (4.19) gives

$$\alpha = 2\sqrt{\pi f \mu \sigma} = \frac{SE_A}{4.34x} \tag{4.30}$$

Using the equation for the skin depth δ (Eq. (4.6)), Eq. (4.30) indicates that

$$\delta = \frac{2}{\alpha} \tag{4.31}$$

In case that the material is not magnetic, with $\mu = \mu_o$, Eq. (4.30) can be used to calculate the conductivity σ as a function of the frequency f from the frequency dependence of α. Moreover, Eq. (4.30) indicates that $\sqrt{\sigma}$ is inversely related to $\sqrt{f}$. The frequency dependence of α, δ, σ and $\sqrt{\sigma}$ is shown in Fig. 4.18 for aluminum foil.

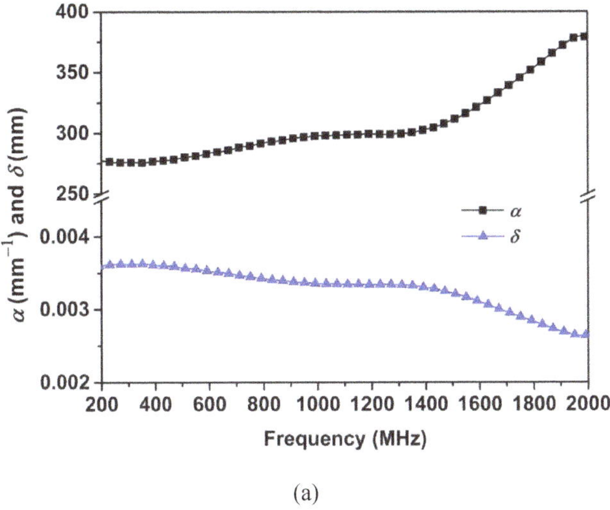

(a)

Fig. 4.18 Linear absorption coefficient α (obtained from SE_A and thickness x), skin depth δ given by Eq. (4.31), and electrical conductivity given by Eq. (4.30) [7]. The material is aluminum foil, which is not magnetic. The α increases with increasing frequency, while δ decreases with increasing frequency; $\sqrt{\sigma}$ is inversely related to $\sqrt{f}$, i.e., it is directly related to $f^{-1/2}$. Due to the pitfall described in Eq. (4.20) and involved in Ref. [7], the curve for α shown in (a) is in error and should be corrected by doubling the values of α. The δ and conductivity values shown are correct.

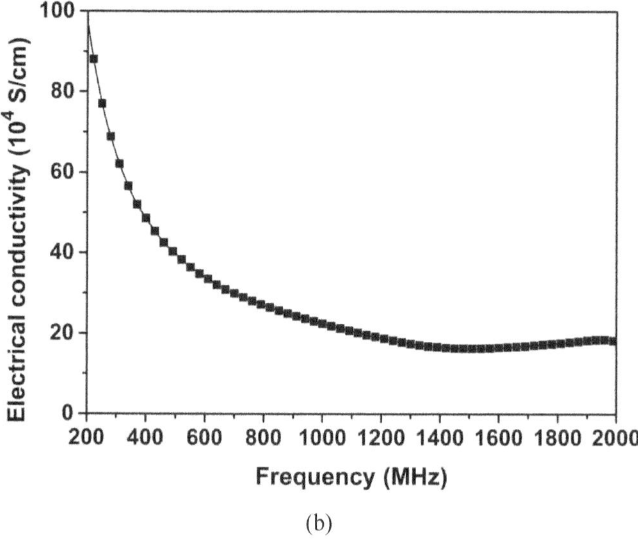

(b)

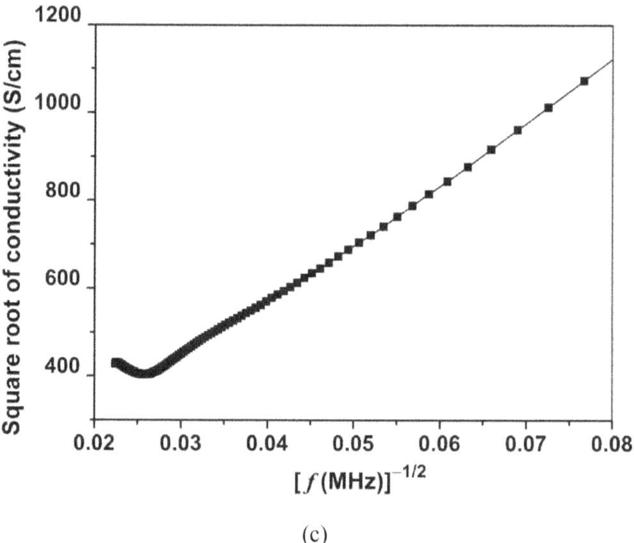

(c)

Fig. 4.18 (*Continued*)

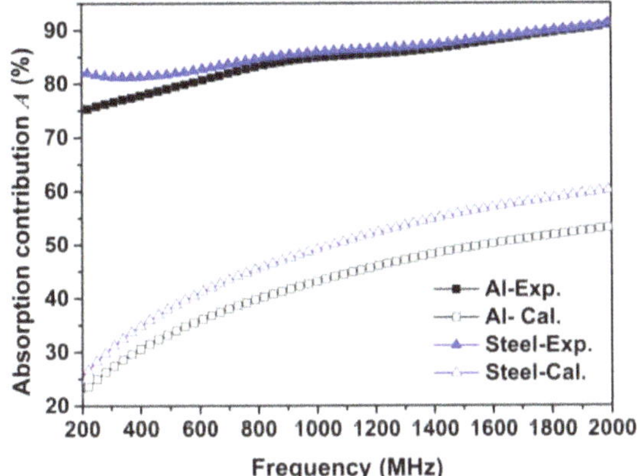

Fig. 4.19 Absorption contribution (absorption loss divided by the total loss) of alumium foil (16 μm thick) and steel foil (25 μm thick), with comparison of experimental (Exp.) and calculated (Cal.) values. [8]

For metals (aluminum and steel) and flexible graphite, it has been shown that the loss is dominated by the absorption loss, as shown by high values of the absorption contribution (defined as the absorption loss divided by the total loss) [8]. As shown in Fig. 4.19, the absorption contribution of the metals reaches 90%. However, the majority of the incident power is reflected. This is because the incident radiation is first reflected and the remaining radiation is then absorbed. The absorption loss is relative to the power that remains after the reflection.

The experimental value of the absorption loss is compared to the value calculated using Eq. (4.28), as shown in Fig. 4.19. The experimental value exceeds 75% (i.e., absorption dominance) in the entire frequency range for both aluminum and steel. The calculated values are much below the experimental values. The difference between the experimental and calculated values is attributed to the permittivity of the materials (Sec. 3.2). The permittivity's contribution to absorption is not considered in Eq. (4.28), which assumes that the conductivity effect overshadows the permittivity effect.

According to electromagnetic theory, the reflection loss SE_R is given by

$$SE_R = 20 \log \frac{\eta_0}{4\eta_s}, \tag{4.32}$$

where η_0 is the intrinsic impedance of free space (with $\eta_0 = 120\pi \ \Omega = 377 \ \Omega$), and η_s is the impedance of the material, as given by the equation

$$\eta_s = \sqrt{2\pi f \mu / \sigma} \qquad (4.33)$$

The reflection loss decreases with increasing frequency (Eq. (4.32 and 4.33)), if μ and σ are frequency independent. This trend is opposite to that for the absorption loss (Eq. (4.29)).

Table 4.2 Electrical conductivity relative to copper (σ_r) and relative magnetic permeability (μ_r) of selected materials.

Material	σ_r	μ_r	$\sigma_r\mu_r$	σ_r/μ_r
Silver	1.05	1	1.05	1.05
Copper	1	1	1	1
Gold	0.7	1	0.7	0.7
Aluminum	0.61	1	0.61	0.61
Brass	0.26	1	0.26	0.26
Bronze	0.18	1	0.18	0.18
Tin	0.15	1	0.15	0.15
Lead	0.08	1	0.08	0.08
Nickel	0.2	100	20	2×10^{-3}
Stainless steel (430)	0.02	500	10	4×10^{-5}
Mumetal (at 1 kHz)	0.03	20,000	600	1.5×10^{-6}
Superpermalloy (at 1 kHz)	0.03	100,000	3,000	3×10^{-7}

The absorption loss is a function of the product $\sigma_r\mu_r$ (Eq. (4.29)), whereas the reflection loss is a function of the ratio σ_r/μ_r (Eq. (4.33)). Table 4.2 shows these factors for various materials. Silver, copper, gold and aluminum are excellent for reflection, due to their high conductivity. Magnetic metals are excellent for shielding through absorption. Examples of magnetic alloys are mu-metal (a nickel–iron soft ferromagnetic alloy with relative magnetic permeability 20,000 at 1 kHz), superpermalloy (a nickel–iron alloy with about 80 wt.% nickel and 20 wt.% iron, and relative permeability 100,000 at 1 kHz), and stainless steel (No. 430, ferritic, with relative permeability 500). However, these alloys are heavy, due to their high densities. Nickel is also magnetic, though the relative permeability (100) is low. However, nickel is attractive because it can be deposited by electroplating to form a coating.

The effectiveness of a magnetic filler for enhancing shielding is greater when it is present with an electrically conductive material [6]. This can be obtained, for example, by using both a magnetic filler and an electrically conductive filler in the same composite. This requirement relates to the eddy current effect of a magnetic field.

The measurement of the EMI shielding effectiveness may be conducted by using the coaxial cable method illustrated in Fig. 4.20. In this method, the specimen of annular shape (like a donut) is sandwiched by two mating fixtures that serve as an expanded coaxial cable. A piece of electronic equipment known as a network analyzer is commonly used to provide the input wave and to analyze the output wave. There are two types of network analyzer, namely the scalar network analyzer (SNA) and the vector network analyzer (VNA). The SNA measures only the amplitude, whereas the VNA measures both the magnitude and phase. The SNA is not suitable for scientific investigation.

The scattering parameters (abbreviated S-parameters) describe the response of a network with N ports to signal(s) incident to one or more of the ports. In the abbreviation for the S-parameters, the first number in the subscript specifies the responding port, while the second number specifies the incident port. Hence, the abbreviation S_{21} refers to the response at port 2 resulting from a signal at port 1.

The VNA has two ports, labeled 1 and 2. The incident and transmitted waves can be expressed by complex scattering parameters (or S-parameters) S_{11} (or S_{22}) and S_{12} (or S_{21}), respectively. The reflectivity R is given by

$$R = |S_{11}|^2 = |S_{22}|^2. \tag{4.34}$$

The transmittivity T is given by

$$T = |S_{12}|^2 = |S_{21}|^2. \tag{4.35}$$

Both T and R should be reported as fractions or percentages, and should not be reported in the unit of dB.

The quantities SE_T, SE_R and SE_A are obtained from the S-parameters through the equations

$$SE_{\text{T}} = -10\log(T) = -10\log|S_{21}|^2 \tag{4.36}$$

$$SE_{\text{R}} = -10\log(1-R) = -10\log\left(1-|S_{11}|^2\right) \tag{4.37}$$

$$SE_{\text{A}} = SE_{\text{T}} - SE_{\text{R}} - SE_{\text{M}} \tag{4.38}$$

In Eq. (4.38), SE_{M} refers to the loss due to multiple reflections within the specimen. Eq. (4.37) stems from Eq. (4.16).

In practice, SE_{M} is typically negligible if the total shielding SE_{T} exceeds 15 dB. Under this situation, Eq. (4.38) becomes

$$SE_{\text{A}} = SE_{\text{T}} - SE_{\text{R}} = -10\log\left(T/(1-R)\right) \tag{4.39}$$

Equation (4.39) is consistent with Eq. (4.28).

The absorption contribution refers to the ratio of SE_{A} to SE_{T}. It is to be distinguished from the absorbance A.

Due to the limited volume in a device such as a cellular phone, functional shielding materials need to be effective at a small thickness. Furthermore, for reducing the weight of the device, functional shielding materials that are low in density are desirable. However, being effective at a low density is not meaningful in practice unless there is also effectiveness at a small thickness. A pitfall is that the shielding effectiveness per unit thickness or per unit density is considered more technologically important than the shielding effectiveness itself. This is not justified in view of the technological needs. A high value of the shielding effectiveness per unit thickness allows a very thin material to provide adequate shielding. However, very thin materials are limited to applications that involve radiation that is essentially at normal incidence, so that the electric field is in the plane of the material. Resiliency is valuable for EMI gaskets, but very thin materials tend not to exhibit adequate resiliency. Furthermore, the shielding effectiveness (better described in terms of the absorption loss) per unit thickness tends to decrease with increasing thickness, primarily due to the Skin Effect. Theoretically, the absorption loss in dB is proportional to the thickness. However, in reality, the simple proportionality may not occur. Assuming the proportionality without adequate experimental support is a pitfall. In addition, the material manufacturability may limit the thickness, so that a high value of the shielding effectiveness cannot be attained in spite of a high value of the shielding effectiveness per unit thickness. On the other hand, due to the large size of structures, shielding materials in the form of structural materials do not need to be effective at a small thickness.

Both the transmitted wave (i.e., the transmission loss, which is the shielding effectiveness) and the reflected wave (i.e., $P_i - P_r$ in Fig. 4.15) may be evaluated experimentally using either of the methods illustrated in Fig. 4.20 and Fig. 4.21. These methods are explained below.

As depicted in Fig. 4.2, the electric field is perpendicular to the direction of propagation of the wave. In both methods (Fig. 4.20 and 4.21), the radiation propagates in the direction perpendicular to the plane of the specimen. In Fig. 4.20, the waveguide is as shown in Fig. 4.6(b), with the electric field being in the plane of the specimen, as in a coaxial cable. The radial direction of the electric field is supported by the presence of the inner and outer conductors, which are concentric circles. In Fig. 4.21, the waveguide is as shown in Fig. 4.6(a), with no inner conductor. Due to the absence of the inner conductor, the radiation bounces back and forth off the interior surface of the waveguide, thus causing the electric field at a given point along the length of the waveguide to be at an angle to the plane of the specimen. If the waveguide is long enough (more exactly, if the aspect ratio of the waveguide is large enough), the radiation primarily propagates along the axis of the waveguide, with the consequence that the angle is small and the electric field is radial.

If the incident radiation is polarized, the electric field is in a well-defined direction that is perpendicular to the direction of propagation. However, the incident radiation is typically unpolarized, so that the electric field direction is two-dimensionally random in a plane perpendicular to the direction of propagation. Under this situation, the electric field can be resolved into two components with in-plane directions that are 90° apart.

In case that the radiation is unpolarized and the specimen is strongly anisotropic in this plane (with the conductivity or permittivity being high in a particular in-plane direction of the specimen), the specimen may not be able to interact with half of the incident radiation. Thus, under this situation, a composite with unidirectional conductive fibers is less effective for shielding than one with conductive fibers in two orthogonal directions in the plane of the specimen. Moreover, a composite with conductive fibers in a planar coil configuration would not be effective for interacting with the radial electric field from the incident radiation. An example of an anisotropic material is a carbon fiber composite, the fibers of which can be arranged differently. As shown in Fig. 4.17(a), the crossply fiber configuration gives higher SE_A compared to the unidirectional configuration, which in turn gives higher SE_A than the planar coil configuration.

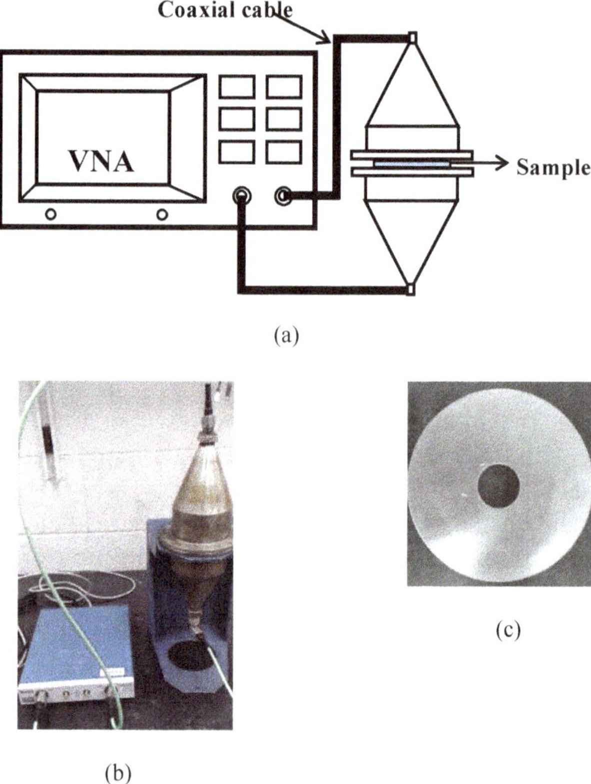

Fig. 4.20 Set-up for measurement of the EMI shielding effectiveness using the Coaxial Cable Method. (a) Schematic illustrations of the EMI shielding testing set-up involving a vector network analyzer (VNA). (b) Photograph of the set-up, showing the specimen fixture on the right and the VNA on the left. (c) Photograph of a metal sample used for the testing. [4]

As illustrated in Fig. 4.2, the electric field and magnetic field in an electromagnetic wave are perpendicular to one another, with both fields being perpendicular to the direction of propagation of the wave. In both methods (Fig. 4.20 and 4.21), the radiation propagates in the direction perpendicular to the plane of the specimen, so the electric field is radial while the magnetic field is circular in the plane of the specimen. In case that the specimen is magnetically anisotropic, the specimen may not be able to interact with a part of the incident radiation. An example is a composite with nickel-coated carbon fiber, in which the nickel is ferromagnetic (Sec. 6.3) and both the nickel and carbon fiber are conductive. As shown in Fig. 4.17(b), the planar coil configuration of the fiber gives SE_A that is similar to that provided by the unidirectional configuration, in contrast to Fig. 4.17(a), where the fiber is not nickel-coated and the unidirectional configuration gives greater SE_A than the planar coil configuration.

The relative magnitudes of the electric field wave and magnetic field wave in an electromagnetic wave are variable. By increasing the magnetic field, the interaction of a magnetic material with the radiation is enhanced. By increasing the electric field, the interaction of a conductive but nonmagnetic material with the radiation is enhanced.

The method in Fig. 4.20 is known as the Coaxial Cable Method. A coaxial cable has an inner conductor with a small diameter and an outer conductor with a larger diameter, such that the two conductors are concentric. as the specimen is sandwiched by two horn antennas, which serve as the waveguide and are akin to two enlarged coaxial cables. The two ends of the antennas are connected to a VNA.

Alternatively, a waveguide (commonly having a rectangular cross section) is used to direct the incident radiation on the specimen and to direct the reflected radiation in the opposite direction (Fig. 4.21). The waveguide is connected to a specimen holder, which typically has a rectangular through-hole for placing the specimen, as shown in the photo in Fig. 4.21. An EMI gasket is not usually used at the interface between the specimen and the interior surface of the hole in the specimen holder, as the specimen is typically just tightly fitted to this hole. The interface between the specimen and specimen holder is prone to result in radiation leakage, though this leakage is commonly ignored. The degree of tightness of the fit depends on the roughness of the edge surface of the specimen, in addition to depending on the dimensions of the specimen.

Different specimens of the same type may have different degrees of roughness and different spatial distributions of the hillocks associated with the roughness at their edge surface, even if the specimen dimensions are the same. The quality of the interface between the specimen and specimen holder may not be adequately controlled, thus reducing the repeatability in the EMI testing results obtained for different specimens of the same type. The output after transmission through the specimen is connected to a VNA. The input end at the front end of the waveguide is connected to the other port of the VNA.

Electromagnetic radiation leakage around the edges of the specim encauses the loss of a part of the incident radiation, so that this part of the radiation does not interact with the specimen. In spite of the leakage, the totality of the incident radiation is used in the data analysis associated with the VNA. A reduction of the amount of radiation that is actually received by the specimen would cause a reduction of the amount of radiation that is transmitted through the specimen. The leakage typically results in a decrease in the measured shielding effectiveness.

In order to reduce the leakage, the specimen should be electrically connected to the waveguide, such that there is no gap at the interface. This connection may be attained by using an EMI gasket as the interface material. An EMI gasket is a shielding material that is resilient. It is commonly in the form of an elastomer (e.g., silicone) that is filled with conductive particles (e.g., silver particles). The resiliency allows the gasket to serve as a cushion that is compressible, thereby reducing the chance of having gaps or voids at the interface.

Mechanical polishing of the proximate surfaces is not adequate for smoothening the interface completely. Even after polishing, there are still microscopic hills and valleys in the topography. The application of silver conductive paint in the wet state at one of the two proximate surfaces of the interface helps the electrical connection. The wetness allows the flow of the paint to fill the valleys. However, the paint after drying is not resilient and is small in thickness, so it is limited in the effectiveness for filling the valleys in the topography associated with the interface.

In the method of Fig. 4.20, the specimen is in the form of an annular ring (like a donut, as shown in Fig. 4.20(c)). The inner edge of the annular ring should be electrically connected to the inner conductor of each of the two horns, while the outer edge of the annular ring should be electrically connected to the outer conductor of each of the two horns. EMI gaskets (commonly made from silicone that is filled with conductive particles) may be used at the interfaces to provide the electrical connection and avoid the radiation leakage. In the case of the Coaxial Cable Method, the EMI gasket may be an O-ring. O-rings of two diameters are needed to fit the inner and outer rims of the annular-ring specimen. Gaskets of two diameters are used to improve the interface between the specimen and one of the two horns and another pair of gaskets is used to improve the interface between specimen and the other horn. The application of pressure in the direction perpendicular to the plane of the gasket helps improve the interface. For reliable comparative evaluation of various specimens, this pressure should be controlled, say, by controlling the torque on the screws used for the fastening at the joint. In case that the specimen is not highly conductive, silver paint may be applied to the inner and outer rims of the specimen in order to improve the electrical contact between the specimen and the gaskets.

An alternate method of EMI testing uses no chamber, but involves placing the specimen symmetrically between two horn antennas, which are connected to a VNA (Fig. 4.22). This method is inferior to the methods that involve chambers (Fig. 4.20 and 4.21), because a part of the incident radiation goes around the specimen rather than going through the specimen.

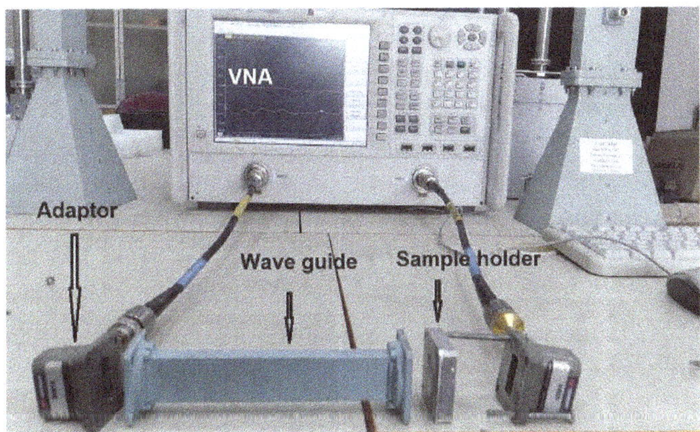

Fig. 4.21 Set-up for measurement of the EMI shielding effectiveness using the Waveguide Method. With courtesy from Murat Ozturk, Turkey.

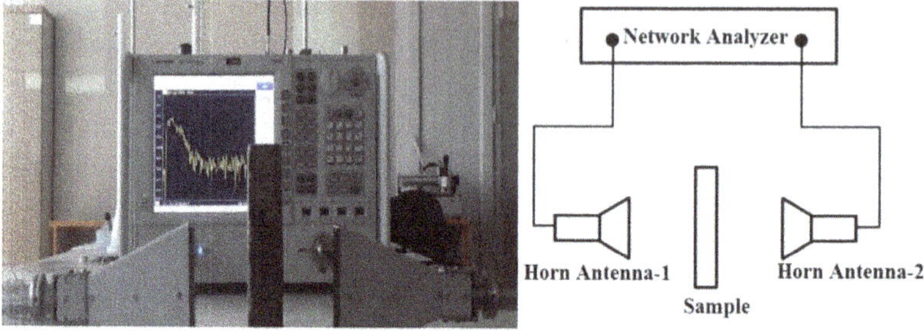

Fig. 4.22 Set-up for measurement of the EMI shielding effectiveness using the Antenna Method. The photo at the left corresponds to the schematic drawing at the right. Reproduced from the following publication with permission. Murat Ozturk, O. Akgol, U.K. Sevim, M. Karaaslan, M. Demirci, E. Unal. Experimental work on mechanical, electromagnetic and microwave shielding effectiveness properties of mortar containing electric arc furnace slag. Constr. Building Mater. 165, 58-63 (2018).

The electromagnetic absorption behavior can be assessed by measuring the "reflection loss"', which is not SE_R, but refers to the loss in reflection due to the effect of an absorbing specimen placed on a highly reflecting plane (Fig. 4.23). In other words, the reflection is diminished due to the presence of the specimen and this effect is assumed to be completely due to the absorption by the specimen. However, reflection from the surface of the specimen would decrease the "reflection loss". As a result, the measured "reflection loss" is not equal to the absorption loss, unless the specimen does not reflect at all. Hence, the absorption loss based on the "reflection loss" is not equal to SE_A. In spite of this, the "reflection loss" obtained by using this method is commonly referred to as the absorption loss. Furthermore, this method suffers from the fact that it does not provide a measurement of SE_R.

In the method of Fig. 4.23, the specimen, reflecting plane and two horn antennas (connected to a VNA) are contained in an anechoic chamber, which is a chamber designed to completely absorb the electromagnetic waves incident on the walls of the chamber. The pyramidal absorber shown in Fig. 4.23 serves to render this characteristic to the chamber, due to the jagged shape of the absorber causing numerous reflections. This method of EMI testing is known as the Arch Reflection Method.

Example problems 11–20 at the end of the chapter are relevant.

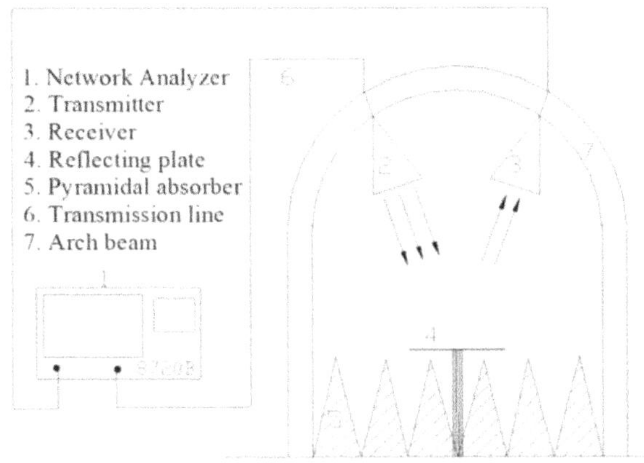

Fig. 4.23 Set-up for measurement of the EMI shielding effectiveness using the Arch Reflecting Method. Reproduced from the following publication with permission. Hongtao Guan, Shunhua Liu, Yuping Duan, Yanbo Zhao. Investigation of the electromagnetic characteristics of cement based composites filled with EPS. Cem. Concr. Compos. 29(1), 49-54 (2007).

4.5 Low observability

The difficulty of low observability relates to the high reflectivity of metals and carbon fiber composites used in making aircraft. A Stealth aircraft, such as the F-117A military aircraft of USA, is rendered low observability mainly by the shaping of the aircraft. The shape is jagged, involving flat surfaces and sharp edges. Due to the shape, the radiation from a radar is reflected by the aircraft at an angle, so that it does not return to the radar that emits the radiation (Fig. 4.24(a)). In contrast, a conventional aircraft has rounded shapes, so the radiation from a radar is reflected by the aircraft back to the radar (Fig. 4.24(b)).

If the entire aircraft is electromagnetically transparent, it cannot be detected by radars. However, this is impossible, since there are people and equipment in the aircraft. Therefore, the aircraft material needs to absorb radiation to a certain degree, as enabled by materials that have electric or magnetic dipoles, such as ferrite (iron oxide, which is magnetic) that is added to the paint or to the structural composite.

The enhancement of multiple reflections can be used to promote absorption. The use of honeycomb materials (with a high internal surface area) in the wings enhances multiple reflections. The use of lightly

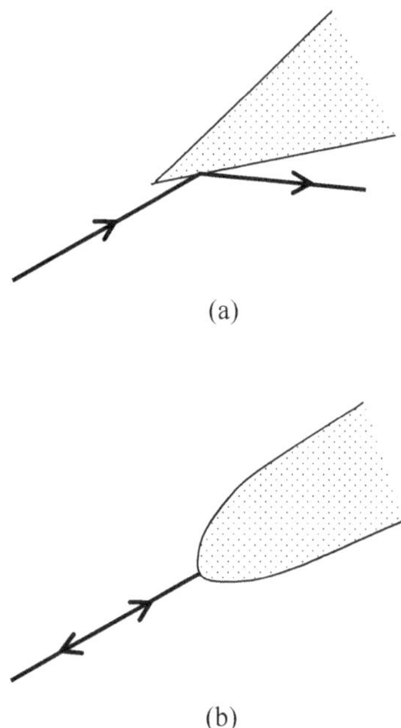

(a)

(b)

Fig. 4.24 Reflection of radar signal by an aircraft. (a) A Stealth aircraft. (b) A conventional aircraft.

activated carbon fiber in the structural composite enhances the fiber-matrix interface area in the composite, thereby enhancing multiple reflections also.

4.6 Composite materials for electromagnetic functions

EMI shielding materials include metals, carbons, ceramics, cement, polymers, hybrids and composites. The hybrids and composites are the dominant avenues used to tailor the structure for enhancing the shielding effectiveness. Metals and carbons are the main functional materials for providing shielding ability. Ceramics, cement and polymers are typically not very effective, unless they are combined with a functional component. Nevertheless, some ceramics (such as MXene, Ti_3AlC_2 and SiC) and intrinsically conductive polymers are conductive and effective for shielding to various degrees. Due to the availability of numerous types of microcarbons and nanocarbons, shielding materials in the form of metal-carbon, ceramic-carbon, cement-carbon and polymer-carbon combinations have received much attention.

Composite materials for shielding have received much attention through the use of microscopic fillers or constituents. Much less attention has been given to the shielding effectiveness of macroscopic structures, such as steel structures [9], which are important in construction. For buildings, shielding is desirable for electronic protection, but is undesirable for television reception. Therefore, there is a need for the ability to tailor structures in terms of the degree of shielding.

Composites science is relevant to all of the electromagnetic applications mentioned in Sec. 4.1, except for radomes. This is because polymers are the best materials for electromagnetic transparency and the addition of a filler to a polymer will make the material less transparent. The permittivity and conductivity promote the interaction of a material with the electric field in the electromagnetic radiation. Ceramics tend to have higher permittivity and higher conductivity than polymers.

Due to the Skin Effect, a composite material having a conductive filler with a small unit size of the filler is more effective than one having a conductive filler with a large unit size of the filler. For effective use of the entire cross-section of a filler unit for shielding, the unit size of the filler should be comparable to or less than the skin depth. Therefore, a filler of unit size 1 μm or less is typically preferred, though such a small unit size is not commonly available for most fillers and the dispersion of the filler is more difficult when the filler unit size decreases. For examples, carbon nanofibers tend to be intertwined, so that their dispersion is more challenging than the dispersions of carbon fibers. Nevertheless, nanocarbons such as graphene, carbon nanofiber and carbon nanotube are effective for shielding. For a matrix that is essentially transparent to the radiation, a nanofiber composite (with the nanofiber adequately dispersed) is superior to a microfiber composite in shielding, when the nanofiber and microfiber are at the same volume fraction. This is because of the Skin Effect and the higher interface area (more exactly, a higher specific interface area, which refers to the interface area per unit volume or per unit mass) in the nanofiber composite. For the same reason, multi-scale composites (hierarchical composites) that involve the combination of fillers of different scales can be attractive. For example, the fillers in a multi-scale composite are microcarbon (e.g., carbon fiber) and nanocarbon (e.g., carbon nanotube). In case that the microcarbon is continuous while the nanocarbon is discontinuous, the addition of the nanocarbon to the microcarbon may not result in a substantial gain in the shielding effectiveness, due to the dominating shielding effectiveness of the continuous filler. Since metal-coated polymer fibers or particles are used as fillers for shielding, but they suffer from the fact that the polymer interior of each fiber or particle does not contribute to shielding.

For EMI gaskets, the filler content must be kept low, since the resiliency is derived from the polymer matrix and it decreases with increasing filler content. Thus, for EMI gasketing, the filler must be effective even at a low volume fraction. This makes the nanofillers attractive for EMI gaskets.

4.6.1 *Composite materials with discontinuous fillers*

Polymer-matrix composites containing conductive fillers are attractive for shielding, due to their processability (e.g., moldability), which helps to reduce or eliminate the seams in the housing that is the shield. The seams are commonly encountered in the case of metal sheets as the shield and they tend to cause leakage of the radiation and diminish the effectiveness of the shield. In addition, polymer-matrix composites are attractive in their low density. The polymer matrix is commonly electrically insulating and does not contribute to shielding, though the polymer matrix can affect the connectivity of the conductive filler and connectivity enhances the shielding effectiveness. In addition, the polymer matrix affects the processability.

Electrically conducting polymers are becoming increasingly available, but they are not common and tend to be poor in the processability and mechanical properties. Nevertheless, electrically conducting polymers do not require a conductive filler in order to provide shielding, so that they may be used with or without a filler. In the presence of a conductive filler, an electrically conducting polymer matrix has the added advantage of being able to electrically connect the filler units that do not touch one another, thereby enhancing the connectivity.

Cement is slightly conducting, so the use of a cement matrix also allows the conductive filler units in the composite to be electrically connected, even when the filler units do not touch one another. Thus, cement-matrix composites have higher shielding effectiveness than corresponding polymer-matrix composites in which the polymer matrix is insulating. Moreover, cement is less expensive than polymers and cement-matrix composites are useful for the shielding of rooms in a building. Similarly, carbon is a superior matrix than polymers for shielding due to its conductivity, but carbon-matrix composites are expensive.

A seam in a housing that serves as an EMI shield needs to be filled with an EMI gasket (i.e., a resilient EMI shielding material), which is commonly a material based on an elastomer, such as rubber and silicone. An elastomer is resilient, but is itself not able to shield, unless it is coated with a conductor (e.g., a metal coating called metallization) or is filled with a conductive filler (typically metal particles such as Ag-Cu). The coating suffers from its poor wear resistance due to the tendency for the coating to debond from the elastomer. The use of a conductive filler suffers from the resulting decrease in resilience, especially at a high filler volume fraction that is usually required for sufficient shielding effectiveness. As the decrease in resilience becomes more severe as the filler concentration increases, the use of a filler that is effective even at a low volume fraction is desirable. Therefore, the development of EMI gaskets is more challenging than that of EMI shielding materials in general.

Flexible graphite (compressed exfoliated graphite, Sec. 1.3) is a particularly effective EMI gasket material, due to its combination of electrical conductivity, high surface area ($15 \ m^2/g$) and resiliency in the direction perpendicular to the sheet. The shielding effectiveness is up to 130 dB at 1 GHz. This performance is comparable to the best conductor filled elastomer-matrix composite materials.

For a general EMI shielding material in the form of a composite material, a filler that is effective at a low concentration is also desirable, although it is not as critical as for EMI gaskets. This is because the strength and ductibility of a composite tend to decrease with increasing filler content when the filler-matrix bonding is poor. Poor bonding is quite common for thermoplastic polymer matrices. Furthermore, a low filler content is desirable due to the greater processability, which decreases with increasing viscosity. In addition, a low content of the filler is desirable, due to the cost saving and weight saving.

In order for a conductive filler to be highly effective, it preferably should have a small unit size (due to the skin effect), a high conductivity (for shielding by reflection and absorption) and a high aspect ratio (for connectivity). Metals are more attractive for shielding than carbons due to their higher conductivity, though carbons are attractive in their oxidation resistance and thermal stability. Fibers are more attractive than particles due to their high aspect ratio. Thus, metal fibers of a small diameter are desirable. Nickel nanofibers of diameter 0.4 µm (as made by electroplating carbon nanofibers of diameter 0.1 µm) are particularly effective. Nickel is more attractive than copper due to its superior oxidation resistance. The oxide film is poor in conductivity and is thus detrimental to the connectivity among filler units.

4.6.2 *Composite materials with continuous fillers*

Continuous fiber polymer-matrix structural composites that are capable of EMI shielding are needed for aircrafts and electronic enclosures. The fibers in these composites are typically carbon fibers (Fig. 4.17(a)), which are attractive for their low density, high strength, high modulus, high electrical conductivity and high thermal conductivity. The carbon fibers which may be coated with a metal (e.g., nickel, Fig. 4.17(b)). Less commonly, the carbon fibers are intercalated (i.e., doped) to increase the conductivity.

An alternate design involves the use of glass fibers (not conductive) together with conductive interlayers, which render conductivity to the composite. Yet another design involves the use of polyester fibers (not conductive) and a conducting polymer (e.g., polypyrrole) matrix, which render conductivity to the composite. Still another design involves the use of moderately activated carbon fibers, the moderately high specific surface area (90 m^2/g) of which results in extensive multiple reflections while the tensile properties are maintained. Heavily activated carbon fibers would not be suitable for structural applications, due to the low strength and modulus resulting from the heavy activation.

Carbon fibers are electromagnetically reflective and absorptive, so they are undesirable for low observability. To alleviate this problem, an electromagnetically absorbing layer is attached to the carbon fiber polymer-matrix composite substrate. The absorbing layer material can be a ferrite particle epoxy-matrix composite, a carbonyl-iron particle polymer-matrix composite, a conductive polymer or other related materials which absorb through the interaction of the electric and magnetic dipoles in the absorbing layer with the electromagnetic radiation. However, the attached layer may fall off due to degradation of the bond between the layer and the substrate. Thus, it is desirable to decrease the observability of the carbon fiber polymer-matrix structural composite itself.

A way to decrease the observability is to decrease the reflectivity. The reflectivity is related to the impedance mismatch, which in turn relates to the electrical conductivity. Hence, a way to decrease the reflectivity is to decrease the conductivity. The conductivity of a carbon fiber polymer-matrix depends not only on that of the fibers themselves, but also depends on the electrical connectivity among the fibers. This is particularly true for the conductivity in the transverse direction (direction perpendicular to the fiber direction). The electrical connectivity can be decreased by using epoxy-coated carbon fibers. In this way, the observability is decreased.

Another way to decrease the reflectivity is to use coiled carbon nanofibers as a filler. The coiled configuration, with the pitch of the coil appropriately chosen, helps interaction with the electromagnetic radiation, thereby decreasing the reflectivity.

Continuous carbon fiber polymer-matrix and carbon-matrix composites are highly effective for shielding. The latter is also known as carbon-carbon composites. The effectiveness is partly due to the continuity of the carbon fiber in both composites. Electrical continuity is valuable, even though the frequency is high and the excursion of the charges is not large. Due to the conductivity of the carbon matrix and the non-conductivity of the polymer matrix, the carbon-matrix composite is more effective than the corresponding polymer-matrix composite for shielding. The carbon matrix enhances the electrical continuity, in addition to contributing to the shielding. Also due to the electrical continuity, continuous carbon fibers are more effective than discontinuous carbon fibers for providing shielding.

4.6.3 *Relationship of EMI shielding effectiveness, electrical resistivity and surface area*

The electrical conductivity is one of the main parameters that affect SE_T (the total loss, i.e., the shielding effectiveness). Figure 4.25 shows the SE_T and DC electrical resistivity for each of a variety of composites that contain various conductive fillers in a polymer or slightly conductive cement matrix.

The correlation of SE_T and the electrical resistivity applies to composites that have the same matrix, but different volume fractions of the same filler. When different fillers in the same matrix are considered or when different matrices are considered, this correlation does not work well. Figure 4.25 shows that materials that exhibit the same resistivity

                                Functional Materials

can have very different levels of SE_T. At essentially the same resistivity of 2×10^{-4} Ω.m, the SE_T of PES-matrix composite with nickel nanofiber (0.4 μm diameter) is 40 dB higher than that of PES-matrix composite with nickel fiber (2 μm diameter), which is in turn superior to PES-matrix composite with nickel fiber (20 μm diameter) by 40 dB. This suggests that a large geometric surface area of the filler helps shielding. At essentially the same resistivity of 10 Ω.m, cement-matrix composite with steel fiber (8 μm diameter) gives more shielding than that with carbon fiber (15 μm diameter), due to the smaller diameter and higher conductivity of steel fiber compared to carbon fiber.

In Fig. 4.25, cement-matrix composites tend to have higher resistivity than polymer-matrix composites at the same SE_T. As a consequence, the highest SE_T attained by cement-matrix composites is lower than that attained by polymer-matrix composites. Nevertheless, at essentially the same resistivity of 1 Ω.m, cement-matrix composite with steel fiber

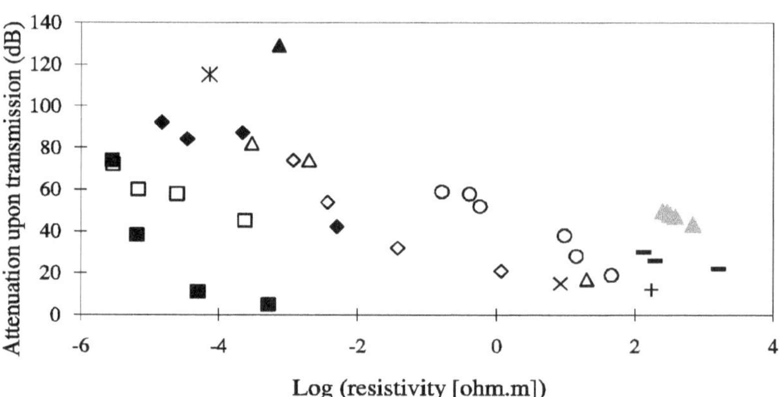

Fig. 4.25 Correlation of the EMI shielding effectiveness (attenuation upon transmission, i.e., the total loss) and the DC electrical resistivity. ▲: Flexible graphite (thickness 3.1 mm). ◇: PES/carbon nanofibers (diameter 0.1 μm, length 0.1 mm). ◆: PES/Ni nanofibers (diameter 0.4 μm, length > 0.1 mm, i.e., nickel coated carbon nanofibers). □: PES/Ni fiber (diameter 2 μm, length 1 mm),[7] ■: PES/Ni fiber (diameter 20 μm, length 1 mm).[7] △: Silicone/Ni nanofibers (diameter 0.4 μm, length > 0.1 mm, i.e., nickel coated carbon nanofibers). ×: Cement/carbon fiber (diameter 15 μm). —: Cement/C nanofibers (diameter 0.1 μm, length 100 mm). ○: Cement/steel fiber (diameter 8μm, length 6 mm). +: Cement/steel fiber (diameter 60 μm). *: Continuous carbon fiber/epoxy composite (diameter 7 μm). ▲: Cement/calcined petroleum coke powder (particle size < 75 μm). All data were obtained using the same equipment at a frequency around 1 GHz.

(8 μm diameter) gives more shielding than PES-matrix composite with carbon nanofiber. This means that cement is a superior matrix to polymer such as PES, due to its slight electrical conductivity and the consequent greater degree of electrical connectivity in the composite.

Figure 4.25 shows that flexible graphite and continuous carbon fiber epoxy-matrix composite are outstanding in the SE_T, as they are superior to any of the polymer-matrix and cement-matrix composites. In particular, flexible graphite is superior to the silicone-matrix composite with nickel nanofiber, though both are resilient (which helps to avoid electromagnetic radiation leakage from the testing fixture, thereby promoting the measured SE_T) and their resistivity values are comparable. The superiority of flexible graphite and continuous carbon fiber epoxy-matrix composite is due to their high degrees of electrical connectivity. The resiliency is valuable for EMI gasketing, which serves to provide electromagnetic sealing. Coke as a filler in cement gives higher SE_T than carbon nanofibers, as shown in Fig. 4.25, in spite of the high resistivity of the composite and the large particle size of the coke (<75 μm, with 50 μm taken as the mean particle size). The cause of the high SE_T is presently not clear.

In order to compare the SE_T of composites with different combinations of matrix and filler, the geometric surface area Q of a unit of the filler (e.g., a fiber) is calculated. This geometric area does not take into consideration the microscopic pores, if any, on the surface of the filler.

$$Q = D\pi L + \pi(D/2)^2 \qquad (4.40)$$

where D is the diameter, and L is the length of the fiber. Since the aspect ratio of the fiber is high, the second term $\pi(D/2)^2$ is negligible. The volume of fibers (in m^3) in 100 m^3 of the volume (V) of the composite is given by

$$V = (D/2)^2\pi L \qquad (4.41)$$

$$L = f(2/D)^2/\pi \qquad (4.42)$$

where f is the fiber volume fraction. Combination of Eq. (4.40) and Eq. (4.42) gives

$$Q = D\pi L = 4\,f/D \qquad (4.43)$$

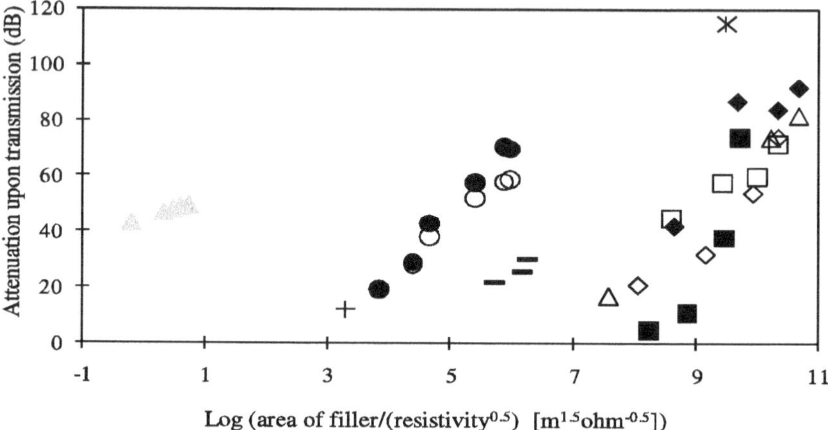

Log (area of filler/(resistivity$^{0.5}$) [m$^{1.5}$ohm$^{-0.5}$])

Fig. 4.26 Relationship between EMI shielding effectiveness (attenuation upon trans-
mission, i.e., the total loss) and a figure of merit, which is the surface area of the filler
divided by the square root of the electrical resistivity. The symbols are as defined in
Fig. 4.23. •: Cement/steel fiber (diameter 8 μm, length 6 mm) at 1.5 GHz.

Figure 4.26 shows that both the resistivity and the geometric surface
area of the filler (Eq. (4.38)) are important factors that affect the *S.E.*
Combination of these two factors gives a figure of merit, which is shown
in the horizontal axis of Fig. 4.24 as the logarithm of the geometric
surface area of the filler divided by the square root of the resistivity
of the composite. This figure of merit reflects the concepts behind of
Eq. (4.28) and (4.29). The SE_T correlates well with this figure of merit
for the variety of polymer-matrix composites. The correlation is also
good for the variety of cement-matrix composites, except for the carbon
nanofiber cement-matrix composites, the data points of which are in
the center portion of the graph. The difference between cement-matrix
and polymer-matrix composites is probably due to the slight electrical
conductivity of cement (10^3 Ω.m) and very slight conductivity of PES
(10^8 Ω.m). The data points for polymer-matrix composites with micro-
fibers are in the right part of Fig. 4.26. Moreover, the data points for the
continuous carbon fiber epoxy-matrix nanofibers and microfibers all fall
in the same band in the right part of in Fig. 4.26. However, the data
points for cement-matrix composites with nanofibers fall outside the
band (to their left) for cement-matrix composites with microfibers and
also fall outside the band (to their right) for polymer-matrix composites

with discontinuous fillers. This means that the electrical connectivity enhanced by either the cement matrix or the continuous fiber can make the figure of merit, which does not take into account the electrical connectivity, inadequate. For coke powder, SE_T is high at exceptionally low values of the figure of merit; the cause of this exceptional behavior is not clear.

4.6.4 *Summary on composite materials*

Composite materials for electromagnetic functions (particularly EMI shielding) are mainly electrically conducting materials with polymer, cement and carbon matrices. Among multifunctional structural materials, polymer-matrix composites with continuous carbon fibers and cement-matrix composites with discontinuous submicron-diameter carbon filaments are attractive. Among non-structural materials, polymer-matrix composites with submicron-diameter nickel-coated carbon filaments (i.e., nanofibers) are particularly effective.

The electrical conductivity, conductive component geometric surface area and electrical connectivity govern the shielding effectiveness of a material. Due to their electrical connectivity, flexible graphite and continuous carbon fiber epoxy-matrix composites are more effective than any of the composites with discontinuous conductive fillers. As a conductive filler, nickel nanofiber (nickel coated carbon nanofiber) is more effective than nickel microfiber or carbon nanofiber, as shown for a polymer matrix. Steel fiber (8 μm diameter) is more effective than carbon nanofiber, carbon fiber (15 μm diameter) or steel fiber (60 μm diameter), as shown for a cement matrix. Cement-matrix with coke powder shows high SE_T, in spite of the large particle size and high electrical resistivity.

Review questions

1. Why are radio waves used in broadcasting?
   Answer: Radio waves are at frequencies at which the atmosphere is essentially transparent to the electromagnetic radiation.

2. How does the Doppler effect enhance the function of a radar?
   Answer: The Dopplere effect enables a radar to detect the speed of movement of an object.

3. What are the main limitations of eddy current inspection?
    Answer: Eddy current inspection is limited to materials that are highly conductive electrically. The suitable materials are mainly metals.

4. Why is it that a metal detector can detect metals but not polymers or ceramics?
    Answer: A metal detector involves eddy current flowing in the material. Polymers are not conductive, so eddy current cannot flow. Metals are conductive.

5. What is the principle behind the function of a step-up transformer?
    Answer: A step-up transformer converts AC voltage of one voltage amplitude to another one of a higher voltage amplitude.

6. What is the principle behind how the rotation of the blades of a windmill generates electricity?
    Answer: The conversion of the mechanical energy to electrical energy is provided by the presence of a magnet and Faraday's Law.

7. What is the main aspect of aircraft design that renders low electromagnetic observability?
    Answer: The design involves a jagged shape, so that the radar signal is not reflected back to the radar.

8. What are the main criteria that govern the effectiveness of a material for low-observable aircraft?
    Answer: The material should have low reflectivity, but high absorptivity.

9. What are the two required properties of an EMI gasket material?
    Answer: An EMI gasket material should have high EMI shielding effectiveness and resilience.

10. What are the two main mechanisms of EMI shielding?
    Answer: Reflection and absorption are the two main mechanisms.

11. Why are nickel nanofibers of diameter 0.4 μm more effective than nickel fibers of diameter 2 μm for EMI shielding?
    Answer: The Skin Effect causes the nanofibers to be more effective.

12. Why are carbon nanofibers more effective than short pitch-based carbon fibers as a filler in cement for providing EMI shielding?
    Answer: The Skin Effect causes the nanofibers to be more effective.

13. Why is cement a more effective matrix than polymer for providing composite materials for EMI shielding? The polymer is a conventional one – not an inherently conductive polymer.
    Answer: Cement is slightly conductive, thus promoting electrical connectivity in the composite.

Example problems

Problem 1 (Frequency of electromagnetic radiation)

What is the frequency of electromagnetic radiation of wavelength 100 m? The speed of propagation of the radiation is 3×10^8 m/s.

Solution:

From Eq. (4.2), the frequency f is given by

$$f = s/\lambda = (3 \times 10^8 \text{ m/s})/(100 \text{ m})$$

$$= 3 \times 10^6 \text{ s}^{-1}$$

$$= 3 \times 10^6 \text{ Hz}$$

$$= 3 \text{ MHz}$$

This answer is consistent with Fig. 4.4.

Problem 2 (Wavelength of electromagnetic radiation)

What is the wavelength of electromagnetic radiation with photon energy 1.1 eV? The speed of propagation of the radiation is 3×10^8 m/s.

Solution:

According to Eq. (4.1) and Eq. (4.2), the wavelength λ is given by

$$\lambda = hs/E,$$

where s is the speed of propagation of the radiation, h is the Planck's constant and E is the photon energy. Hence,

$$\lambda = (6.626 \times 10^{-34} \text{ J.s}) (3 \times 10^8 \text{ m/s})/[(1.1) (1.9 \times 10^{-19} \text{ J})]$$

$$= 9.5 \times 10^{-7} \text{ m} = 0.95 \text{ } \mu\text{m}$$

This answer is consistent with Fig. 4.4.

Problem 3 (Frequency, wavelength and photon energy)

For electromagnetic radiation at a frequency of 1.5 GHz.

(a) Calculate the wavelength (in m).

(b) Calculate the photon energy (in both Joules and electron-volts).

Solution:

(a) Based on Eq. (4.2), the wavelength λ is given by

$$\lambda = s/f,$$

where s is the speed of propagation of the radiation, and f is the frequency. Hence,

$$\lambda = (3 \times 10^8 \text{ m/s})/(1.5 \times 10^9/\text{s}) = 0.2 \text{ m}$$

Note that the wavelength corresponding to a frequency of 1.5 GHz is long (0.2 m).

(b) According to Eq. (4.1), the photon energy E is given by

$$E = hf,$$

where f is the frequency and h is the Planck's constant $= 6.626 \times 10^{-34}$ m^2.kg/s. Hence,

$$E = (6.626 \times 10^{-34} \text{ m}^2.\text{kg/s}) (1.5 \times 10^9/\text{s})$$

$$= 9.9 \times 10^{-25} \text{ J}$$

$$= (9.9 \times 10^{-25} \text{ J}) (6.242 \times 10^{18} \text{ eV/J})$$

$$= 6.2 \times 10^{-6} \text{ eV}$$

$$= 6.2 \ \mu\text{eV}$$

Problem 4 (Induction)

In the conversion of mechanical energy to electrical energy using Faraday's Law (i.e., the rotation of a loop in the presence of a magnetic field in a fixed direction), the voltage V generated is given by

$$V = -\omega B A_o \cos \omega t,$$

where ω is the angular frequency of the rotation of the loop, t is the time, and A_o is the area of the loop. Derive this equation.
Hint: $A = A_o \sin \omega t$.

Solution:

$$\Delta\Phi/\Delta t = B\Delta A/\Delta t = \omega B A_o \cos \omega t.$$

$$V = -\Delta\Phi/\Delta t = -\omega B A_o \cos \omega t.$$

Problem 5 (Induction)

A circular conductor coil that has diameter 6.5 cm and consists of 165 turns encounters an AC magnetic flux density B of amplitude 3.6×10^{-3} T and frequency 1.5 GHz. The B is perpendicular to the plane of the turns in the coil. Calculate the amplitude of the AC voltage difference between the two ends of the coil.

Hint:

> T is the abbreviation for Tesla. It is the unit for the magnetic flux density B.
> $1\ T = 1\ Wb/m^2$
> Weber (Wb) is the unit for the magnetic flux Φ.
> $Wb = V.s$
> $\Phi = \Phi_o \sin \omega t$,
> where ω is the angular frequency, i.e., $\omega = 2\pi f$, where f is the frequency, and t is the time.
> Magnetic flux $\Phi = BA$, where A is the area.
> Voltage $V = -N\Delta\Phi/\Delta t$, where N is the number of turns in the coil.

Solution:

Area $A = \pi\,(0.065\ \text{m}/2)^2$

Magnetic flux $\Phi = \Phi_o \sin \omega t$

Amplitude of magnetic flux $= \Phi_o$

Amplitude of the magnetic flux $\Phi_o = BA$

$$= (3.6 \times 10^{-3}\ \text{T})\,[\pi\,(0.065\ \text{m}/2)^2]$$

$\Delta\Phi/\Delta t = \omega\Phi_o \cos \omega t$

Amplitude of $\Delta\Phi/\Delta t = \omega\Phi_o$

$\omega = 2\pi v = 2\pi\,(1.5 \times 10^9/\text{s})$

$N = 165$

Amplitude of voltage $V = N\omega\Phi_o$

$$= 165\ [2\pi\ (1.5 \times 10^9/\text{s})]\ (3.6 \times 10^{-3}\ \text{T})\ [\pi\ (0.065\ \text{m}/2)^2]$$

$$= 1.86 \times 10^7\ \text{V}$$

Problem 6 (Voltage induced by a magnetic flux density)

Calculate the voltage induced by a magnetic flux density (i.e., magnetic inductance) of 0.3 T (T = Tesla = wb/m^2) in a coil with three turns, such that the area of the coil is changed at the rate of 0.2 m^2/s. Note: A magnetic flux density of 1 T is quite strong, as the earth's magnetic flux density is of the order of 0.0001 T.

Solution:

$$\text{Rate of change of magnetic flux} = (0.3\ \text{T})\ (0.2\ \text{m}^2/\text{s})$$

From Eq. (4.8),

$$\text{Induced voltage} = -\,(3)\ (0.3\ \text{T})\ (0.2\ \text{m}^2/\text{s}) = -\,0.18\ \text{V}$$

Problem 7 (Induction)

A rectangular coil consisting of 76 turns and having dimensions 18.9 cm $\times$ 25.8 cm is exposed to a DC magnetic flux density B of 1.28×10^{-2} T. The coil is rotated along an axis perpendicular to B at a frequency of 115 revolutions per second. Calculate the amplitude of the voltage generated. Hint: $V = -NB\Delta\Phi/\Delta t = -NB\omega A_o \cos \omega t$

Solution:

$$\omega = 2\pi f = 2\pi\ (115/\text{s}),$$

where f is the frequency.

$N = 115$

$A_o = (18.9 \text{ cm}) (25.8 \text{ cm}) = (0.189 \text{ m}) (0.258 \text{ m})$

$\Phi = BA$

$V = NB\omega A_o = (76) (1.28 \times 10^{-2} \text{ T}) [2\pi (115/\text{s})] (0.189 \text{ m}) (0.258 \text{ m})$

$\qquad = 34.3 \text{ V}$

Problem 8 (Transformer)

(a) Describe the design of a transformer that increases the voltage from 1 V to 5 V.

(b) The output power may be lower than the input power. Why?

Solution:

(a) Have 10 turns for the primary coil (around the transformer core at the 1V input part of the transformer) and 50 turns for the secondary coil (at the 5 V output part of the transformer). A factor of 5 in the number of turns is what is required.

(b) Eddy current in the transformer core.

Problem 9 (Skin Effect)

What is the electric field at a depth of 6.7 μm for a material of skin depth 4.5 μm. The incident electric field is 2.5 V/mm.
Hint: $E = E_i\, e^{-x/\delta}$

Solution:

Electric field $= (2.5 \text{ V/mm})\, e^{-6.7/4.5} = 0.56 \text{ V/mm}$

Problem 10 (Skin depth)

Calculate the skin depth at a frequency of 1.5 GHz for a material with the following properties.

Electrical resistivity $= 5.87 \times 10^{-5}$ Ω.cm

Relative magnetic permeability $= 113$

Solution:

The skin depth δ is given by

$$\delta = 1/\sqrt{(\pi f \mu \sigma)}$$

where f is the frequency and σ is the electrical conductivity in $\Omega^{-1}.m^{-1}$.

$\sigma = 1/\text{resistivity} = 1/(5.87 \times 10^{-5}$ Ω.cm$) = 1/(5.87 \times 10^{-7}$ Ω.m$)$

Magnetic permeability $\mu = \mu_o\,\mu_r$,

where μ_o is the permeability of free space and μ_r is the relative permeability.

$$\mu_o = 4\pi \times 10^{-7} \text{ H/m}$$

H is abbreviation for Henry $= \Omega$.s; 1 Henry corresponds to the situation in which a current of 1 Ampere flowing through a coil produces a magnetic flux of 1 Weber per turn of the coil.

$\delta = 1/\sqrt{(\pi f \mu \sigma)}$

$\quad = 1/\sqrt{[\pi\ (1.5 \times 10^{9}\text{/s)}\ (113)\ (4\pi \times 10^{-7} \text{ H/m})/(5.87 \times 10^{-7}\ \Omega.\text{m})]}$

$\quad = 9.4 \times 10^{-7}$ m

$\quad = 0.94\ \mu$m

Problem 11 (Shielding effectiveness)

Let P/P_i be 0.02, where P is the transmitted power and P_i is the incident power. What is the shielding effectiveness in dB?
Hint: Shielding effectiveness $= -10 \log P/P_i$.

Solution:

Shielding effectiveness $= -10 \log 0.02$

$$= +10 \log (1/0.02)$$

$$= 17 \text{ dB}$$

Problem 12 (Power loss in dB)

The power loss in electromagnetic radiation propagation is commonly expressed in decibels (abbreviated dB). It is defined as

$$\text{Loss (dB)} = -10 \log (P/P_i),$$

where P is the power output and P_i is the power input. Calculate the ratio P/P_i corresponding to a loss of 40 dB.

Solution:

$$\text{Log } (P/P_i) = 40/(-10) = -4$$

$$P/P_i = 10^{-4} = 0.0001$$

Problem 13 (Absorption loss and linear absorption coefficient)

The fraction of the input electric power in a radio wave radiation that is absorbed upon traversing through a depth of 0.27 mm in a material is 20%. This is due to absorption only.

(a) Calculate the absorption loss in dB.

(b) Calculate the linear absorption coefficient of the material in mm^{-1}.

Solution:

(a) Fraction of power that remains after the absorption = 80%

$$\text{Absorption loss (dB)} = -10 \log (P_a/P_i)$$

$$= -10 \log (0.8)$$

$$= 10 \log (1/0.8)$$

$$= 10 \log (1.25)$$

$$= 0.97 \text{ dB}$$

(b) $\text{Absorption loss (dB)} = (10/2.3) \, \alpha x$

$$\alpha = (0.97) (2.3/10)/(0.27 \text{ mm})$$

$$= 0.83 \text{ mm}^{-1}$$

Problem 14 (Absorption loss)

The absorption loss in dB is proportional to the distance traversed. Derive this relationship.
Hint:

$$P = P_o \, e^{-\alpha x},$$

where α is the absorption coefficient and x is the distance traversed.

Solution:

$$\text{Loss (dB)} = -10 \, \log \, (P/P_o)$$

$$= -10 \, \log \, e^{-\alpha x}$$

$$= (10/2.303) \, \alpha x$$

$$= 4.34 \, \alpha x$$

Problem 15 (Absorption loss)

The linear absorption coefficient of an EMI shield of thickness 1.4 cm is 15 cm^{-1}. What is the absorption loss (in dB) due to the entire thickness of the shield?

Solution:

From Eq. (4.18),

$$\text{Absorption loss (dB)} = (10/2.303) \, \alpha x$$

$$= (10/2.303) \, (15 \text{ cm}^{-1}) \, (1.4 \text{ cm})$$

$$= 47 \text{ dB}$$

Problem 16 (Reflection loss and total loss)

(a) The fraction of the input electric power in a radio wave radiation that is reflected at the surface of a material is 30%. This is due to reflection only. Calculate the reflection loss in dB.

(b) The radiation that remains after reflection then enters the material and undergoes absorption, which gives an absorption loss of 0.97 dB. What is the total loss (due to reflection and absorption combined) in dB?

Solution:

(a) Fraction of power that remains after the reflection $= 70\%$

$$\text{Reflection loss (dB)} = -10 \log (P_r/P_i)$$

$$= -10 \log (0.7)$$

$$= 10 \log (1/0.7)$$

$$= 10 \log (1.43)$$

$$= 1.55 \text{ dB}$$

(b)   Total loss $=$ Reflection loss $+$ absorption loss

$$= (1.55 + 0.97) \text{ dB}$$

$$= 2.52 \text{ dB}$$

Problem 17 (Power of the transmitted beam)

The power of the incident beam is 39 mW. The shielding effectiveness is 22 dB. What is the power of the transmitted beam?

Solution:

$$22 = -10 \log (P/P_i)$$

$$= -10 \log [P/(39 \text{ mW})]$$

$$\log [P/(39 \text{ mW})] = -2.2$$

$$(39 \text{ mW})/P = 158$$

$$P = 0.25 \text{ mW}$$

Problem 18 (Power of the reflected beam)

The reflection loss is 2.4 dB. The incident power is 2.9 mW. What is the power of the reflected beam?

Solution:

$$2.4 = -10 \log (P_r/P_i)$$

$$= -10 \log (P_r/2.9 \text{ mW})$$

$$\log (P_r/2.9 \text{ mW}) = -2.4/10 = -0.24$$

$$\log (2.9 \text{ mW}/P_r) = 0.24$$

$$2.9 \text{ mW}/P_r = 1.74$$

$$P_r = 1.7 \text{ mW}$$

Power of the reflected beam $= P_i - P_r$

$$= (2.9 - 1.7) \text{ mW} = 1.2 \text{ mW}$$

Problem 19 (Absorption loss per unit distance)

(a) A material for electromagnetic interference (EMI) shielding exhibits absorption loss 53 dB and reflection loss 9 dB. What is the EMI shielding effectiveness?

(b) A material exhibits an absorption loss per unit thickness of 18 dB/mm. Calculate the absorption loss for the case that the material has thickness 2.8 mm.

(c) The power P_a after absorption by the path of distance x is given by

$$P_a = P_i \, e^{-\alpha x},$$

where P_i is the incident power and α is the linear absorption coefficient of the material. Derive the mathematical relationship between α and the absorption loss per unit distance. Hint:

$$\text{Absorption loss (dB)} = -10 \log (P_a/P_i)$$

Solution:

EMI shielding effectiveness = (53 + 9) dB = 62 dB

(a) Absorption loss = (18 dB/mm) (2.8 mm) = 50 dB

(b) Absorption loss (dB) $= -10 \log (P_a/P_i)$

$$= -10 \log e^{-\alpha x}$$

$$= (10/2.303)\, \alpha x$$

(c) Absorption loss/x = $(10/2.303)\, \alpha$ = 4.34 α.

Problem 20 (Absorption loss and reflection loss)

Calculate the absorption loss (in dB) and the reflection loss (in dB) for a shield of thickness 2.7 mm, electrical resistivity 3.6×10^{-4} Ω.cm and relative magnetic permeability 12. The frequency is 1.0 GHz. The electrical conductivity of copper is 6.0×10^5 $(\Omega.\text{cm})^{-1}$.

Solution:

The electrical resistivity of the shield = 3.6×10^{-4} Ω.cm

$$= 3.6 \times 10^{-4}\, \Omega\, (10^{-2}\,\text{m})$$

$$= 3.6 \times 10^{-6}\, \Omega.\text{m}.$$

So, the conductivity of the shield = $1/(3.6 \times 10^{-6}\, \Omega.\text{m})$

$$= 2.8 \times 10^5\, (\Omega.\text{m})^{-1}$$

Conductivity of shield relative to copper $= (2.8 \times 10^3)/(6.0 \times 10^5)$

$$= 4.6 \times 10^{-3}$$

From Eq. (4.29),

$$SE_A = 1.31 \; x \; \sqrt{(f\sigma_r\mu_r)}$$

$$= (1.31)(2.7 \times 10^{-3} \text{ m}) \sqrt{[(10^9)(4.6 \times 10^{-3})(12)]}$$

$$= 2.6 \text{ dB}$$

$\mu = \mu_0\mu_r$, where $\mu_0 = 4\pi \times 10^{-7}$ H/m

$$SE_R = 20\log\frac{\eta_0}{4\eta_s}, \tag{4.32}$$

where $\eta_0 = 377 \; \Omega$ and η_s is as given by the equation

$$\eta_s = \sqrt{2\pi f \mu/\sigma} \tag{4.33}$$

Hence,

$$SE_R = 20 \; [\log 377 - \log 4\sqrt{(2\pi f\mu/\sigma)}]$$

$$= 20 \log(377) - 20 \log[4\sqrt{(2\pi f\mu/\sigma)}]$$

$$= 20 \log(377) - 20 \log(4) - 20 \log\sqrt{(2\pi f\mu/\sigma)}$$

$$= 20 \log(377/4) - (20/2) \log(2\pi f\mu/\sigma)$$

$$= 20\log\left(\frac{377}{4}\right) - \frac{1}{2} \times 20\log\left(\frac{2\pi \times 10^9 \times 12 \times 4\pi \times 10^{-7}}{2.8 \times 10^3 \times 100}\right)$$

$$= 20 \times 1.974 - 10 \times (-0.470)$$

$$= 44.2$$

i.e., 44.2 dB

References

1. Chung DDL. Materials for electromagnetic interference shielding (Feature Article). Mater Chem Phys 2020;255:123587.

2. Hong X, Chung DDL. Carbon nanofiber mats for electromagnetic interference shielding. Carbon 2017;111:529-527.

3. Guan H, Chung DDL. Effect of the planar coil and linear arrangements of continuous carbon fiber tow on the electromagnetic interference shielding effectiveness, with comparison of carbon fibers with and without nickel coating. Carbon 2019;152:898-908.

4. Chung DDL, Ozturk M. Radio-wave absorption by aluminum and its dependence on the absorption distance. J Mater Sci 2021; 56(15):9263-9273.

5. Ozturk M, Chung DDL. Enhancing the electromagnetic interference shielding effectiveness of carbon-fiber reinforced cement paste by coating the carbon fiber with nickel. J Building Eng 2021;41:102757.

6. Wu J, Chung DDL. Combined use of magnetic and electrically conductive fillers in a polymer matrix for electromagnetic interference shielding. J Electron Mater 2008;37(8):1088-1094.

7. Guan H, Chung DDL. Radio-wave electrical conductivity and absorption-dominant interaction with radio wave of exfoliated-graphite-based flexible graphite, with relevance to electromagnetic shielding and antennas. Carbon 2020;157:549-562.

8. Guan H, Chung DDL. Absorption-dominant radio-wave attenuation loss of metals and graphite. J Mater Sci 2021;56(13):8037-8047.

9. Ozturk M, Chung DDL. Radio-wave shielding behavior of steel structures. J Electromagnetic Waves Appl 2021;35(11):1407-1419.

Chapter 5

Optical Behavior

This chapter covers the science and applications that pertain to the optical behavior of materials. The applications relate to optical fiber sensors, optical fiber imaging, light sources, light detectors, liquid crystal display, thermal emission and compact discs. However, this chapter does not address polymer-matrix composites for nonlinear optical applications, which include optical data storage, optical image processing, holography, optical computing and pattern recognition, due to the complexity of the physics behind nonlinear optics.

5.1 Basic concepts on the optical behavior of materials

Light is electromagnetic radiation in the visible region, which includes light of colors violet, blue, green, yellow, orange and red, in order of increasing wavelength from 0.4 to 0.7 µm. Optical behavior refers to properties associated with the interaction of electromagnetic radiation in the ultraviolet, visible and infrared regimes (Fig. 4.5) with materials. Please refer to Sec. 4.2.1 for basic concepts on electromagnetic radiation.

The electric and magnetic fields in electromagnetic radiation allow the radiation to interact with nuclei, electrons, molecular vibrations and molecular rotations in a material. Interaction with nuclei causing nuclear reactions requires high energy radiation (γ-rays or x-rays); interaction with inner electrons causing electronic transitions in atoms requires lower energy radiation (x-rays or ultraviolet); interaction with outer electrons causing electronic transitions in atoms requires even lower energy radiation (ultraviolet or visible); interaction with molecular vibrations requires still lower energy radiation (infrared); interaction with molecular rotations requires yet lower energy radiation (infrared or microwave).

Figure 5.1(a) illustrates the excitation of an electron in an isolated atom from energy E_2 to energy E_4 due to absorption of a photon of energy equal to the difference in energy between E_4 and E_2. After the absorption, the excited electron spontaneously decays to a lower energy (say the initial energy E_2) once the incident radiation is removed, thereby emitting a photon of energy equal to the difference in energy between E_4 and E_2 (Fig. 5.1(b)). Instead of decaying to the initial energy E_2 in one step, the electron may decay to an intermediate energy (say E_3), emitting a photon of energy equal to the difference between E_4 and E_3, before continuing to decay to the initial energy E_2 and, in the process, emitting a photon of energy equal to the difference between E_3 and E_2 (Fig. 5.1(c)). In other words, electron decay can cause the emission of one photon (Fig. 5.1(b)) or multiple photons (Fig. 5.1(c)). This emission is called luminescence, if the photons are in the visible region.

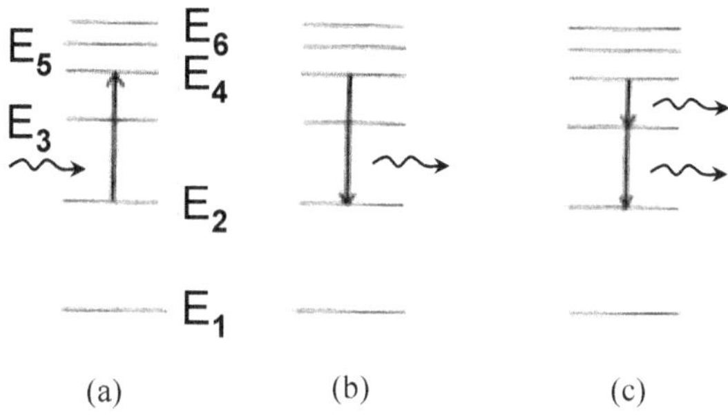

(a)   (b)   (c)

Fig. 5.1 Schematic electron energy level diagram for an isolated atom.
(a) Absorption of incident photon of energy $h\upsilon = E_4 - E_2$ causing electronic transition from E_2 to E_4.
(b) Emission of photon of energy $h\upsilon = E_4 - E_2$ due to electronic transition from E_4 to E_2.
(c) Emission of photon of energy $h\upsilon = E_4 - E_3$ and photon of energy $h\upsilon = E_3 - E_2$ due to electronic transition from E_4 to E_3 and that from E_3 to E_2 respectively.

In the case of a metal instead of an isolated atom, an electron at the Fermi energy (highest energy of the filled electron states in the absence of thermal agitation, i.e., at the temperature 0 K) absorbs a photon and is thus excited to an empty state above the Fermi energy (Fig. 5.2(a)). The change in energy is equal to the photon energy. After that, the electron spontaneously returns to the initial energy, thereby emitting a photon (Fig. 5.2(b)).

In the case of an intrinsic (undoped) semiconductor (Sec. 2.12), an electron at the top of the valence band absorbs a photon of energy exceeding the energy band gap E_g, thereby it is excited to an empty state in the conduction band (Fig. 5.3(a)). The excited electron is a conduction (free) electron. At the same time, a hole is created at the top of the

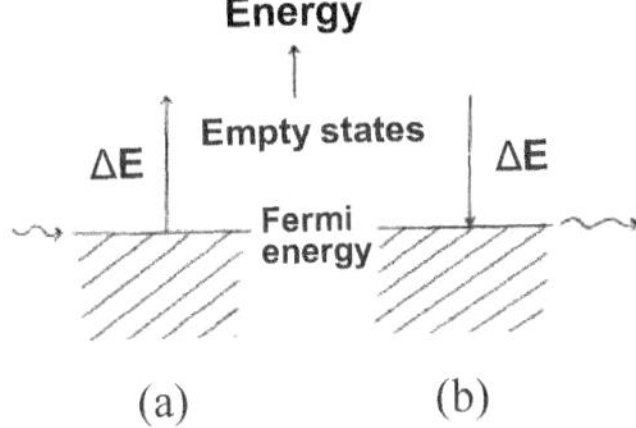

Fig. 5.2 Schematic electron energy diagram for a metal, showing filled states below the Fermi energy and empty states above the Fermi energy.

(a) Absorption of incident photon of energy ΔE causing electronic transition from Fermi energy to an empty state at energy ΔE above the Fermi energy.

(b) Emission of photon of energy ΔE due to electronic transition from energy ΔE above the Fermi energy to the Fermi energy.

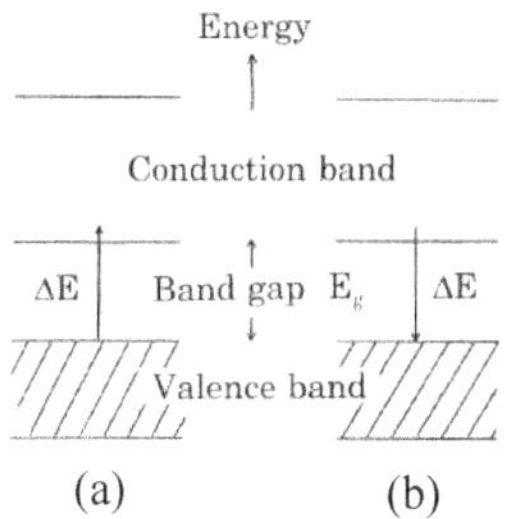

Fig. 5.3 Schematic electron energy diagram for an intrinsic semiconductor, showing filled states in the valence band and empty states in the conduction band. The two bands are separated by band gap E_g.

(a) Absorption of incident photon of energy $\Delta E > E_g$ causing electronic transition from top of the valence band to the conduction band, thereby generating a hole in the valence band and a conduction electron in the conduction band.

(b) Emission of photon of energy $\Delta E > E_g$ due to electronic transition from the conduction band to the top of the valence band. The transition involves recombination of conduction electron and hole.

valence band (where the electron resided). Hence, a pair of conduction electron and hole is created. This results in increases in conduction electron and hole concentrations, so that the electrical conductivity of the semiconductor is increased. This phenomenon in which the electrical conductivity is increased by incident electromagnetic radiation is called photoconduction. Upon removal of the radiation, the excited electron spontaneously returns to the top of the valence band, thus emitting a photon (Fig. 5.3(b)).

In the case of a p-type semiconductor (i.e., a semiconductor that has been doped by an electron acceptor, which results in holes. The holes are positively charged electronic vacant sites, with energy in the valence band (Sec. 2.12). The absorption of a photon of energy E_A causes the transition of an electron from the top of the valence band to the acceptor level a little above the top of the valence band. In the case of an n-type semiconductor (i.e., a semiconductor that has been doped by an electron donor, which results in mobile electrons with energy in the conduction band), the absorption of a photon of energy E_D causes the transition of an electron from the donor level a little below the bottom of the conduction band to the bottom of the conduction band (Fig. 5.4).

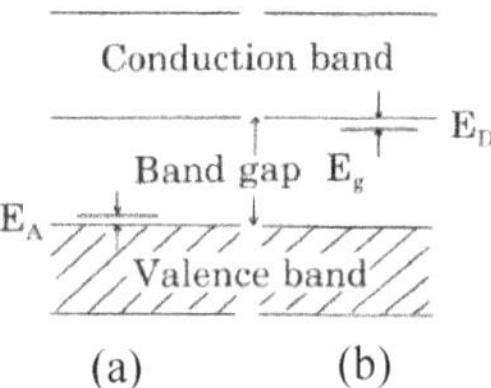

Fig. 5.4 Schematic electron energy diagrams for extrinsic (doped) semiconductors.
(a) p-type semiconductor with the acceptor energy level E_A above the top of the valence band.
(b) n-type semiconductor with the donor level E_D below the bottom of the conduction band.

Since a metal has no energy band gap, it absorbs photons of any energy. An intrinsic semiconductor has an energy band gap E_g, so it absorbs photons of energy exceeding E_g. An extrinsic semiconductor has energy levels within the band gap, so it absorbs photons of energy exceeding E_a (for p-type semiconductors) or exceeding E_d (for n-type semiconductors).

In metals, there is no energy gap, so luminescence (emission of photons in the visible region) does not occur. In intrinsic (undoped) semiconductors, there is an energy gap, so luminescence occurs and is called fluorescence. In extrinsic semiconductors, the donor or acceptor level within the band gap can trap an electron for a limited time when the electron is making a transition downward across the energy band gap. In other words, the donor or acceptor level acts as a stepping stone for the electronic transition across the energy band gap. As a result, two photons are emitted at different times. This emission over a period of time is called phosphorescence. In the case of an n-type semiconductor, the first photon emitted is of energy E_d and the second photon is of energy $E_g - E_d$.

Different materials absorb photons of different energies, so the colors of different materials are different and the measurement of the absorptivity (the fraction of the incident intensity of the electromagnetic radiation that is absorbed) as a function of photon energy provides useful information on the electronic properties of the material. For a highly absorbing material, the reflectivity (the fraction of the incident intensity that is reflected) is usually measured instead of the absorptivity, particularly if the material is opaque. Due to the law of the conservation of energy, the fraction absorbed, the fraction reflected and the fraction transmitted (transmittivity) must add up to 1, i.e.,

$$\text{Absorptivity} + \text{reflectivity} + \text{transmittivity} = 1. \qquad (5.1)$$

The absorptivity is usually not measured directly, but it is obtained from Eq. (5.1) after the reflectivity and transmittivity have been measured. The reflectivity of a semiconductor decreases abruptly as the photon energy is increased beyond the energy band gap (i.e., as the wavelength is decreased below that corresponding to the energy band gap), because of increased absorptivity.

Refer to Example Problems 1–5 at the end of the chapter.

5.2 Absorption, reflection and refraction

The intensity (also known as the irradiance) of light (electromagnetic radiation) is defined as the energy per unit area per unit time, i.e., power per unit area, where the area is measured on the plane perpendicular to the direction of propagation of the energy. The unit of the intensity is W/m^2. Thus, the intensity is directly related to the power P mentioned in Eq. (4.13). By dividing the energy by the photon energy (hf, where f is the frequency and h is the Planck's constant), one obtains the number of photons. Thus, the intensity relates to the number of photons per unit area per unit time. In case of a plane wave (Fig. 4.3), the intensity is proportional to the square of the amplitude E of the electric field wave. The electrical energy density (electrical energy per unit volume) is proportional to E^2.

Upon travelling of light in a material over an incremental distance dx, the intensity I of the light is incrementally reduced by dI, such that the fractional change in I (i.e., dI/I) is proportional to dx. Let the proportionality constant be $-\alpha$, where the minus sign is because dI/I is negative. The quantity α is known as the linear absorption coefficient (a material property). In other words,

$$dI/I = -\alpha\, dx. \tag{5.2}$$

The integration of both sides of Eq. (5.2) from $x = 0$ to $x = x$ (i.e., from $I = I_o$ to $I = I$, where I_o is the intensity at $x = 0$ and I is the intensity at $x = x$) gives

$$\ln I - \ln I_o = -\alpha\,(x - 0) = -\alpha x. \tag{5.3}$$

Equation (5.3) can be written as

$$\ln (I/I_o) = -\alpha x. \tag{5.4}$$

Hence,

$$I = I_o\, e^{-\alpha x}. \tag{5.5}$$

Equation (5.5) is known as the Beer-Lambert law and is often taken as the definition of α. The unit of α is m^{-1}. This definition of α is consistent with Eq. (4.7).

The value of α depends on the material. For a given material, it depends on the wavelength (or photon energy). This is because the mechanism of absorption depends on the material, particularly the electronic energy levels of the material, due to the absorption stemming from electronic transition between energy levels. In addition, different materials tend to absorb light at different wavelengths, due to their difference in the electronic energy levels. Some of the electronic transitions are illustrated in Fig. 5.1–5.4.

Converting natural logarithm in Eq. (5.4) to logarithm to the base 10 gives

$$2.3 \log (I/I_o) = -\alpha x. \tag{5.6}$$

As log (I/I_o) is proportional to x, Eq. (5.6) is more convenient to use than Eq. (5.5). Thus, it is customary to define

$$\text{attenuation loss (in dB)} = -10 \log (I/I_o). \qquad (5.7)$$

In this case, the attenuation is due to absorption, so the attenuation loss in this case is the same as the absorption loss. This definition is the same as that in Eq. (4.6). When $I/I_o = 0.1$, the attenuation loss is 10 dB. When $I/I_o = 0.01$, the attenuation loss is 20 dB. From Eq. (5.6) and (5.7),

$$\text{absorption loss (in dB)} = 10 \ \alpha x/(2.3) = 4.34 \ \alpha x. \qquad (5.8)$$

Hence, the absorption loss in dB is proportional to x, as mentioned in relation to Eq. (4.18).

The speed s of electromagnetic radiation depends on the medium. If the medium is free space (vacuum), $s = c = 3 \times 10^8$ m/s. If the medium is not vacuum, $s < c$. The ratio of c to s is defined as the refractive index (also called the index of refraction, abbreviated n) of the medium, i.e.,

$$n = c/s. \qquad (5.9)$$

Table 5.1 gives the values of n for various materials at the wavelength of 5,893 Å (yellow sodium D line). Silicon has a higher value of n than diamond, through they are in the same group of the Periodic Table. This is because of the larger number of electron shells in silicon than in diamond and the larger number of electrons results in more interaction with the electromagnetic radiation. Borosilicate glass has higher value of n than fused silica, due to the presence of ions that are relatively not strongly held in the borosilicate glass. Among polymers, teflon is relatively low in n, due to the symmetry in its molecular structure.

The refractive index of a material varies with the wavelength. The above values for n are for a wavelength of 5,890 Å (yellow sodium light). Typically, n decreases with increasing wavelength, as shown in Fig. 5.5 for fused quartz. The phenomenon in which n varies with the wavelength is known as dispersion, which is responsible for an optical prism to be able to disperse (divide) white light into rays with different colors.

Table 5.1 Refractive index of various materials at a wavelength of 5,893 Å.

Material	Refractive index
Vacuum	1 (exactly)
Air (standard temperature and pressure)	1.0002926
Water	1.333
Teflon	1.35–1.38
Fused silica	1.458
Borosilicate glass (Pyrex, a tradename)	1.470
Sodium chloride	1.50
Polycarbonate	1.584–1.586
Sapphire	1.762–1.778
Cubic zirconia	2.15–2.18
Strontium titanate	2.41
Diamond	2.419
Gallium arsenide	3.927
Silicon	4.01

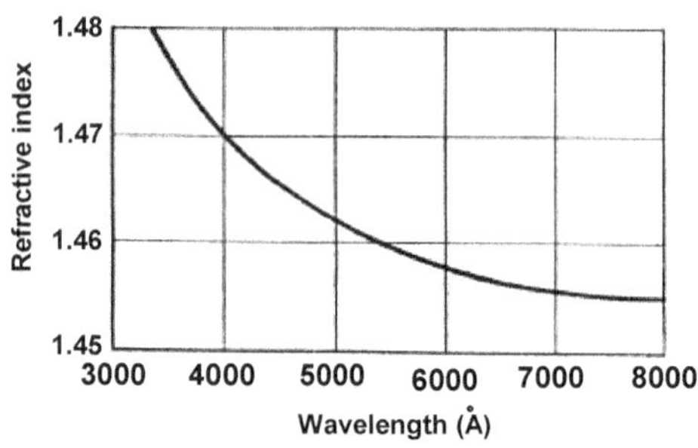

Fig. 5.5 Variation of refractive index with wavelength for fused quartz.

The refractive index n of a material is governed by how the material interacts with the electric field and the magnetic field in the electromagnetic radiation. The interaction of the material with the electric field results in an increase in the electric dipole moment, which relates to the relative dielectric constant (also called the relative permittivity) κ

(Sec. 3.1). The higher is κ, the greater is the interaction with the electric field. The interaction with the magnetic field results in an increase in the magnetic dipole moment, which relates to the relative magnetic permeability μ_r. The higher is μ_r, the greater is the interaction with the magnetic field. Both κ and μ_r are relative to the corresponding values for vacuum. Consideration of the detailed physics gives the relation

$$n = \sqrt{(\kappa\,\mu_r)}. \tag{5.10}$$

Equation (5.10) is consistent with n being equal to 1 for vacuum.

Both κ and μ_r depend on the wavelength. For some materials, there are peaks superimposed on the curve of refractive index versus wavelength (Fig. 5.5), due to corresponding peaks in either κ or μ_r. A peak in κ or μ_r at a certain wavelength is due to wavelength-dependent interaction between the electric/magnetic dipole in the material and the electric/magnetic field in the electromagnetic radiation. A common type of wavelength-dependent interaction is absorption of the radiation by the material.

Just like κ being considered a complex quantity, with real and imaginary parts (Fig. 3.12), n can be considered a complex quantity, with real and imaginary parts. Hence, n can be expressed as

$$n = n' + i\,n'', \tag{5.11}$$

where n' is the real part and n'' (called the extinction coefficient, due to its relation to the absorption loss that occurs when the electromagnetic radiation travels through the material) is the imaginary part of the refractive index. Just like the imaginary part of κ being related to the conduction behavior, the imaginary part of n is related to the conduction behavior, which is responsible for the energy loss of a lossy capacitor (Sec. 3.5). An electrically conductive material, such as a metal, tends to be opaque (with a high absorptivity), due to the interaction of the free electrons in the material with the electromagnetic radiation. As a result, n'' is substantial. In contrast, a glass is not conductive and tends to be transparent (with low absorptivity), with n'' being small. In case that the material does not absorb the radiation, the real part n' decrease with increasing wavelength (Fig. 5.5). In case that the material absorbs the radiation, it is possible for n' to increase with increasing wavelength.

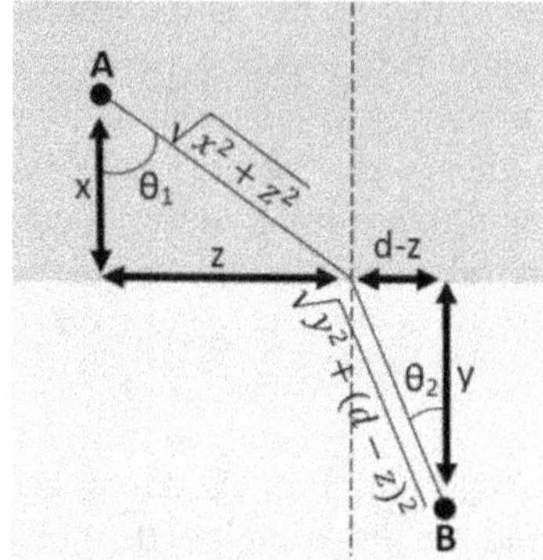

Fig. 5.6 Derivation of Snell's Law. The speed of light is v_1 and v_2 in the upper medium (containing point A) and lower medium (containing point B), respectively, with $v_1 > v_2$. Adapted from https://www.youtube.com/watch?v=8wYkgZKboss.

When an incident ray traveling in a medium of refractive index n_1 encounters the interface with a medium of refractive index n_2, as shown in Fig. 5.6, it is partly reflected and partly transmitted. The reflected ray is symmetric with the incident ray around the normal to the interface, such that the angle of incidence (angle between incident ray and the normal) equals the angle of reflection (angle between reflected ray and the normal), as in the case of mirror reflection. The transmitted ray is not quite in the same direction as the incident ray, if n_1 is not equal to n_2. This phenomenon is known as refraction. The angle of refraction (angle between refracted ray and the normal) is not equal to the angle of incidence. The relationship between the angle of incidence θ_1 and the angle of refraction θ_2 is

$$n_1 \sin \theta_1 = n_2 \sin \theta_2, \tag{5.12}$$

which is known as Snell's Law.

Snell's Law can be derived based on Fermat's principle, which states that, when light travels between two points, it takes the path that requires the least time. Consider point A in a medium with light velocity v_1 and point B in an adjoining medium with light velocity v_2, such that $v_1 > v_2$ (Fig. 5.6). The path corresponding to the straight line from A to B is not the quickest path. Rather, the path shown in Fig. 5.6, since the path length should be relatively short in the medium in which the light travels relatively slowly. The distance z (Fig. 5.6) that minimizes the travel time can be calculated. Since the travel time is given by the distance divided by the speed,

$$\text{the travel time from A to B} = \frac{\sqrt{x^2 + z^2}}{v_1} + \frac{\sqrt{y^2 + (d - z)^2}}{v_2} \tag{5.13}$$

In Eq. (5.13), the first term is the travel time in the first medium and the second term is the travel time in the second medium. The distance z needs to be chosen in order to minimize this travel time. Thus, the differential of this travel time with respect to time needs to equal zero. Hence,

$$\frac{d}{dz}\left(\frac{\sqrt{x^2 + z^2}}{v_1} + \frac{\sqrt{y^2 + (d - z)^2}}{v_2} \right) = 0 \tag{5.14}$$

Differentiation gives

$$\frac{1}{v_1} \cdot \frac{z}{\sqrt{x^2 + z^2}} - \frac{1}{v_2} \cdot \frac{(d - z)}{\sqrt{y^2 + (d - z)^2}} = 0 \tag{5.15}$$

Rearrangement gives

$$\frac{1}{v_1} \cdot \frac{z}{\sqrt{x^2 + z^2}} = \frac{1}{v_2} \cdot \frac{d - z}{\sqrt{y^2 + (d - z)^2}} \tag{5.16}$$

Since each of the angles θ_1 and θ_2 is, by definition, the angle between the path and the normal to the interface between the two media, Fig. 5.6 shows by geometry that

$$sin\theta_1 = \frac{z}{\sqrt{x^2 + z^2}} \tag{5.17}$$

$$sin\theta_2 = \frac{d - z}{\sqrt{y^2 + (d - z)^2}} \tag{5.18}$$

Hence, Eq. (5.16) becomes

$$\frac{sin\theta_1}{v_1} = \frac{sin\theta_2}{v_2} \tag{5.19}$$

However,

$$v_1 = \frac{c}{n_1} \tag{5.20}$$

$$v_2 = \frac{c}{n_2} \tag{5.21}$$

where c is the speed of light in vacuum. Hence, Eq. (5.19) becomes

$$\frac{n_1 sin\theta_1}{c} = \frac{n_2 sin\theta_2}{c} \tag{5.22}$$

Cancelling the factor c on both sides of Eq. (5.22) gives Snell's Law (Eq. (5.12)).

Based on Snell's Law, if $n_1 > n_2$, then $\theta_2 > \theta_1$. Since a larger refractive index means lower speed, $n_1 > n_2$ means $v_2 > v_1$. Thus, the medium with the larger speed is associated with a larger angle between the ray in it and the normal.

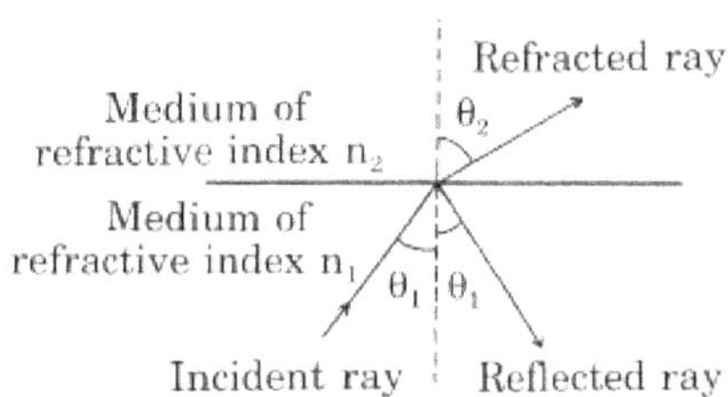

Fig. 5.7 Snell's Law, which governs the geometry of reflection and refraction at the interface between media of different values of the refractive index.

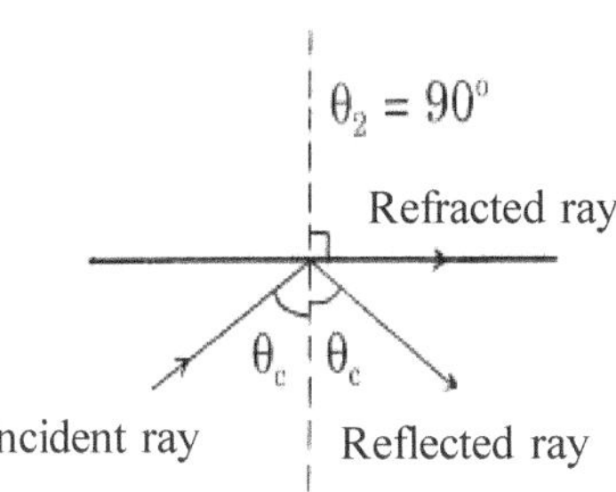

Fig. 5.8 When the angle if incidence θ_1 equals the critical angle θ_c, the refracted ray is along the interface between the media of different values of the refractive index.

When $\theta_2 = 90°$, the refracted ray is along the interface (Fig. 5.8). According to Eq. (5.5), this occurs when

$$\sin \theta_1 = n_2/n_1. \tag{5.23}$$

The value of θ_1 corresponding to $\theta_2 = 90°$ is called θ_c (the critical angle).

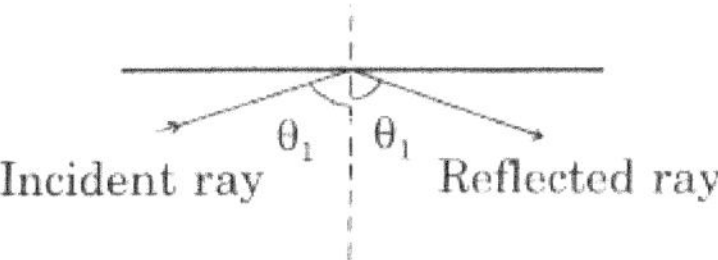

Fig. 5.9 When the angle of incidence θ_1 exceeds the critical angle θ_c, there is no refracted ray and total internal reflection occurs.

When $\theta_1 > \theta_c$, there is no refracted ray and all the incident ray is reflected (Fig. 5.9). This is known as total internal reflection.
Refer to Example Problems 6-9 and 16 at the end of the chapter.

5.3 Optical fiber

An optical fiber (a solid fiber, not a hollow fiber) guides the light in it so that the light stays inside even when the fiber is bent (Fig. 5.10). This is because the fiber has a cladding of refractive index n_2 and a core of refractive index n_1, such that $n_1 > n_2$ and total internal reflection takes place when $\theta_1 > \theta_c$. This means that the incident ray should have an angle of incidence more than θ_c in order to have the light not leak out of the core (Fig. 5.11). Hence, incoming rays that are at too large an angle (exceeding θ_{NA}) from the axis of the fiber leak. The acceptance angle of the fiber is defined as twice θ_{NA}. Rays within the acceptance angle do not leak.

Fig. 5.10 The trapping of light within an optical fiber as the light travels through a bent optical fiber due to total internal reflection.

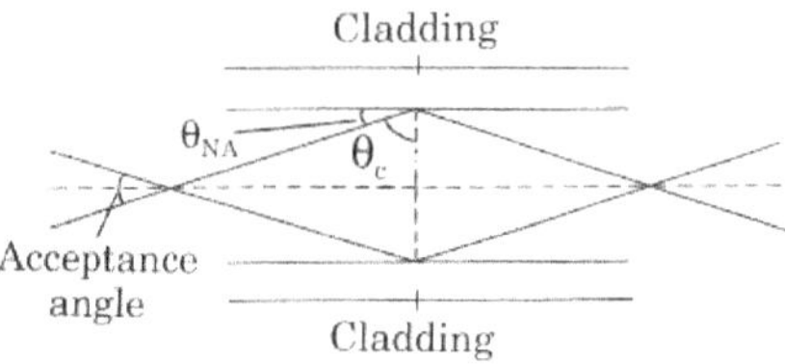

Fig. 5.11 Relationship between the critical angle θ_c, the numerical aperture angle θ_{NA}, and the acceptance angle.

The numerical aperture (NA) of an optical system is a dimensionless number that describes the range of angles over which the system can accept or emit light. The NA of an optical fiber is defined as $n_1 \sin \theta_{NA}$. Since $\theta_{NA} = 90° - \theta_c$,

$$n_1 \sin \theta_{NA} = n_1 \sin (90° - \theta_c)$$

$$= n_1 \cos \theta_c. \qquad (5.24)$$

Since

$$\sin \theta_c = n_2/n_1,$$

$$\cos\theta_c = \sqrt{1-\left(\frac{n_2}{n_1}\right)^2} = \sqrt{\frac{n_1^2 - n_2^2}{n_1^2}} = \frac{\sqrt{n_1^2 - n_2^2}}{n_1} \tag{5.25}$$

Hence,

$$\text{numerical aperture} = n_1\left(\frac{\sqrt{n_1^2 - n_2^2}}{n_1}\right)$$

$$= \sqrt{n_1^2 - n_2^2} \tag{5.26}$$

An optical fiber (or optical wave guide) has a low-index glass cladding and a normal-index glass core. The refractive index may decrease sharply or gradually from core to cladding (Fig. 5.12), depending on how the fiber is made. A sharp decrease in index is obtained in a composite glass fiber; a gradual decrease is obtained in a glass fiber that is doped at the surface to lower the index. A gradual decrease is akin to having a diffuse interface between core and cladding. As a consequence, a ray does not change direction sharply as it is reflected by the interface (Fig. 5.12). In contrast, a sharp decrease in index corresponds to a sharp interface and a ray changes direction sharply upon reflection by the interface. A fiber with a sharp change in index is called a stepped index fiber. A fiber with a gradual change in index is called a graded index fiber. A graded index fiber gives a sharper output pulse (i.e., less pulse distortion) in response to an input pulse, compared to a stepped index fiber.

An optical fiber may have different diameters of the core. A small core (e.g., 3 μm diameter) means that only rays that are essentially parallel to the fiber axis can go all the way through the fiber, as off-axis rays need to be reflected too many times as they travel through the fiber and, as a result, tend to leak. A large core (e.g., 50–200 μm) means that both on-axis and off-axis rays make their way through the fiber. Thus, a fiber with a large core is called a multimode fiber, whereas one with a small core is called single-mode fiber (Fig. 5.13). A single-mode fiber gives less pulse distortion than a multimode fiber, so it is preferred for long-distance optical communication. However, the intensity of light that

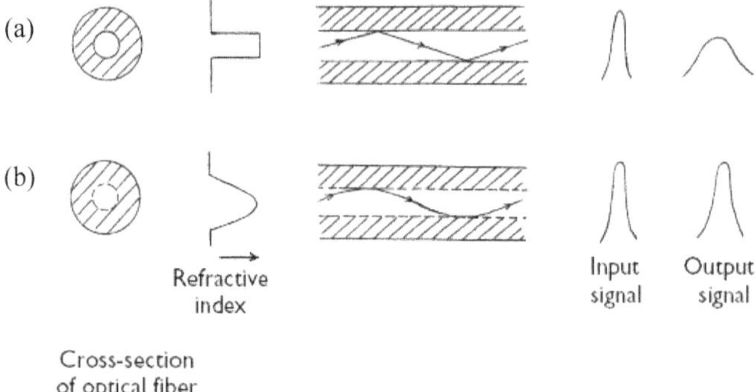

Fig. 5.12 (a) A stepped index optical fiber. (b) A graded index optical fiber.

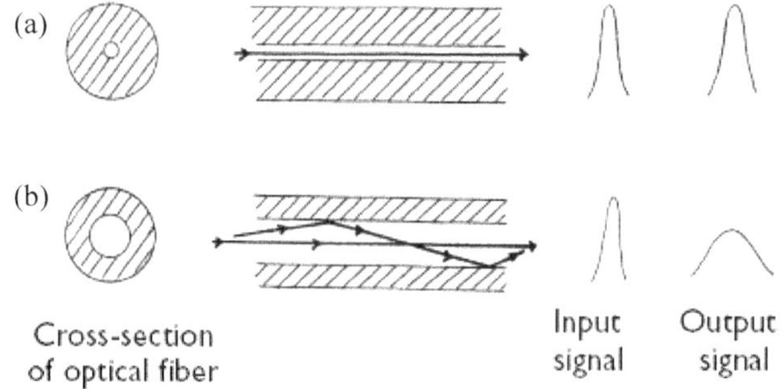

Fig. 5.13 (a) A single-mode optical fiber. (b) A multimode optical fiber.

can go through a single-mode fiber is smaller than that for a multimode fiber. The NA tends to be around 0.1 for a single-mode glass fiber and around 0.2 for a multimode glass fiber.

A single-mode fiber tends to have the cladding thicker than the core, so that the overall fiber diameter is not too small. For example, the cladding may be 70–150 μm thick, while the core diameter is 3 μm. A multimode fiber tends to have the cladding thinner than the core, as the core is already large. For example, the cladding may be 1–50 μm thick, while the core diameter is 50–200 μm.

A single-mode fiber is stepped index, whereas a multimode fiber may be either stepped index or graded index. Thus, there are three basic types of optical fibers: single-mode stepped index, multimode stepped index and multimode graded index. Pulse distortion increases in the order: single-mode stepped index, multimode graded index and multimode stepped index.

The attenuation loss, also called optical loss, per unit length of an optical fiber varies with wavelength, because the absorptivity and light scattering (i.e., scattering of light out of the medium) depend on the wavelength. Absorption losses occur when the frequency of the light is resonant with oscillations in the electronic or molecular structure of the fiber material. Various ions and functional groups have characteristic absorption peaks at well-defined frequencies. Scattering that occurs at inhomogenieties in the fiber is linear, i.e., there is no change of frequency. Scattering due to phonon-phonon interaction or Raman scattering is nonlinear, i.e., there is a change of frequency. A typical loss for glass fibers is around 1 dB/km. Polymers are not as attractive as glass for use as optical fibers because of their relatively high attenuation loss.

The imperfect coupling between the light source and an optical fiber is another source of loss, called coupling loss, which is typically 10–12 dB. This loss is because the light from the source has rays that are at angles greater than the acceptance angle of the optical fiber (Fig. 5.11). Even if the light source (a light emitting diode with rays exiting it within an angle of 100°) is attached directly to the optical fiber, coupling loss still occurs. Less coupling loss occurs if the light source is a laser, since laser light diverges negligibly as it travels.

An optical fiber is most commonly used for the transmission of signals, e.g., for optical communication. Related to this application is the use of an optical fiber as a light guide to connect light to a sensor and to return the light from the sensor to an analyzer. However, an optical fiber can also serve as a sensor. For example, it can be used for sensing the strain and damage in a structure in which the optical fiber is embedded. The amplitude, phase and polarization of light that travels through an optical fiber can be affected by the strain and damage in the fiber. These quantities can be monitored, using appropriate instrumentation. Strain in the structure gives rise to strain or bending in the fiber; damage in the structure gives rise to more strain, more bending or even damage in the fiber. For strain sensing, the optical fiber is preferably a ductile material, so polymers are sometimes used in place of glass, although they suffer from high attenuation loss. The intensity of light (related to the amplitude) that goes through an optical fiber is called the light throughput, which decreases as the fiber decreases in diameter, as the fiber bends (causing leakage through the cladding) and as the fiber is damaged. (Damage increases the attenuation loss, due to increased absorption or light scattering by the fiber.) An optical fiber may contain partially reflecting (partially transmitting) mirrors at certain points along its length within the fiber. In this way, a part of the light is reflected and a part is transmitted. By measuring the time it takes for the reflected light to reach the start of the fiber, information can be obtained concerning the location of the strain or damage. This technique is called time domain reflectometry.

Fiber-optic imaging involves sending an image through a bundle of optical fibers (Fig. 5.14). The extent of light sent by a particular fiber in the bundle depends on the location of the fiber with respect the lit part of the image. With the packing geometry of the fibers in the bundle fixed, the bundle transmits a mosaic image. Fiber optic imaging is valuable for sending an image to an area which is otherwise inaccessible.

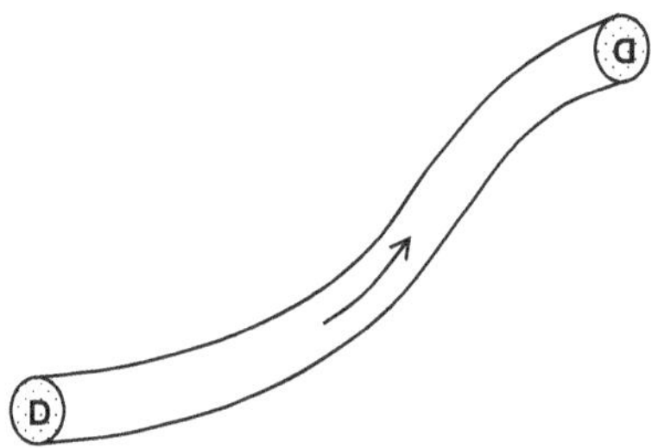

Fig. 5.14 Fiber-optic imaging using a bundle of optical fibers.

An optical fiber sensor (also called a fiber-optic sensor) can be of one of three types. They are known as transmission-gap sensor, evanescent-wave sensor and internal-sensing sensor, as explained below.

A transmission-gap sensor has a gap between the input fiber and the output fiber (which are end-to-end except for the gap) (Fig. 5.15(a)) and the disturbance at the gap affects the output. The disturbance may be pressure, temperature, etc. Thus, the sensor provides pressure or temperature sensing. An application is associated with downhole measurement in oil wells, which are too hot for the operation of semiconductor sensors.

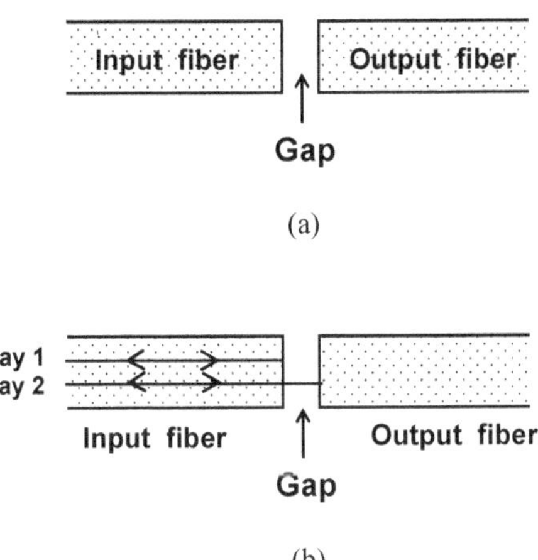

Fig. 5.15 A transmission-gap optical fiber sensor. (a) General configuration. (b) Fabry-Perot fiber optic strain sensor based on the interference between rays 1 and 2.

In case that the ends of the fibers delineating the gap of a transmission-gap sensor are polished to enhance light reflection, a slight change in the gap distance causes a change in the optical path length and hence change in the phase difference between the light rays (rays 1 and 2 in Fig. 5.15(b)) reflected from the adjacent ends of the two fibers and travelling in the same direction back toward the light source. This is the principle behind a Fabry-Perot fiber optic strain sensor. A special application relates to hydrophones (devices for converting sound to electricity in a liquid environment) and microphones (similar devices that operate in air). The acoustic vibrations associated with sound can be detected by using a Fabry-Perot fiber optic strain sensor. The light output of the sensor can be converted to electricity by using an LED receiver (which operates like a solar cell).

An evanescent-wave sensor has a part of the length of an optical fiber stripped of its cladding (Fig. 5.16). The stripped part is the sensor, since the light loss from the stripped part is affected by the refractive index of the medium around the stripped part. Hence, a change in medium is detected by this sensor. For example, an evanescent-wave sensor is used for monitoring the curing of the polymer during polymer-matrix composite fabrication, since the refractive index of the polymer changes as it cures.

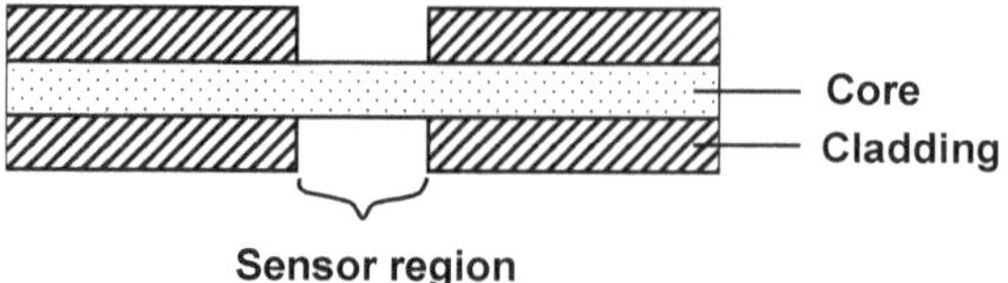

Fig. 5.16 An evanescent-wave optical fiber sensor.

An internal-sensing sensor is just an unmodified optical fiber. The amplitude and phase of light going through the fiber is affected by the disturbance encountered by the fiber.

An optical fiber's sensing region may be coated with special materials to enhance the response to certain disturbance. A coating in the form of a magnetostrictive material (Sec. 6.11) may be used to enhance the response to magnetic fields and one in the form of a piezoelectric material may be used to enhance the response to electric fields. The special material may be in the cladding. For example, the special material is a sensor material that reacts with a chemical species and changes its refractive index, thereby changing the light loss from the fiber and enabling the fiber to sense the chemical species.

Optical fibers are often embedded in a structure to form a grid, i.e., a number of fibers in the x direction and a number of fibers in the y direction. The grid allows information to be obtained concerning the location of strain or damage.

In order for an optical fiber embedded in a structure to be truly sensitive to the strain/damage of the structure, the optical fiber must be well adhered to the structure. Thus, the surface of an optical fiber may be treated or coated so as to promote the adhesion. The long-term durability of the fiber in the structure should also be considered. In particular, glass slowly dissolves in concrete, which is alkaline.

In case of embedding an optical fiber between the layers of reinforcing fibers (not optical fibers) in a structural composite material, the large diameter of the optical fiber compared to the reinforcing fibers causes the reinforcing fibers that are perpendicular to the optical fiber to be bent or distorted in the vicinity of an optical fiber (Fig. 5.17). This distortion degrades the effectiveness of the reinforcing fibers for reinforcing the composite. Furthermore, the optical fiber is akin to a flaw in the composite; it causes stress concentration and weakens the composite. Thus, the optical fiber is an intrusive sensor.

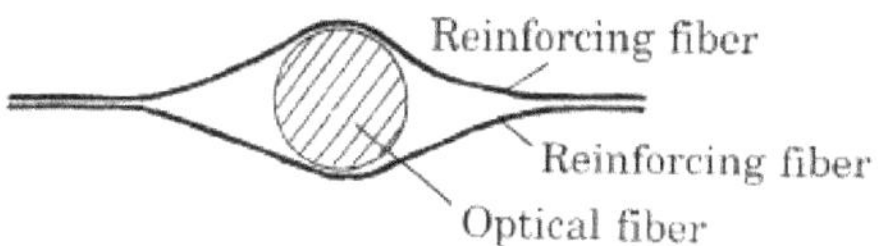

Fig. 5.17 Distortion of reinforcing fibers in the vicinity of an optical fiber perpendicular to them.

A fiber-optic sensor system includes a light source, a fiber-optic sensor, a light detector and electronic processing equipment, which are connected in the order given.
Refer to Example Problems 10–15 at the end of the chapter.

5.4 Light sources

Light sources include the sun, lamps, light emitting diodes and lasers. For use with optical fibers, which are small in size, light emitting diodes and lasers are appropriate.

5.4.1 *Light emitting diodes*

A light emitting diode (abbreviated LED) is a *pn*-junction (a junction between a *p*-type semiconductor and an *n*-type semiconductor) under forward bias (i.e., the applied voltage having its positive end at the *p*-side). The majority holes in the *p*-side move toward the n-side while the majority electrons in the *n*-side move toward the *p*-side (Fig. 5.18). The holes meet the electrons at the junction, resulting in recombination. As a result, light of photon energy equal to the energy band gap is emitted from the junction. By making one of the sides (say, the *p*-side) thin and essentially not covered by electrical contacts, light can come out from the junction through the thin side (Fig. 5.19). By using semiconductors of different band gaps, light of different wavelengths (different colors) can be obtained. This phenomenon of light emission is known as electro-luminescence, since an applied electrical signal is used to cause photon emission.

An LED lamp is a type of solid state lighting that uses LEDs as the source of light. In contrast to incandescent and fluorescent lamps, solid state lighting does not use any gas. An LED lamp is a cluster of LEDs in

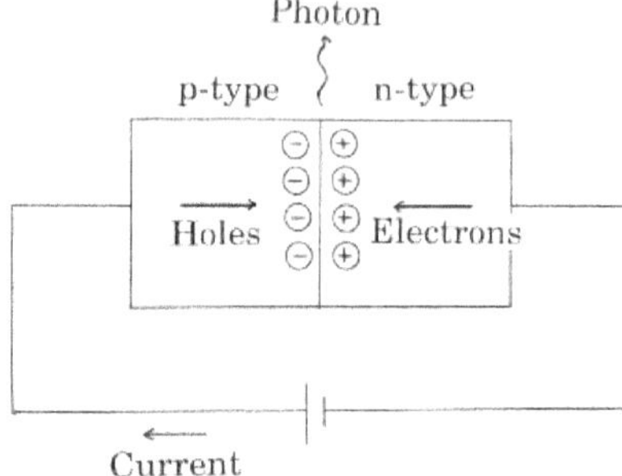

Fig. 5.18 A light emitting diode made from a *pn*-junction under forward bias.

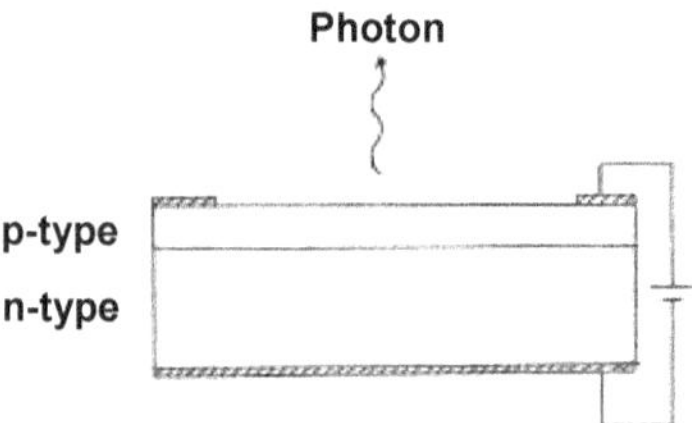

Fig. 5.19 A light emitting diode with light coming out from the exposed thin p-side of the *pn*-junction. The shaded regions are metal thin films serving as electrical contacts.

a housing. LEDs come in various colors, which are produced without the use of filters. Their attractions include high efficiency, small size, high durability, full dimmability and being free from mercury. Disadvantages include lighting directionality and the need for special circuitry for operation with the power from mains AC (due to the LEDs being low-voltage DC devices).

5.4.2 *Lasers*

Laser is abbreviation for light amplification by stimulated emission of radiation. Stimulated emission, to be distinguished from spontaneous emission, refers to the downward transition of an electron (say, from the conduction band to the valence band or from an excited state to the ground state) occurring due to another electron doing the same thing. In other words, the downward transition of one electron stimulates another electron to do the same thing. In spontaneous emission, no stimulation

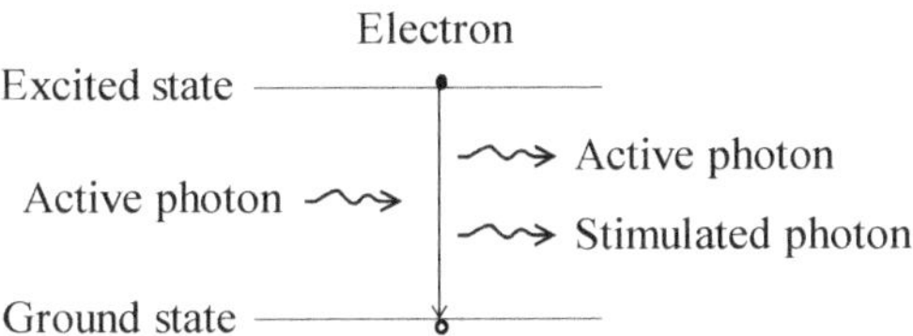

Fig. 5.20 Schematic electron energy level diagram illustrating the emission of a stimulated photon by an active photon of the same energy.

occurs. Stimulated emission causes a stimulated photon, which exists along with the original photon (called the active photon), as illustrated in Fig. 5.20.

A helium-neon (He-Ne) laser involves a plasma tube containing a mixture of He and Ne gases (90:10 ratio, 3×10^{-3} atm pressure), such that a DC voltage (~ 2 kV) is applied along the length of the tube (Fig. 5.21). The ground state electronic configuration of He is $1s^2$ (with no occupancy of the 2s level) that of Ne is $1s^2 2s^2 2p^6$ (with no occupancy of the 3s level) The applied electric field causes ionization of the atoms, thus forming a plasma in the plasma tube. This excitation causes the originally empty 2s energy level of He to be occupied (Fig. 5.22). Because the He 2s level is at essentially the same energy as the Ne 3s level (also empty in the ground state of Ne), thermal collision causes the transfer of the electron from He 2s to Ne 3s. The Ne 3s electron stays in the Ne 3s level for $\sim 6 \times 10^{-6}$ s before relaxing to lower levels. As a result, the Ne 3s level is more populated with electrons than the Ne 2p level – a situation known as population inversion (a necessary situation for stimulated emission). The population inversion causes stimulated emission to occur as the electrons relax from the Ne 3s level to the Ne 2p level. The resulting laser light is red, with wavelength 6,328 Å. Outside the plasma tube and at its two ends are two parallel mirrors, each with a reflectivity of 0.9999. The emitted light is reflected by each mirror, except for a fraction of 0.0001, which is transmitted through the mirror. The reflected portion reenters the plasma tube and stimulates other electrons to relax from the Ne 3s level to the Ne 2p level. By reflecting back and forth between the two mirrors, the number of stimulated photons becomes higher and higher. An intense laser beam is then transmitted through either mirror.

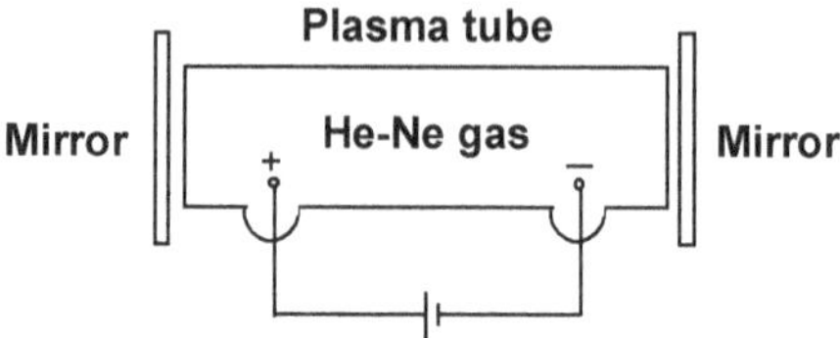

Fig. 5.21 He-Ne laser comprising a plasma tube, two mirrors and a DC power supply.

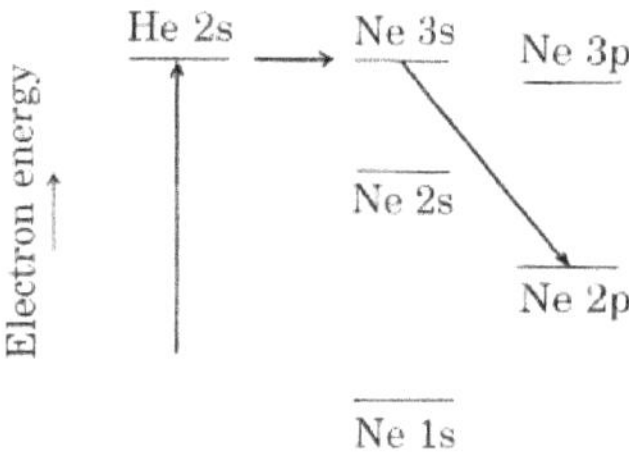

Fig. 5.22 Energy level diagram of the He-Ne system. Arrows indicate electronic transitions.

In case of a semiconductor laser, electrons are excited into the conduction band by an applied voltage. An electron recombines with a hole to produce a photon, which stimulates the emission of a second photon by a second recombination. The ends of the semiconductor are mirrored, such that one mirror is totally (100%) reflective and the other mirror is partially reflective (e.g., 99% reflective). The reflection of the photons by the mirrors back into the semiconductor allows these photons to stimulate even more photons. A fraction of the photons is emitted through the partially reflective mirror as a laser beam.

A particularly effective semiconductor laser is a *pn*-junction with very heavily doped *n*- and *p*-type regions. Furthermore, the semiconductor is a direct gap semiconductor, i.e., a semiconductor in which the momentum of the electrons at the bottom of the conduction band and the momentum of the holes at the top of the valence band are equal. The tendency for recombination is enhanced by having the momenta of the electron and hole equal. Examples of such semiconductors are GaAs and InP, which are compound semiconductors exhibiting the zinc blende crystal structure. In each compound, one element is from Group III of the Periodic Table whereas the other element is from Group V. Hence, these semiconductors are called III-V compound semiconductors. Silicon, an elemental semiconductor, is an indirect gap semiconductor, i.e., the momenta of the electrons and holes are not equal, so silicon is not used for lasers. The doping is so heavy that the Fermi energy is above the bottom of the conduction band in the *n*-side and is below the top of the valence band in the *p*-side (Fig. 5.23) in equilibrium (i.e., without bias). Upon sufficient forward bias, population inversion occurs at the junction, i.e., the electron-rich conduction band of the *n*-side overlaps the hole-rich valence band of the *p*-side. Recombination occurs and light is emitted from the junction (the light confinement region).

An improved form of semiconductor laser involves the use of semi-conductors in a multiplayer form, as illustrated in Fig. 5.23, where an undoped GaAs layer is sandwiched by *p*-type (heavily doped) GaAlAs and *n*-type GaAlAs (heavily doped). The GaAlAs is a ternary semi-conductor compound with energy band gap higher than that of GaAs and refractive index lower than that of GaAs. The difference in energy gap causes population inversion in the GaAs layer in the middle, as shown in Fig. 5.24 and Fig. 5.25. The difference in refractive index causes the light to be trapped in the GaAs layer (with the GaAlAs acting like the cladding of an optical fiber).

A laser beam is characterized by its being parallel (not diverging), nearly monochromatic (of nearly one wavelength) and coherent (with any two points in the laser beam having a predictable phase relationship).

By "nearly monochromatic", one means that the range or band of frequencies (called the frequency bandwidth Δf) is narrow. A He-Ne laser (stabilized) has $\Delta f = 10^4$ Hz, whereas spectral lines emitted by gas discharge tubes have $\Delta f = 10^9$ Hz. White light ranges in frequency from 4×10^{14} Hz to 7×10^{14} Hz, so its $\Delta f = 3 \times 10^{14}$ Hz. Light emitting diodes have much larger Δf than lasers.

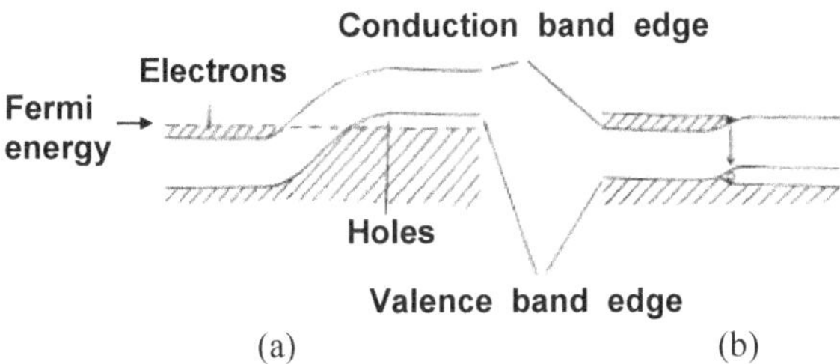

Fig. 5.23 Semiconductor laser. (a) *pn*-junction without bias. (b) *pn*-junction under forward bias. The shaded regions are occupied by electrons.

 Functional Materials

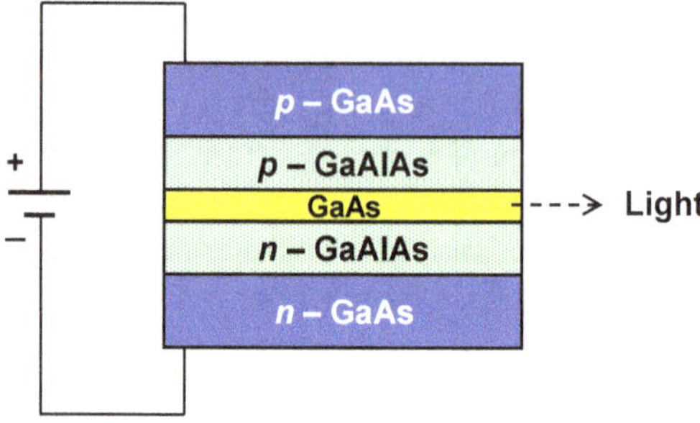

Fig. 5.24 Semiconductor laser involving a semiconductor multilayer.

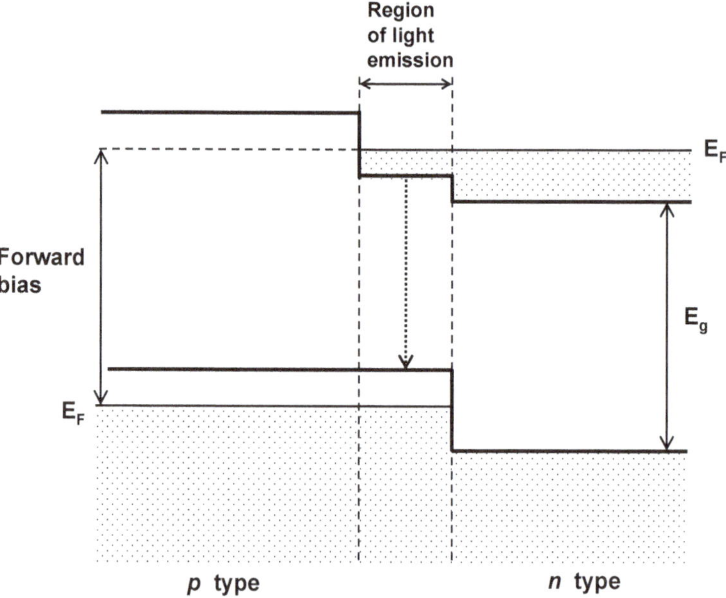

Fig. 5.25 Population inversion and light confinement in a semiconductor multilayer laser.

To understand coherence, consider predicting the relationship between the phase of a light wave at two different times at the same point in space. Suppose that at time t_o, the wave is at a peak. At time $t_o + \Delta t$, the wave would have gone through $f\Delta t$ cycles, where f is the frequency (cycles per second), if the light were truly monochromatic. From the number of cycles, the phase at $t_o + \Delta t$ is obtained. If the light is not monochromatic, but has frequency range from f to $f + \Delta f$, the number of cycles is between $f\Delta t$ and $(f + \Delta f)\Delta t$. If Δt is small, $\Delta f\Delta t \ll 1$ and the number of cycles is $f\Delta t$, just like the monochromatic case. The condition for coherence (i.e., having predictable phase relationship) is

$$\Delta f\Delta t \ll 1, \tag{5.27}$$

or

$$\Delta t \ll 1/\Delta f \tag{5.28}$$

In a time interval Δt, a wave propagates by a distance Δx, where $\Delta x = c\,\Delta t$ and c is the speed of light. Thus, comparing the phases at two points a distance of Δx apart at a fixed time is the same as comparing the phases at the same point in space over a time interval Δt. Hence, Eq. (5.28) can be written as

$$\Delta x \ll c/\Delta f \tag{5.29}$$

The distance $c/\Delta f$ is called the coherence length (x_c). In other words,

$$x_c = c/\Delta f \tag{5.30}$$

and the condition for coherence is

$$\Delta x \ll x_c. \tag{5.31}$$

The coherence length is the distance within which the light has a predictable phase relationship, i.e., the distance within which the light is coherent.

Refer to Example Problems 18 and 19 at the end of the chapter.

5.4.3 *Gas-discharge lamps*

A gas-discharge lamp is a lamp based on the use of an electric field to ionize a gas (the ionized gas being known as a plasma) and the electrons resulting from the ionization being excited to a higher energy state through collision with the gas (e.g., a noble gas such as argon) and other materials present (e.g., mercury vapor). Upon subsequent return of an electron to the original state, a photon (visible or ultraviolet) is emitted.

The ultraviolet radiation is converted to visible light by a fluorescent coating (a phosphor) on the inside of the lamp's glass surface, thus resulting in the widely used fluorescent lamp. Large fluorescent lamps in the form of long tubes have long been used. Compact fluorescent lamps (abbreviated CFLs) in a form akin to incandescent light bulbs are now increasingly used in homes in place of incandescent light bulbs for the purpose of saving energy. Compared to incandescent light bulbs, CFLs use less power for the same amount of light, generate less heat and generally last longer, though they are more expensive and more bulky. However, energy is lost in converting ultraviolet light to visible light. Furthermore, to prevent mercury release, recycling requires specialist routes.

5.5 Light detection and photocopying

The sensing or detection of light is relevant to smart structures. For example, a laser beam hitting an aircraft in a war is sensed as soon as it hits, and the aircraft is then directed to respond to the threat in an appropriate fashion. As another example, a machine component rotates when light is directed at it and stops when the light is off. In the latter example, light serves as a switch. A light detector (also called an optical detector) is usually a semiconductor, which can be an elemental semiconductor (e.g., Si) or a compound semiconductor (e.g., GaAs, InP, etc.).

Semiconductors (preferably intrinsic) are effective for detecting light of wavelength exceeding the energy band gap, as the light excites electrons (since light is electromagnetic radiation and electrons are charged) from the valence band to the conduction band, thus generating conduction electrons and holes and causing the electrical conductivity to increase. The increase in conductivity results in an increase in current in the circuit. The increased current causes an increased voltage across a resistor in series with the semiconductor in the circuit. The increase in voltage is called the photovoltage. This phenomenon is called photoconduction or photoconductivity. It is a photovoltaic effect, which is any effect in which radiation energy is converted to electrical energy. By monitoring the conductivity, light can be detected.

Photoconduction is valuable not only for light detection, but also for photocopying (xerography), where light is used to dissipate the charge in the part of the photoconductor layer that corresponds to the white region of an image (Fig. 5.26). The photoconductor layer is on a cylindrical drum. Subsequently, toner particles are applied to the photoconductor surface. Due to their positive charge, the toner particles are attracted to the regions of the photoconductor layer where negative charges are present (i.e., where the image is black). The toner image is then transferred to a piece of paper which is more negatively charged than the drum.

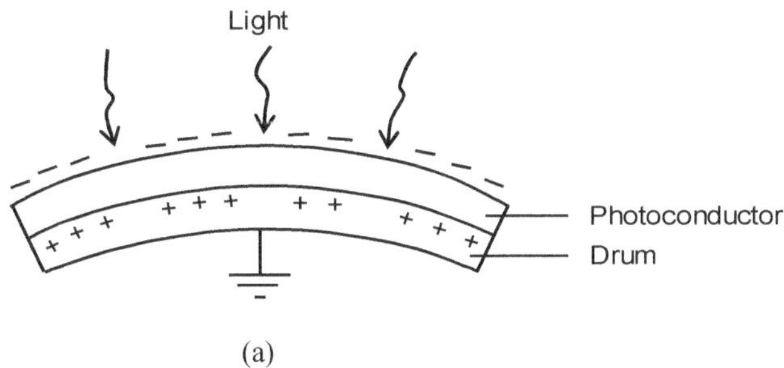

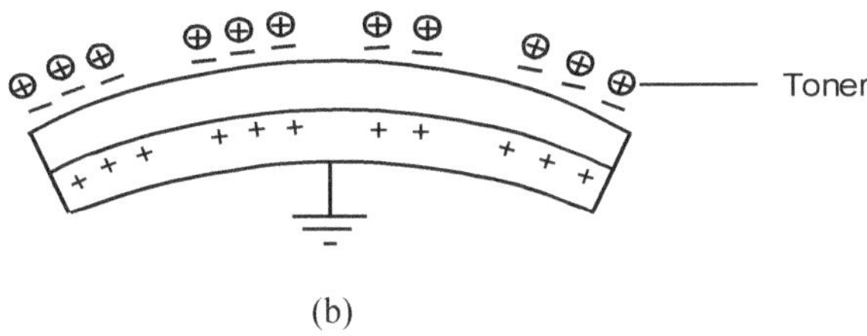

Fig. 5.26 Photocopying involving a photoconductor layer on a drum. (a) Light exposure to form an image on the photoconductor surface. (b) The application of charged toner particles on the photoconductor surface to form a toner image, which is then transferred to a piece of negatively charged paper.

The electrical conductivity σ of a semiconductor is given by

$$\sigma = qn\mu_n + qp\mu_p, \tag{5.32}$$

where n = conduction electron concentration,

      p = hole concentration,

      μ_n = conduction electron mobility, and

      μ_p = hole mobility.

For an intrinsic (undoped) semiconductor, $n = p$. Thus, Eq. (5.32) becomes

$$\sigma = qn\,(\mu_n + \mu_p). \tag{5.33}$$

For an intrinsic semiconductor, the conductivity σ_1 in darkness (i.e., no incident light) is given by

$$\sigma_1 = qn_1\,(\mu_n + \mu_p), \tag{5.34}$$

where n_1 is the conduction electron (or hole) concentration in darkness. The conductivity σ_2 in light (i.e., with incident light) is given by

$$\sigma_2 = qn_2\,(\mu_n + \mu_p), \tag{5.35}$$

where n_2 is the conduction electron (or hole) concentration in light. Implicit in Eq. (5.35) is the notion that light affects the mobilities negligibly. The ratio of the light conductivity σ_2 to the dark conductivity σ_1 is called the photoresponse, which describes the effectiveness of the light detector. Hence, from Eq. (5.34) and (5.35),

$$\text{photoresponse} = \sigma_2/\sigma_1 = n_2/n_1. \tag{5.36}$$

A good light detector has photoresponse as high as 10^3. For a given light detector, the photoresponse increases with the light intensity, since n_2 increases with the light intensity.

Photoconduction is the basis of detecting electromagnetic radiation of a large range of wavelength. X-ray is electromagnetic radiation of a small wavelength (Fig. 4.5). It is emitted from a solid material that has been excited by having a core electron removed from an atom in the material. The emission occurs upon the transition of an electron in the atom from a higher energy level to the core level with an electron removed, although the emission of another electron (called the Auger electron) at a high energy level is a process that competes with the x-ray emission. Because the x-ray emitted has a photon energy that is equal to the difference in energy between the energy level from which the electron comes and the core energy level at which the electron lands, analysis of the x-ray wavelength (or energy) gives information on the energy levels, which can be used to identify the element which gives the x-ray emission. This technique of elemental composition analysis is called x-ray spectroscopy (or x-ray spectrometry).

A widely used detector of x-ray is an intrinsic semiconductor such as silicon. The detector is known as the semiconductor detector. As the photon energy of the x-ray is much greater than the energy band gap of a semiconductor, a photon of x-ray hitting the semiconductor causes the excitation of a large number of electrons of the semiconductor from the valence band to the conduction band. Each electronic transition causes the creation of a hole in the valence band, in addition to a conduction electron in the conduction band, i.e., an electron-hole pair. In the presence of an electric field applied to the semiconductor, the conduction electrons move toward the positive end of the voltage gradient while the holes move toward the negative end of the voltage gradient. This results in a current. The conventional direction of the current is the direction of the flow of holes. The electrons and holes contribute additively to the current. Thus, each photon of x-ray results in a current pulse. The height of each pulse increases with increasing photon energy. The number of pulses over a period of time increases with increasing intensity of the x-ray. Hence, both photon energy and intensity can be determined.

In order to decrease the dark current (current in the absence of photons hitting the detector), which is due to thermal excitation of electrons from the valence band to the conduction band, the semiconductor detector is cooled to 77 K (boiling point of nitrogen) by using liquid nitrogen. As a result, the detector is housed in a cryostat which has a window (typically made of beryllium) for x-ray to enter.

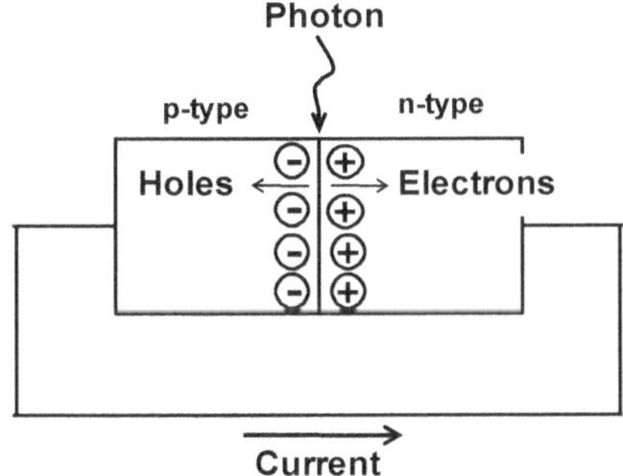

Fig. 5.27 A *pn*-junction solar cell. The principle is the same for an LED receiver.

A short-circuited *pn*-junction is particularly effective as a light detector (Fig. 5.27). When light is directed at the junction, electrons are excited across the band gap and conduction electrons and holes are generated. The potential gradient in the depletion region sweeps the conduction electrons toward the *n*-side and the holes toward the *p*-side, thus resulting in a current from the *n*-side to the *p*-side within the semiconductor. This is the basis of a *pn*-junction solar cell (the most important photovoltaic device). The same principle applies to an LED receiver, which is also a short-circuited *pn*-junction.

Refer to Example Problems 17 and 20–26 at the end of the chapter.

5.6 Liquid crystal display

A liquid crystal display (abbreviated LCD) is a thin, flat display device that is based on the effect of an electric field on a liquid crystal. The device uses very little electric power, so it is valuable for battery-powered electronic devices.

A usual liquid has no positional order or orientational order among the molecules in the liquid. For example, liquid water consists of H_2O molecules that are randomly oriented and arranged. A liquid crystal, on the other hand, is a liquid with some degree of order. In particular, a nematic liquid crystal is a liquid crystal that has some degree of orientational order, though there is no positional order. In other words, the molecules of a nematic liquid crystal have a certain tendency to orient in a particular direction, though not all the molecules are oriented exactly along the direction associated with the order. The degree of order decreases with increasing temperature, such that the order vanishes at a critical temperature (T_c), above which there is no order and the liquid crystal becomes a usual liquid. A type of nematic liquid crystal is known as a twisted nematic (or a chiral nematic), due to a linear array of molecules tending to twist in the orientation, so that the molecules spiral along an axis, as shown in Fig. 5.28(a), where the axis is vertical. This twisted configuration is the natural configuration. In the presence of an applied electric field along this axis, the molecules tend to orient along this axis, so that the twist disappears, as shown in Fig. 5.28(b).

A polarizer is a sheet that converts an unpolarized light beam into a linearly polarized beam (Sec. 4.2). An unpolarized beam incident on a polarizer becomes a linearly polarized beam upon exit from the polarizer.

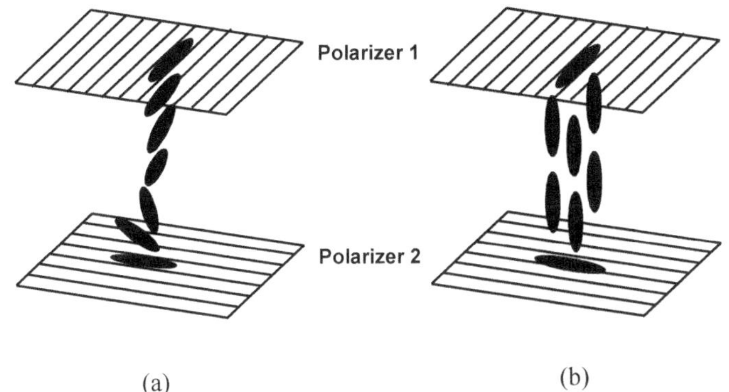

Fig. 5.28 A pixel of a liquid crystal display. (a) Voltage off – light transmitted. (b) Voltage on – light not transmitted.

The function of a polarizer is based on the selective absorption of light of a particular polarization (i.e., a particular direction of the electric field) by the polarizer material, so that only light of polarization perpendicular to this direction can pass through the material. This ability stems from the electrons in the material being able to move easily in the direction corresponding to the polarization that gets absorbed and being not able to move in the perpendicular direction. The direction of easy electron movement is the direction of the aligned molecules in the material. The alignment can be obtained by stretching the polymer film after the film has been softened by heating. In general, a material that absorbs light rays of different polarizations by different amounts is known as a dichroic material. Polaroid (the tradename for the most commonly used dichroic material) is iodine-impregnated polyvinyl alcohol, with the iodine impregnation used to provide electrons that can move easily along the polyvinyl alcohol molecules.

By using two polarizers with their transmission directions oriented perpendicular to one another, the ray that emerges from the first polarizer has a polarization for absorption by the second polarizer. As a result, total absorption is provided by the set of two polarizers, which are called crossed polarizers.

When a twisted nematic liquid crystal is positioned between two crossed polarizers, light is transmitted when the voltage is off (i.e., in the absence of an applied electric field), due to the natural twist of the molecules in the liquid crystal and the consequent twist of the polarization of the light travelling from the top polarizer to the bottom polarizer, as shown in Fig. 5.28(a). However, no light is transmitted when the voltage is on (i.e., in the presence of an applied electric field), due to the voltage disturbing the twist and the total absorption of the light by the crossed polarizers, as shown in Fig. 5.28(b).

The voltage can be applied by using transparent electrodes. The transparency is necessary for light to get through. The electrodes obviously must be electrically conductive. Metals are conductive, but they are not transparent. The choice of transparent electrical conductors is limited. The most common transparent electrode material is indium tin oxide (ITO), which is a mixture of indium oxide (In_2O_3, typically 90 wt.%) and tin oxide (SnO_2, typically 10 wt.%). A set of electrodes (one at the front and the other at the exit) is necessary for each pixel in the display. An electrode array involving crossed row and column strip electrodes at the front and exit glass plates (say, rows at the front and columns at the exit) can be used to decrease the number of required electrodes for a large number of pixels. In this design, each pixel corresponds to one crossing point and can be addressed by applying a voltage between the corresponding row and column. In color displays, each pixel is subdivided into three or four subpixels, which are covered with color filters. The use of three subpixels allows for the three basic colors (red, green and blue), which may be mixed. The use of four subpixels allows for the three basic colors and grey to improve brightness control.

5.7 Thermal emission

Thermal emission refers to the emission of photons from a material due to excitation of the material by heat. The yellow-orange glow of a hot piece of metal is an example of thermal emission. At a particular temperature, the emitted photons have a range of wavelength, such that the emission is greatest at an intermediate wavelength (Fig. 5.29). In general, as the temperature increases, the wavelength of the strongest emission decreases and the intensity of the emission increases (Fig. 5.29). The emitted photons are mainly in the infrared. However, as the temperature increases, more photons that are in the visible range are emitted.

Thermal emission is utilized for sensing warm objects, including people. This form of person detection is particularly valuable at night,

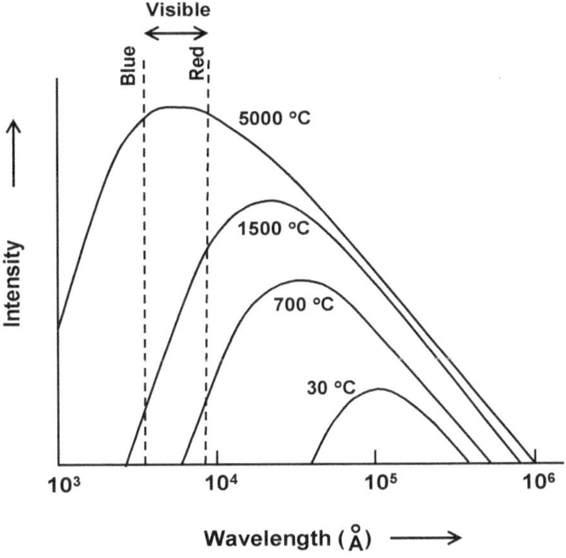

Fig. 5.29 Intensity vs. wavelength of thermally emitted electromagnetic radiation.

when the darkness does not allow visual observation. It can also be used for imaging the thermal energy distribution of a surface, such as the surface of Mars. With the help of a spectrometer for distinguishing different wavelengths, images can be obtained at different wavelengths.

Thermal emission is also used for temperature measurement, though the temperatures suitable for measurement are high, typically above 1,000°C. At temperatures below about 1,000°C, the emission is too weak for effective temperature measurement. A non-contact device based on thermal emission and designed for temperature measurement is known as a pyrometer. In contrast, contact is necessary for thermocouples and thermometers.

Incandescence refers to the emission of visible light from a hot object. The commonly used incandescent light bulb is based on thermal emission from the heated filament (typically tungsten, which is a refractory metal) in the bulb. The heating is provided by the passing of an electric current through the filament, i.e., resistance heating. The glass bulb prevents oxygen from the air to reach the hot filament, which may be destroyed by oxidation. The majority of radiation emitted by an incandescent light bulb is in the infrared (not visible), so these light bulbs are inefficient.

5.8 Compact disc

A compact disc (abbreviated CD) is an optical disc for storing digital data. It is a disc with a spiral array of indentations (called pits) that constitute the stored information (Fig. 5.30). The data track (i.e., the series of pits) is about 0.5 μm wide, with 1.6 μm between the centers of adjacent tracks (Fig. 5.31). The pits are indeed pits if they are viewed from the label side of the CD. If they are viewed from the polycarbonate side of the CD, they are bumps rather than pits. The areas between pits are known as lands. Each bump is at least 0.8 μm long (in the spiral direction) and at least 0.1 μm high.

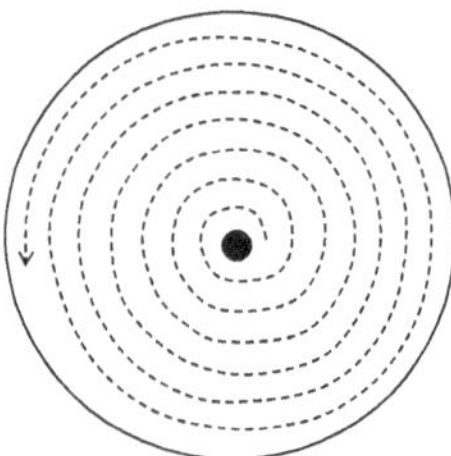

Fig. 5.30 A compact disc showing the spiral array of stored information.

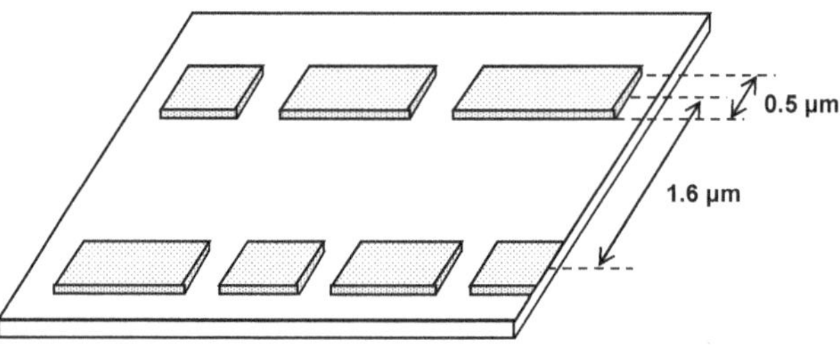

Fig. 5.31 Two tracks of stored information, each track being in the form of an array of bumps (red regions). The blue region is the polycarbonate base that is covered with aluminum.

The pits are made of an aluminum film, which is on a polycarbonate substrate (Fig. 5.32). Aluminum is used because it reflects light, thus allowing the analysis of bumps and lands by using a laser light that is

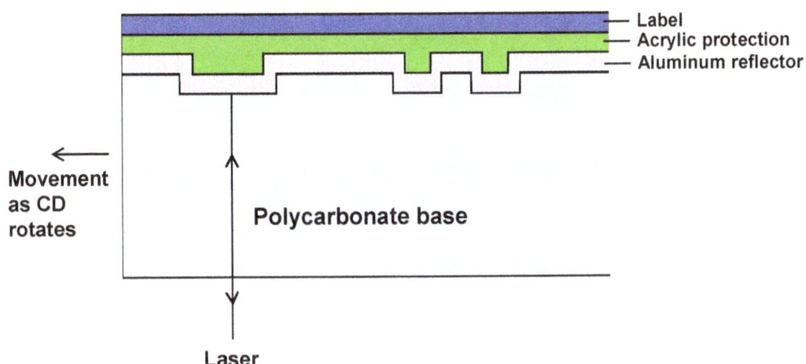

Fig. 5.32 Cross section of a compact disc showing the various layers on a polycarbonate base and the laser incident from below the base.

directed through the transparent polycarbonate substrate and is reflected by the aluminum layer. The laser light is scanned along the spiral path of the disc as the disc rotates in order to read the stored information. The laser is a semiconductor laser of wavelength 7,800 Å (near infrared).

The manufacturing of a CD involves impressing the undulations on the polycarbonate substrate, followed by the sputtering of a thin layer of aluminum on the surface of the polycarbonate substrate. After this, a thin layer of acrylic is sprayed on the aluminum for protection. Finally, a label is affixed to the acrylic protection.

The rate of reading of a CD by the laser is kept constant as the disc rotates, regardless of the position of the laser detector relative to the center of the disc. For this purpose, the disc rotation speed is decreased as the laser detector traverses from the center toward the periphery of the disc. In order to keep the laser focused on the reflective surface as the disc rotates, the position of a focusing lens is adjusted up or down as inadequate focusing is sensed.

5.9 Composite materials for optical applications

Composites science provides an avenue for tailoring the optical behavior of materials. Through the combining of constituents with different optical properties and/or different mechanical properties, materials that are suitable in terms of the optical and/or mechanical properties for specific optical applications are obtained.

5.9.1 *Composite materials for optical waveguides*

Optical waveguides can be in the form of fibers, slabs, thick films and thin films. A requirement for waveguide materials concerns the refractive index. Another requirement is transparency (i.e., low optical loss) at the relevant wavelength. Composite material design provides a route for tailoring the refractive index. For attaining transparency in the direction of wave propagation, different components in a composite should exhibit similar values of the refractive index. Otherwise, one component should be nano-sized. In the case of an optical fiber, the core and cladding are different in the refractive index, by design, while the core must be sufficiently transparent. Examples of transparent composites are polyimide containing nano-sized titanium dioxide particles, a TiO_2-SiO_2 composite film, and glass particle epoxy-matrix composites.

5.9.2 *Composite materials for optical filters*

Optical filters are for transmission or reflection of selected wavelengths. They are usually in the form of thin films, particularly multilayers. The different layers are designed by consideration of the refractive index and thickness (in relation to the wavelength) of each layer. Materials with both high and low values of the refractive index are needed for this purpose. Glasses and polymers tend to have low values of the refractive index. Composite engineering provides a route to obtain materials with high values of the refractive index. Examples of such composites are poly(ethylene oxide) (PEO) containing iron sulfides, and a composite of tantalum oxide and hafnium oxide. Related applications of high refractive index materials pertain to antireflective coatings and lenses.

5.9.3 *Composite materials for lasers*

Solid state lasers involve laser active materials. A laser active material is conventionally used by itself in a monolithic form. However, a composite containing the laser active material (preferably in the form of nanoparticles to increase its surface area) and a refractive-index-matched polymer matrix can be used instead. The attractions of a composite are processability, such as castability to form thick films, and low cost.

Review questions

1. What is the mechanism behind photoconduction?
   Answer: Photoconduction involves the excitation of electrons by the photon of light so that there are more electrons available for conduction. In case of an intrinsic semiconductor, the excitation is from the top of the valence band to the bottom of the conduction band. In the conduction band, the electrons are known as conduction electrons.

2. How does the absorptivity of a semiconductor vary with the photon energy in the range from one below the energy band gap to one above the energy band gap?
   Answer: The absorptivity of a semiconductor increases with the photon energy in this range.

3. How is the refractive index of a medium related to the speed of electromagnetic radiation in the medium?
   Answer: The refractive index is inversely related to the speed.

4. What is the main advantage of a graded index optical fiber compared to a stepped index optical fiber?
   Answer: A graded index fiber gives a sharper output pulse (i.e., less pulse distortion) in response to an input pulse, compared to a stepped index fiber.

5. How can an optical fiber function as a damage sensor?
   Answer: The amplitude, phase and polarization of light that travels through an optical fiber can be affected by the damage in the fiber. Damage in the structure gives rise to more strain, more bending or even damage in the fiber. The intensity of light (related to the amplitude) that goes through an optical fiber is called the light throughput, which decreases as the fiber bends (causing leakage through the cladding) and as the fiber is damaged.

6. Both light emitting diode and solar cell involve a *pn*-junction. What is the difference in operation condition?
   Answer: A light emitting diode is a *pn*-junction under forward bias. A solar cell is a short-circuited *pn*-junction.

7. What are the main advantages of a laser compared to a light emitting diode as a light source?

   Answer: A laser gives light that is more collimated (hence less coupling loss), more monochromatic and more coherent.

8. Give an example of a direct band gap semiconductor and an example of an indirect band gap semiconductor.

   Answer: GaAs is an example of a direct band gap semiconductor. Silicon is an example of an indirect band gap semiconductor.

9. Give an example of a III-V compound semiconductor.

   Answer: GaAs is an example of a III-V compound semiconductor.

10. What is the principle behind x-ray spectroscopy?

    Answer: X-ray is emitted from a solid material that has been excited by having a core electron removed from an atom in the material. The x-ray emission occurs upon the transition of an electron in the atom from a higher energy level to the core level with an electron removed. Because the x-ray emitted has a photon energy that is equal to the difference in energy between the energy level from which the electron comes and the core energy level at which the electron lands, analysis of the x-ray wavelength (or energy) gives information on the energy levels, which can be used to identify the element which gives the x-ray emission. This technique of elemental composition analysis is called x-ray spectroscopy

11. Why should a semiconductor detector be kept cold during use?

    Answer: Keeping a semiconductdor detector cold reduces the background signal due to thermal excitation, thereby improving the signal-to-noise ratio.

12. What are the main requirements for an optical waveguide material?

    Answer: A requirement for waveguide materials concerns the refractive index, as needed for total internal reflection. Another requirement is transparency (i.e., low optical loss) at the relevant wavelength.

Example problems

Problem 1 (Wavelength and frequency)

What are the wavelength and frequency of electromagnetic radiation of photon energy 0.02 eV in free space?

Solution:

The frequency f is given by

$$f = \frac{E}{h} = \frac{0.02 \text{ eV}}{4.14 \times 10^{-15} \text{ eV.s}}$$

$$= 5 \times 10^{12} \text{ s}^{-1}$$

$\lambda = c/f,$

where c is the speed of propagation, i.e., 3×10^8 m/s. Hence,

$\lambda = (3 \times 10^8 \text{ m/s})/(5 \times 10^{12} \text{ s}^{-1}) = 6 \times 10^{-5} \text{ m} = 60 \text{ μm}.$

Problem 2 (Photon energy)

What is the photon energy (in eV) for electromagnetic radiation of wavelength 1.67 μm?
Hint: The speed of propagation of the radiation is the speed of light (3×10^8 m/s).
1 eV $= 1.6 \times 10^{-19}$ J

Solution:

Photon energy $= hf,$
where f is the frequency and h is the Planck's constant (6.6262×10^{-34} J.s).
$f = c/\lambda,$

where c is the speed and λ is the wavelength.

$c = 3 \times 10^8$ m/s

$$f = c/\lambda = (3 \times 10^8 \text{ m/s})/(1.67 \times 10^{-6} \text{ m})$$

$$= 1.8 \times 10^{14}/\text{s}$$

$$= 1.8 \times 10^{14} \text{ Hz}$$

$$hf = (6.6262 \times 10^{-34} \text{ J.s}) (1.8 \times 10^{14}/\text{s})$$

$$= 12 \times 10^{-20} \text{ J}$$

$$= 1.2 \times 10^{-19} \text{ J}$$

$$= (1.2 \times 10^{-19} \text{ J})/(1.6 \times 10^{-19} \text{ J/eV})$$

$$= 0.75 \text{ eV}$$

Problem 3 (Wavelength)

What is the wavelength of a photon with photon energy 1.10 eV.

Hint: $E = hf = hc/\lambda$,

where $c = 3 \times 10^8$ m/s, and $h = 6.626 \times 10^{-34}$ m^2 kg/s.

$1 \text{ eV} = 1.6 \times 10^{-19}$ J

Solution:

$\lambda = hc/E$

$$= (6.626 \times 10^{-34} \text{ m}^2 \text{ kg/s}) (3 \times 10^8 \text{ m/s})/[(1.10) (1.6 \times 10^{-19} \text{ J})]$$

$$= 11.3 \times 10^{-7} \text{ m}$$

$$= 1.13 \text{ } \mu\text{m}.$$

Problem 4 (X-ray detector)

Silicon is a semiconductor with a band gap of 1.1 eV. It can be used as an X-ray detector. X-ray of photon energy 5 keV (5000 eV) is shown on silicon.

(a) What is the number of electron-hole pairs that result from one photon?

(b) The intensity of the x-ray is such that 125 photons arrive at the silicon per second. What is the number of electron-hole pairs that result from 1 second of the x-ray irradiation?

Solution:

(a) No. of electron-hole pairs $= (5000 \text{ eV})/(1.1 \text{ eV})$

$$= 4.5 \times 10^3$$

(b) No. of electron-hole pairs $= 4545\,(125)$

$$= 5.7 \times 10^5$$

Problem 5 (Energy band gap)

The wavelength of visible light ranges from 0.4 μm to 0.7 μm.

(a) In order for a semiconductor not to absorb visible light at all, what is the minimum energy band gap required for the semiconductor?

(b) In order for a semiconductor to absorb the entire range of visible light, what is the maximum energy band gap required for the semiconductor?

Solution:

(a) For no absorption, the energy band gap must be larger than the highest photon energy in the range for visible light. The highest

  Functional Materials

photon energy corresponds to the smallest wavelength. Thus, the minimum energy band gap corresponds to a wavelength of 0.4 μm.

$$E_g = hc/\lambda = (6.62 \times 10^{-34} \text{ J.s}) (3 \times 10^8 \text{ m/s})/(4 \times 10^{-7} \text{ m})$$

$$= 5.0 \times 10^{-19} \text{ J}$$

Since 1 eV = 1.6×10^{-19} J,

$$E_g = [(5.0 \times 10^{-19})/(1.6 \times 10^{-19})] \text{ eV}$$

$$= 3.1 \text{ eV}$$

(b) For complete absorption, the energy band gap must be smaller than the smallest photon energy in the range for visible light. The smallest photon energy corresponds to the longest wavelength. Thus, the maximum energy band gap corresponds to the wavelength of 0.7 μm.

$$E_g = hc/\lambda = (6.62 \times 10^{-34} \text{ J.s}) (3 \times 10^8 \text{ m/s})/(7 \times 10^{-7} \text{ m})$$

$$= 2.8 \times 10^{-19} \text{ J}$$

Since 1 eV = 1.6×10^{-19} J,

$$E_g = [(2.8 \times 10^{-19})/(1.6 \times 10^{-19})] \text{ eV}$$

$$= 1.8 \text{ eV}$$

Problem 6 (Refraction)

The index of refraction of a material is 2.79. What is the speed of light in this material?

Solution:

$n = c/v$, where v is the speed of light in the material, c is the speed of light in vacuum and n is the index of refraction.

$$n = 2.79$$

$$c = 3 \times 10^8 \text{ m/s}$$

Hence,

$$v = c/n = (3 \times 10^8 \text{ m/s})/2.79$$

$$= 1.1 \times 10^8 \text{ m/s}$$

Problem 7 (Refraction)

Light travels from a material with index of refraction 1.76 toward an adjoining material with index of refraction 1.49. What is the critical angle (i.e., the minimum incident angle relative to the normal to the interface between the two materials) so that light does not escape from the material with index of refraction 1.76?

Solution:

$$n_1 \sin \theta_1 = n_2 \sin \theta_2 \text{ (Snell's Law)}$$

$$\theta_2 = 90°, \text{ so}$$

$$\sin \theta_2 = 1$$

$$n_1 = 1.76$$

$$n_2 = 1.49$$

$$\text{Critical angle} = \theta_1$$

$$\sin \theta_1 = n_2/n_1$$

$$= 1.49/1.76$$

$$= 0.8466$$

$$\theta_1 = 57.8°$$

Problem 8 (Refractive index)

Calculate the refractive index of a material with relative dielectric constant 23 and relative magnetic permeability 3.5.

Solution:

From Eq. (5.10),

$$n = \sqrt{(\kappa\, \mu_r)}$$

$$= \sqrt{[(23)\,(3.5)]}$$

$$= 9.0$$

Problem 9 (Refractive index)

The refractive index of a typical soda-lime glass is 1.5. What is the speed of light in this material? The speed of light in vacuum is 3×10^8 m/s.

Solution:

From Eq. (5.9),

$$\text{Speed} = (3 \times 10^8 \text{ m/s})/(1.5) = 2 \times 10^8 \text{ m/s}$$

Problem 10 (Optical fiber)

An optical fiber has refractive index 1.56 and 1.35 for its core and cladding, respectively. Calculate the critical angle.
Hint: $n_1 \sin \theta_1 = n_2 \sin \theta_2$ (Snell's Law)

Solution:

$$\theta_2 = 90°, \text{ i.e., } \sin \theta_2 = 1$$

Let the critical angle be θ_c.

$$\sin \theta_c = \sin \theta_1$$

$$= n_2 \sin \theta_2 / n_1$$

$$= n_2 / n_1$$

$$= 1.35/1.56$$

$$= 0.865$$

$$\theta_c = 60°$$

Problem 11 (Optical fiber)

An optical fiber consists of a cladding of refractive index 1.43 and a core of refractive index 1.67. Calculate the acceptance angle. Hint: Fig. 5.11.

Solution:

$$n_1 \sin \theta_1 = n_2 \sin \theta_2$$

$$1.67 \sin \theta_C = 1.43 \sin 90° = 1.43$$

$$\sin \theta_C = \frac{1.43}{1.67} = 0.856$$

$$\theta_C = 58.9°$$

$$\theta_{NA} = 90° - \theta_C = 90° - 58.9° = 31.1°$$

$$Acceptance\ Angle = 2\theta_{NA} = 2(31.1°) = 62.2°$$

Problem 12 (Optical fiber)

(a) By using a sketch, explain the meaning of the critical angle of an optical fiber.

(b) An optical fiber consists of a core with refractive index 1.58. In order to achieve a critical angle of 70°, what should be the refractive index of the cladding?

Solution:

Refer to Fig. 5.8.

Critical angle θ_c is given by

$$\sin \theta_c = \sin 70°$$

$$= n_2/n_1$$

$$= n_2/1.58$$

$$n_2 = (1.58) \sin 70°$$

$$= 1.48$$

Problem 13 (Optical fiber)

An optical fiber of length 1.8 km has an attenuation loss of 1.5 dB/km.

(a) What is the ratio of the intensity of light coming out of the fiber to that going into the fiber at the other end of the fiber?

(b) What is the absorption coefficient of the optical fiber material?

Solution:

(a) Attenuation loss = (1.5 dB/km) (1.8 km)

$$= 2.7 \text{ dB}$$

From Eq. (5.7),

$$2.7 = -10 \log \frac{I}{I_o} = 10 \log \frac{I_o}{I}$$

$$\frac{I_o}{I} = 1.86$$

$$\frac{I}{I_o} = 0.54$$

(b) From Eq. (5.8), with $x = 1$ km,

$$1.5 = \frac{10\alpha}{2.3}$$

Hence,

$$\alpha = 0.35 \text{ km}^{-1}$$

Problem 14 (Optical fiber)

An optical fiber has a core of refractive index 1.48 and a cladding of refractive index 1.35.

(a) What is the critical angle?

(b) What is the acceptance angle?

(c) What is the numerical aperture?

Solution:

(a) $\sin \theta_c = n_2/n_1$

$\qquad = 1.35/1.48$

$\theta_c - 66°$

(b) Acceptance angle $= 2\,(90° - \theta_c)$

$$= 2\,(90° - 66°)$$

$$= 48°$$

(c) From Eq. (5.25),

$$\text{numerical aperture} = \sqrt{n_1^2 - n_2^2}$$

$$= \sqrt{1.48^2 - 1.35^2}$$

$$= 0.61$$

Problem 15 (Optical loss)

An optical fiber has an optical loss of 1.28 dB/km. What is the ratio of the transmitted intensity to the incident intensity after light has traveled for a distance of 2.75 km in the fiber?

Hint: The attenuation loss in dB is given by Eq. (5.7) as

$$-10 \log \frac{I}{I_o}.$$

Solution:

$$Loss = \left(1.28\ \frac{dB}{km}\right)(2.75\ km) = -10\ log\ \frac{I}{I_0}$$

$$log\ \frac{I}{I_0} = -0.352 = -1 + 0.648$$

$$\frac{I}{I_0} = 4.4 \times 10^{-1} = 0.44$$

Problem 16 (Absorption)

The linear absorption coefficient of a material is 61.7 m^{-1}. Light of intensity I_o enters this material.

(a) What is the intensity (in terms of I_o) of the light after it has travelled in the material by a distance of 15 cm?

(b) What is the attenuation in dB after the light has travelled in the material by a distance of 15 cm?

Solution:

(a) $I/I_o = e^{-\alpha x}$, where α is the linear absorption coefficient.

$x = 15$ cm $= 0.15$ m.

$\ln I/I_o = -\alpha x$, where ln is natural logarithm.

$\ln I_o/I = +\alpha x = (61.7 \text{ m}^{-1}) (0.15 \text{ m}) = 9.26$

$I_o/I = 10457$

$I/I_o = 1/10457$

$\quad = 9.56 \times 10^{-5}$

(b) Attenuation in dB $= -10 \log (I/I_o)$

$\quad\quad\quad\quad = +10 \log (I_o/I)$

$\quad\quad\quad\quad = 10 \,(4.02) = 40.2$

Problem 17 (Photoresponse)

A semiconductor has a photoresponse of 145. Its carrier concentration in darkness is 5.4×10^{15} cm^{-3}. What is the carrier concentration in light?

Solution:

From Eq. (5.36),

$$n_2 = n_1 \text{ (photoresponse)}$$

$$= (5.4 \times 10^{15} \text{ cm}^{-3})\,(145)$$

$$= 7.8 \times 10^{17} \text{ cm}^{-3}$$

Problem 18 (Coherence length)

What is the coherence length of a laser with frequency bandwidth

(a) $\Delta f = 10^4$ Hz,

(b) $\Delta f = 10^8$ Hz?

Solution:

(a) From Eq. (5.30), the coherence length x_c is given by

$$x_c = c/\Delta f$$

Hence,

$$x_c = (3 \times 10^8 \text{ m/s})/(10^4 \text{ Hz}) = 3 \times 10^4 \text{ m}$$

(b)

$$x_c = (3 \times 10^8 \text{ m/s})/(10^8 \text{ Hz}) = 3 \text{ m}$$

Problem 19 (Coherence)

The coherence length of light travelling in a medium is given by the speed of light in the medium divided by the frequency bandwidth of the light. For the coherence length to be at least 10 m, what is the maximum frequency bandwidth in Hz?

Solution:

Frequency bandwidth Δf = speed of light/coherence length

$$= (3 \times 10^8 \text{ m/s})/(10 \text{ m}) = 0.3 \times 10^8/\text{s}$$

$$= 0.3 \times 10^8 \text{ Hz} = 3 \times 10^7 \text{ Hz}$$

Problem 20 (Light absorptivity)

(a) For a semiconductor with energy band gap 1.1 eV, what is the minimum photon energy for the semiconductor to be able to absorb the photon(s).

(b) The light absorptivity of a semiconductor depends on the wavelength of the light. Describe this dependence.

Solution:

(a) The minimum photon energy is 1.1 eV.

(b) The absorptivity abruptly decreases when the wavelength is increased to that corresponding to the energy band gap.

Problem 21 (Photoconduction)

Ultraviolet light of photon energy 52.6 eV is incident on an intrinsic semiconductor with energy band gap 1.1 eV.

(a) How many electron-hole pairs are generated by one photon of the incident light?

(b) How many electron-hole pairs are generated by 10 photons of the incident light?

(b) How many electron-hole pairs are generated by 10 photons of the incident light?

Solution:

(a) No. of electron-hole pairs generated by 1 photon

$$= (52.6 \text{ eV})/(1.1 \text{ eV})$$

$$= 48$$

(b) No. of electron-hole pairs generated by 10 photons $= 48 \times 10$

$$= 480$$

Problem 22 (Light detection)

How many electron-hole pairs are generated in a semiconductor (silicon) detector by a photon of x-ray of energy 1.49 keV? The energy band gap of silicon is 1.1 eV.

Solution:

$$\text{No. of electron-hole pairs} = \frac{1.49 \text{ keV}}{1.1 \text{ eV}} = \frac{1490 \text{ eV}}{1.1 \text{ eV}} = 1355$$

Problem 23 (Light detection)

An n-type semiconductor has energy band gap 1.1784 eV and the energy between the donor level and the bottom of the conduction band is 0.0482 eV. It is used as a detector of electromagnetic radiation

Hints:

$E = hf$, where E is the photon energy and f is the frequency.

h is the Planck's constant (4.1357×10^{-15} eV.s).

$f = c/\lambda$, where c is the speed of light (2.998×10^8 m/s) and λ is the wavelength.

More than one electron can be excited by a single photon of the light.

(a) What is the maximum wavelength (if any) of electromagnetic radiation that can be detected by this detector?

(b) What is the minimum wavelength (if any) of electromagnetic radiation that can be detected by this detector?

(c) What are the wavelengths at which the detection is most sensitive?

(d) Photons are emitted upon the downward transition of the excited electrons. What are the dominant wavelengths of the emitted photons?

(e) The refractive index of the semiconductor is 4.475. Calculate the speed of light in the semiconductor.

Solution:

(a) The minimum photon energy E is 0.0482 eV. This corresponds to maximum wavelength λ.

$$\lambda = hc/E = (4.1357 \times 10^{-15} \text{ eV.s}) (2.998 \times 10^{8} \text{ m/s})/(0.0482 \text{ eV})$$

$$= 2.57 \times 10^{-5} \text{ m}$$

(b) None

(c) 2.57×10^{-5} m from (a) and the following.

$$\lambda = hc/E = (4.1357 \times 10^{-15} \text{ eV.s}) (2.998 \times 10^{8} \text{ m/s})/(1.1784 \text{ eV})$$

$$= 1.05 \times 10^{-6} \text{ m}$$

(d) Same as (c).

(e) Speed of light in the semiconductor $= c/n$

$$= (2.998 \times 10^{8} \text{ m/s})/4.475$$

$$= 6.70 \times 10^{7} \text{ m/s}$$

$$= 6.70 \times 10^7 \text{ m/s}$$

Problem 24 (Optical behavior of a semiconductor)

An intrinsic semiconductor has energy band gap $E_g = 0.9$ eV and refractive index 2.58.

(a) What is the maximum wavelength (in μm) of electromagnetic radiation that can be absorbed by this material?

(b) What is the speed of electromagnetic radiation in this material?

(c) What is the critical angle θ_c (in any) for electromagnetic radiation traveling in air and incident on this material?

(d) The intensity of electromagnetic radiation incident on the semiconductor is I_o and that transmitted through the back side of the semiconductor is I, such that $I/I_o = 0.38$. What is the attenuation loss in dB?

(e) This semiconductor is used to make a *pn*-junction to serve as a light emitting diode. What is the frequency of light emitted?

(f) The frequency bandwidth of the light emitting diode is 10^7 Hz. What is the coherence length of light generated by this diode?

(g) The electrical conductivity of this semiconductor when light is incident on it is 250 times that when it is dark. What is the photoresponse?

Solution:

(a) $E = hf = hc/\lambda$

$\lambda = hc/E$

$$= \frac{\left(4.14 \times 10^{-15}\,\text{eV.s}\right)\left(3 \times 10^8\,\text{m/s}\right)}{0.9\,\text{eV}} = 1.4 \times 10^{-6}\,\text{m}$$

(b)

$$n = c/v$$

$$v = c/n = (3 \times 10^8 \text{ m/s})/2.58 = 1.2 \times 10^8 \text{ m/s}$$

(c)

$$\sin \theta_c = \frac{n_2}{n_1} = \frac{2.58}{1} \Rightarrow \theta_c \text{ does not exist.}$$

(d) Attenuation loss $= -10 \log \dfrac{I}{I_o} = -10 \log 0.38 = 4.2 \text{ dB}$

(e) $f = E/h = (0.9 \text{ eV})/(4.14 \times 10^{-15} \text{ eV.s}) = 2.2 \times 10^{14} \text{ s}^{-1}$

(f) $x_c = c/\Delta f = (3 \times 10^8 \text{ m/s})/(10^7 \text{ Hz}) = 30 \text{ m}$

(g) Photoresponse $= \dfrac{\sigma_2}{\sigma_1} = 250$

Problem 25 (pn-junction)

(a) What is meant by a *pn*-junction?

(b) Where is the positive end of the applied bias voltage in case of a forward-based *pn*-junction?

(c) A *pn*-junction needs to have a forward bias in order for a substantial current to flow through it. Explain from the viewpoint of the scientific origin.

Solution:

(a) A *pn*-junction is the junction between a *p*-type semiconductor and an *n*-type semiconductor.

(b) The positive end is at the *p*-side.

(c) There is a potential drop at the junction impeding the current flow.

Problem 26 (LED and solar cell)

(a) An LED is a *pn*-junction under forward bias (i.e., positive at the *p*-side). Why is forward bias suitable?

(b) A solar cell is a *pn*-junction under reverse bias (i.e., negative at the *p*-side). Why is reverse bias suitable?

(c) Ternary compound semiconductors (e.g., AlGaAs, as opposed to binary compounds such as GaAs) are used for LEDs. Why?

(d) The mobility of the holes of silicon decreases with increasing temperature. Explain from the viewpoint of the scientific origin.

Solution:

(a) Under forward bias, the majority holes flow from the *p*-side to the *n*-side while the majority electrons flow from the *n*-side to the *p*-side. They recombine at the junction.

(b) Under reverse bias, the electrons generated at the junction flow to the *n*-side (positive), while the holes generated flow to the *p*-side (negative).

(c) Ternary compound semiconductors are used to adjust the energy band gap so as to adjust the color of the light.

(d) The thermal vibration amplitude increases upon the temperature increase.

References

1. https://www.youtube.com/watch?v=8wYkgZKboss

2. https://www.fiberoptics4sale.com/blogs/archive-posts/95146054-optical-fiber-tutorial-optic-fiber-communication-fiber

3. https://www.corning.com/worldwide/en/products/advanced-optics/product-materials/optical-materials.html

4. https://www.data-connect.com/fiber_tutorial.htm

5. https://www.newport.com/t/fiber-optic-basics

6. https://www.optcore.net/fiber-optic-cable-basics-tutorial/

7. https://www.youtube.com/watch?v=P2yYkv2MQw4

Chapter 6

Magnetic Behavior

This chapter addresses the science and applications that pertain to the magnetic behavior of materials. The applications include magnetic recording, computer memories, magnetic field sensing (magnetoresistance, Hall effect magnetometer), fluid pumping, stress sensing (Villari effect), actuation (magnetostriction, ferromagnetic shape memory effect), magnetic levitation (magnetic repulsion), vibration damping (magnetorheology), loudspeakers, nondestructive evaluation (magnetic particle inspection), chemical analysis (mass spectrometry), energy conversion, and mechanical energy storage. Electromagnetic radiation has both electric field and magnetic field and is addressed in Sec. 4.2.

6.1 Force generated by the interaction of a magnetic field with moving charged particles

An applied magnetic field interacts with moving charged particles, such as electrons, thereby resulting in a force known as the Lorentz force, which is given by

$$F = qv \times B, \qquad (6.1)$$

where q is the charge of the particle, v is the velocity of the particle and B is the magnetic flux density. As a result, a current carrying wire in the presence of an applied magnetic field experiences a force. Due to the cross vector product in Eq. (6.1), the force is perpendicular to the velocity, as illustrated in Fig. 6.1.

The magnetic flux density (also known as the magnetic induction, with the unit Tesla, abbreviated T) is the number of lines of force passing through a unit area of material. The magnetizing force, which induces the lines of force through a material, is called the magnetic field intensity or H, which has the unit ampere per meter (A/m). The relationship between B and H is discussed in Sec. 6.3 (particularly Eq. (6.4)).

Example Problem 6 at the end of the chapter is relevant.

457

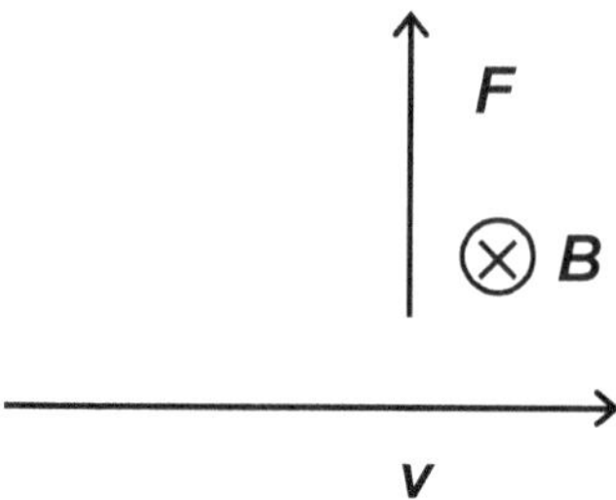

Fig. 6.1 The force *F* on a positively charged particle moving with velocity *v* in a magnetic flux density *B* that goes into the page. The direction of *F* would be opposite to the indicated direction if the particle were negatively charged.

6.1.1 *Hall effect*

The Hall effect refers to the deflection of charged particles by the Lorentz force and the consequent generation of a voltage, called the Hall voltage, in the direction after the deflection. In Fig. 6.1, the Hall voltage gradient is in the direction of the Lorentz force *F*, such that the polarity of the Hall voltage depends on the sign of the charge in the particles. If the charge is positive, the Hall voltage has its positive end above the original path prior to deflection. If the charge is negative, the Hall voltage has its positive end below the original path prior to deflection. Thus, the sign of the charge can be discerned experimentally from the polarity of the Hall voltage. The concentration of the charged particles relates to the magnitude of the Hall voltage. Resistivity measurement (Sec. 2.9.1) does not allow determination of the sign of the carrier, but Hall measurement can.

The Hall effect provides a method a measuring *B*. The instrument, which is known as a Hall effect magnetometer, provides a voltage that is proportional to *B* and that indicates the polarity of *B* as well.

6.1.2 *Motor and flywheel*

A motor is a device that converts electrical/magnetic energy to mechanical energy. Figure 6.2 shows a motor that involves a magnetic field (dashed curves) that is generated by the electric current in windings located at the protruded parts of the frame, which provides a path for the magnetic flux lines. The arc shape of the surface of the protruded parts causes the magnetic field to be radial in the gap between the frame and the armature. At the center of the frame circle is a cylindrical armature that is free to rotate around its axis. Electrical conductor wire is wound on the armature, so that it is embedded at opposite slots on the cylindrical surface of the armature. The current in the wire is into (indicated by a cross) and out of (indicated by a dot) the page, as shown in Fig. 6.2. The Lorentz force resulting from this current and the magnetic field causes the armature to rotate counterclockwise, as shown. By using a commutator (an automatic switching mechanism that is not shown), the current in the wire is maintained in the directions shown, regardless of the position of the armature as the armature rotates. As a result, the armature is always rotating counterclockwise.

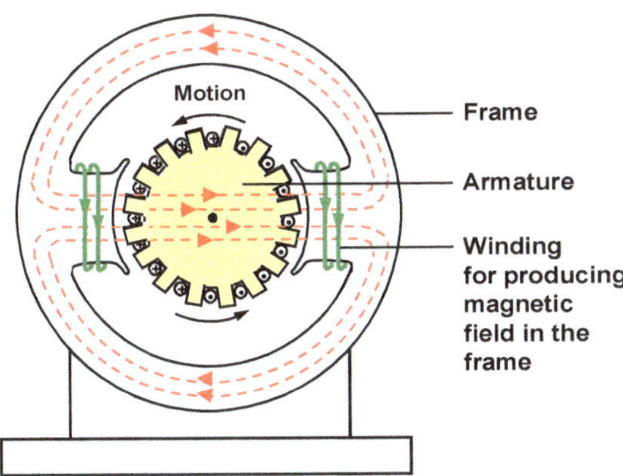

Fig. 6.2 A motor utilizing the Lorentz force for converting electrical/magnetic energy to mechanical energy.

The principle of a motor can be extended to a flywheel, which is a device for storing mechanical energy. The flywheel involves a rotor (i.e., the armature) which rotates at speeds as high as 50,000 rpm. The rotor is suspended by the using of non-contact magnetic bearings. In order to reduce friction, the wheel operates in vacuum. In order to increase the angular momentum, the diameter of the armature needs to be large, although a substantial mass also helps the angular momentum. A balance between mass and volume is obtained by using a carbon fiber polymer-matrix composite, which is low in density, to make the rotor. The mechanical energy can be converted back to electrical energy by slowing down the rotation and using an electric generator (Sec. 4.3.4) to perform the energy conversion.

6.1.3 *Magnetic pump for electrically conductive fluids*

Electrically conductive fluids include blood (which is an ionic conductor) and liquid metals (which are used to help transfer heat from a nuclear reactor core). As illustrated in Fig. 6.3, a conductive fluid is contained in a pipe. The combination of a current applied in the vertical direction through the conductive fluid and a magnetic field in the horizontal direction results in a Lorentz force in the direction perpendicular to these two directions, thus causing the conductive fluid to move in the pipe along the force direction. Magnetic pumps are used for pumping blood, as needed for artificial kidney machines and heart-lung machines.

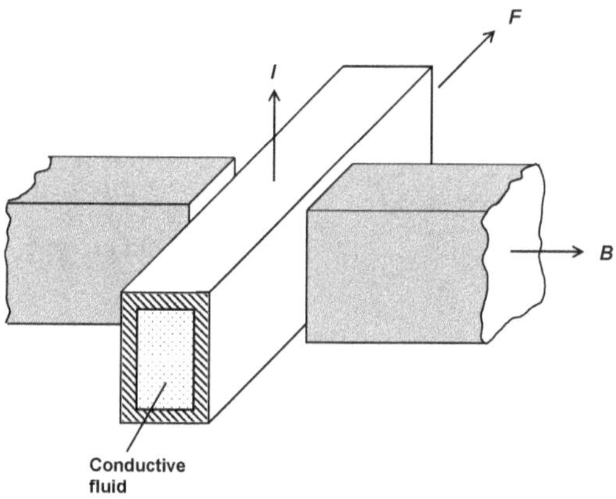

Fig. 6.3 Magnetic pump.

6.1.4 *Loudspeaker*

The Lorentz force is utilized in loudspeakers. A typical loudspeaker consists of a circular magnet surrounded by a freely moving coil that is attached to a cone shaped diaphragm. Electric current sent through the coil, which is immersed in a magnetic field, causes the coil to experience a radial force, which is transferred to the diaphragm. Because the current is AC, the force is alternating in sign, thus causing the diaphragm to vibrate radially, as illustrated in Fig. 6.4. The vibration of the diaphragm causes air molecules to vibrate and hence sound wave generation. By controlling the magnitude and frequency of the AC current, the amplitude and frequency of the vibration is controlled, thereby controlling the volume and pitch of the resulting sound.

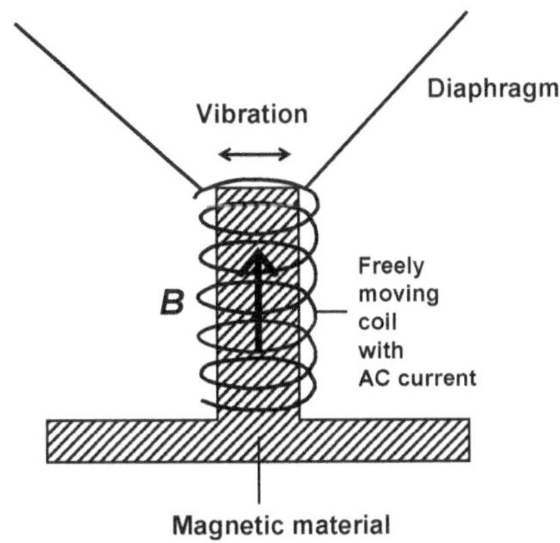

Fig. 6.4 The Lorentz force is utilized in a loudspeaker, which uses magnetic flux density B along the axis of a coil with a flowing AC current to cause the coil to vibrate radially, thereby causing the diaphragm connected to the coil to also vibrate radially.

6.1.5 *Mass spectrometer*

The Lorentz force is also utilized in a type of mass spectrometer, which is an instrument for chemical analysis in the molecular level. For example, a mass spectrometer is used to analyze the urine of an athlete in order to check if the athlete has taken certain drugs. This spectrometer, as illustrated in Fig. 6.5, is based on the Lorentz force on molecular ions (as obtained by fragmentation of a molecule by, say, electron bombardment) that are moving at controlled velocities. The force, which is perpendicular to the velocity, deflects the ions. The path of the deflected ion is in equilibrium when the centripetal force (which depends on the ion mass) is balanced by the Lorentz force, i.e.,

$$qv \times B = mv^2/r, \tag{6.2}$$

where r is the radius of the path and m is the mass of the ion.

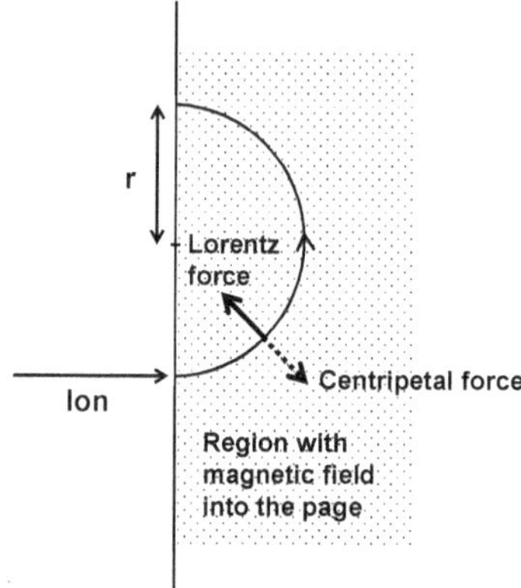

Fig. 6.5 The Lorentz force is utilized in a mass spectrometer that uses a magnetic field to deflect ions according to their mass-to-charge ratio.

Rearrangement of Eq. (6.2) gives

$$m/q = Br/v. \tag{6.3}$$

Thus, the path, as described by r, depends on the ratio of the ion mass to the ion charge, so that ions of different values of the mass-to-charge ratio are deflected by different degrees. Hence, the ions are separated according to their mass-to-charge ratio.

Example Problem 7 at the end of the chapter is relevant.

6.2 Magnetic moment

The magnetic moment (also called the magnetic dipole moment) describes the strength of a magnetic source, which involves a pair of north and south poles. There are two sources of magnetic moment, namely the motion of electric charges and the intrinsic magnetism of elementary particles (e.g., electrons). The orbiting of an electron around the nucleus of an atom (i.e., the motion of an electric charge) gives rise to a magnetic moment. In addition, the spin of an electron (i.e., the intrinsic magnetism of the electron) produces a magnetic moment.

The magnetic moment associated with the spin of an electron is 9.27×10^{-24} A.m², or 1 Bohr magneton (β). A filled orbital contains two electrons of opposite spin, so the net spin magnetic moment of a filled orbital is zero.

Example 1: O (oxygen atom)

Atomic oxygen has an electronic configuration of

$$1s^2\, 2s^2\, 2p^4$$

The 1s orbital is filled; so is the 2s orbital. The 2p subshell (with 3 orbitals) contains two unpaired electrons. Therefore, the magnetic moment of one oxygen atom is 2 (9.27×10^{-24} A.m²), or 2 Bohr magnetons, or $\beta = 2$.

2p | ↑↓ | ↑ | ↑ |

Example 2: O^{2-} (oxide ion)

An O^{2-} ion has an electronic configuration of $1s^2\, 2s^2\, 2p^6$. The 1s, 2s, as well as all three 2p orbitals are full, so the magnetic moment of an O^{2-} ion is zero.

Example 3: O_2 (oxygen molecule)

An O_2 molecule consists of two atoms covalently bonded by the sharing of two pairs of electrons (i.e., a double bond). Each oxygen atom in O_2 has two unpaired electrons. The $2p_y$ electron cloud of one atom overlaps end to end with the $2p_y$ cloud of the other atom, forming a σ bond (Fig. 6.6(a)). The $2p_z$ electron cloud of one atom overlaps side to side with the $2p_z$ electron cloud of the other atom, forming a π bond (Fig. 6.6(b)). The combination of a σ bond and a π bond is a double bond, i.e., O = O. The electron sharing is such that the two electrons being shared have opposite spins. Therefore, an O_2 molecule has a magnetic moment of zero.

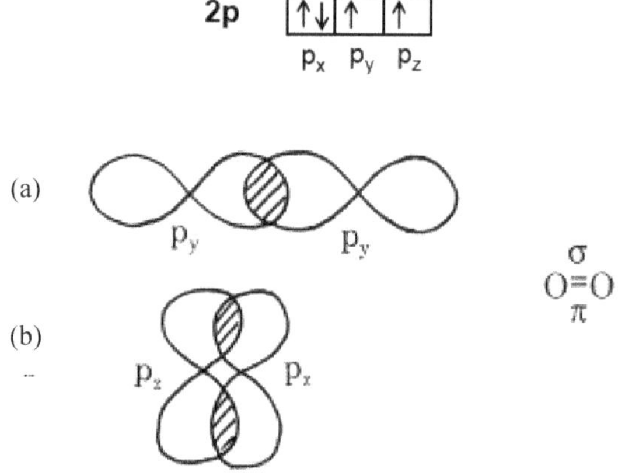

Fig. 6.6 (a) Overlap of $2p_y$ electron clouds of adjacent atoms end to end to form a σ bond. (b) Overlap of $2p_z$ electron clouds of adjacent atoms side to side to form a π bond.

Example 4: Fe (iron atom)

Iron (Fe) solid consists of Fe atoms, each having an electronic configuration of ...$3d^64s^2$. The 4s orbital is full, but the 3d subshell (with 5 orbitals) has 4 unpaired electrons, so $\beta = 4$, or the magnetic moment per Fe atom is 4 (9.27×10^{-24} A.m^2).

Example 5: Co (cobalt atom)

Cobalt (Co) solid consists of Co atoms, each having an electronic configuration of ... $3d^74s^2$. The 4s orbital is full, but the 3d subshell has 3 unpaired electrons, so $\beta = 3$.

Example 6: Ni (nickel atom)

Ni is a solid with electronic configuration ... $3d^84s^2$, so $\beta = 2$.

3d | ↑↓ | ↑↓ | ↑↓ | ↑ | | ↑ |

Example 7: Nd (neodymium atom)

Nd is a solid with electronic configuration ...$4f^4 5d^0 6s^2$. The 4f subshell has 7 orbitals, with four of them being half filled with a single electron. Thus, $\beta = 4$. The large number (7) of orbitals of each of the rare earth elements (elements in the same row as Nd in the Periodic Table) make these elements attractive for magnetic materials. Samarium (Sm) is another example of an attractive rare earth element.

4f | ↑ | ↑ | ↑ | ↑ | | | |

Example 8: Zn (zinc atom)

Zn is a solid with electronic configuration ... $3d^{10} 4s^2$. All orbitals are full, so $\beta = 0$.
   Example Problems 8–14 at the end of the chapter are relevant.

6.3 Ferromagnetic behavior

Since the magnetic moment of an electron is in a certain direction, the magnetic moment of an atom or ion is also in a certain direction. Atoms or ions that are neighboring or close by in a solid can communicate with each other so that, even in the absence of an applied magnetic field, atoms or ions within a limited volume have all their magnetic moments oriented with respect to one another. This communication is a form of quantum-mechanical coupling called the exchange interaction. A material that exhibits this behavior in such a way that the communication causes the magnetic moments to be oriented in the same direction is called a ferromagnetic material. Examples of ferromagnetic materials include iron (Fe), cobalt (Co), nickel (Ni) and gadolinium (Gd). A region in which all atoms have magnetic moments in the same direction is called a ferromagnetic domain. Because there is a tendency for the magnetic moment to be along a certain crystallographic direction, a domain is usually smaller than a grain in a polycrystalline material. In other words, a grain can contain one or more domains.

When no external magnetic field has been applied to a ferromagnetic material (i.e., when the magnetic field strength $H = 0$; the unit for H is A/m), the magnetic moments of various domains cancel one another, so the magnetic induction (also called the magnetic flux density) B is zero. In the presence of an external magnetic field ($H > 0$), the domains with magnetic moments in the same direction as H grow while those with magnetic moments in other directions shrink. This process, which involves the shift of the domain boundaries and the rotation of the domains, results in $B > 0$. The unit for H is A/m. The unit for B is Tesla = wb/m^2, where wb = V.s.

The curve of B versus H during the first application of H is known as the magnetization curve. The slope of this curve is known as the magnetic permeability (μ), as shown in Fig. 6.7(a). Hence,

$$B = \mu H. \tag{6.4}$$

The magnetic permeability is to be distinguished from the carrier mobility in relation to electrical conduction. The mobility is also abbreviated μ (Eq. (2.1). The value of the permeability μ depends on H, since the curve in Fig. 6.7 is not linear. It is common to consider the average permeability, which is the average slope of the curve. It is also common to consider the maximum permeability, which is the maximum slope, as shown in Fig. 6.7(b). However, in applications involving low H (as in magnetic shielding), it is suitable to consider the initial permeability, which is the slope at the low H extreme of the curve.

Consider that H varies at a certain frequency, which corresponds to the angular frequency $\omega = 2\pi\upsilon$, where υ is the frequency in Hz. If the time dependence is sinusoidal, H can be expressed as

$$H = H_\mathrm{o}\, e^{i\omega t}, \tag{6.5}$$

where t is the time and H_o is the amplitude of the H wave. In case that the frequency is high, B does not respond to H fast enough. As a consequence, there is be a time lag of the B wave relative to the H wave, as expressed by the equation

$$B = B_\mathrm{o}\, e^{i(\omega t - \delta)}, \tag{6.6}$$

where δ is the phase lag. The combination of Eq. (6.5) and (6.6) gives

$$\mu = B/H = (B_o/H_o)\, e^{-i\delta}. \tag{6.7}$$

In Eq. (6.7), μ is a complex quantity, with

$$\mu = \mu' + j\mu'', \tag{6.8}$$

where μ' and μ'' are the real and imaginary parts of μ, respectively. Based on Eq. (6.7),

$$\mu' = (B_o/H_o)\cos\delta, \tag{6.9}$$

and

$$\mu'' = -(B_o/H_o)\sin\delta. \tag{6.10}$$

The loss tangent is defined as $\tan\delta$, which is given by

$$\tan\delta = (\sin\delta)/(\cos\delta) = -\mu''/\mu', \tag{6.11}$$

where Eq. (6.9) and Eq. (6.10) are used. The $\tan\delta$ is akin to that in Eq. (3.32), which pertains to the dielectric behavior.

A magnetic dipole is a pair of north and south poles that are separated from one another. The direction of the magnetic flux density B in a magnet is in the direction from the south pole to the north pole. However, outside the magnet, B is in the direction from the north pole to the south pole, due to the looping of the magnetic flux lines.

A magnetic dipole is associated with potential energy, which is the magnetic energy. Thus, magnetization requires energy input, which is given by the shaded area in Fig. 6.7(a), i.e.,

$$\text{Energy per unit volume for magnetization} = \int H\,\Delta B \tag{6.12}$$

The unit of the product of H and ΔB is $(A/m^3)\,(V.s) = J/m^3$, since $V.A = W$ and $W.s = J$. This is consistent with the notion that the integral in Eq. (6.12) describes energy per unit volume. This is analogous to the notion that mechanical energy for deforming a material equals the area under a stress-strain curve. In other words, H is analogous to stress and B is analogous to strain.

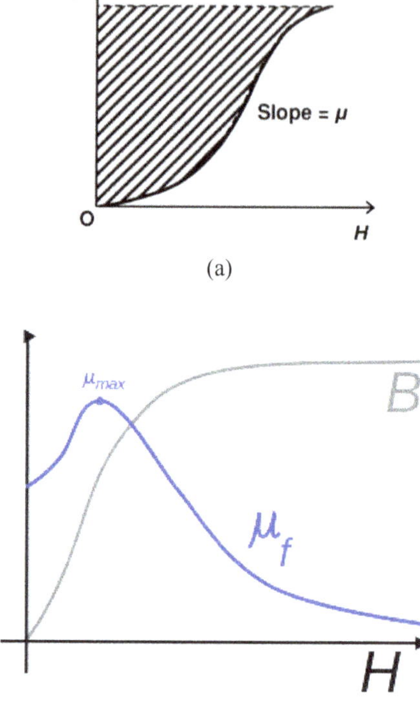

Fig. 6.7 (a) Variation of the magnetic flux density B with the magnetic field strength H during first application of H (virgin curve). The slope of the curve is the permeability μ, which depends on H. The energy (per unit volume) spent in magnetization is given by the shaded area. (b) The slope varies with H, so μ varies with H, having a maximum value μ_{max} and decreasing with increasing H when H exceeds the value corresponding to μ_{max}. (https://en.wikipedia.org/wiki/Permeability_(electromagnetism), public domain)

When B lags H due to the high frequency, a part of the input magnetic energy is lost, typically as heat. This energy loss is related to tan δ, which is a measure of how much power is lost versus how much is stored by the magnetization.

Equation (6.4) can be written as

$$B = \mu_o (H + M), \tag{6.13}$$

where μ_o = permeability of free space (vacuum) = $4\pi \times 10^{-7}$ H/m, H (Henry) = wb/A and wb means weber, and M = magnetization $\equiv$ magnetic moment per unit volume. The unit for the magnetic moment is A.m^2, so the unit for M is A.m^2/m^3 = A/m, which is the same as the unit for H. The quantity $\mu_o H$ is the magnetic flux density (magnetic induction) in vacuum (i.e., no material), so $\mu_o M$ is the extra magnetic induction present due to having a material rather than vacuum as the medium.

The magnetization M is also expressed as

$$M = \chi_m H, \tag{6.14}$$

where χ_m is the magnetic susceptibility. Hence,

$$B = \mu_o \, (1 + \chi_m) \, H = \mu_o \, \mu_r \, H = \mu H, \tag{6.15}$$

where $\mu_r = \mu/\mu_o = 1 + \chi_m$ is the relative permeability of the material. For a ferromagnetic material, $\mu_r \gg 1$ (up to 10^6). For a ferromagnetic material, μ_r is much greater than 1, so μ_r is approximately equal to χ_m. For example, μ_r is 5,000 for iron (99.95 Fe), 10^5 for Permalloy (79Ni-21Fe), and 10^6 for Supermalloy (79Ni-15Fe-5Mo).

The saturation magnetization is reached when the magnetic moments of all domains are in the same direction. Subsequent decrease of H back to zero does not remove the magnetization totally, because the domain walls tend to resist moving. As a result, a remanent magnetization (or remanence, abbreviate M_r) or equivalently a remanent magnetic flux density (or remanent magnetic induction, abbreviated B_r), remains at $H = 0$ (Fig. 6.6). In order to bring the magnetization back to zero, H must be applied in the reverse direction, i.e., $H < 0$. At $H = -H_c$, the magnetization is zero; H_c is called the coercive field (or coercivity). The coercivity is 4 kA/m for carbon steel (0.9C-1Mn alloy), 44 kA/m for Cunife (20Fe-20Ni-60Cu alloy), 123 kA/m for Alnico V (50Fe-14Ni-25Co-8Al-3Cu alloy), 600 kA/m for samarium cobalt (SmCo$_5$ compound) and 900 kA/m for Nd$_2$Fe$_{14}$B compound. The application of an even more negative H causes magnetic alignment in the reverse direction, eventually causing saturation of the magnetization in the reverse direction. The cycling of H between positive and negative values after the initial magnetization results in a B-H hysteresis loop (Fig. 6.8).

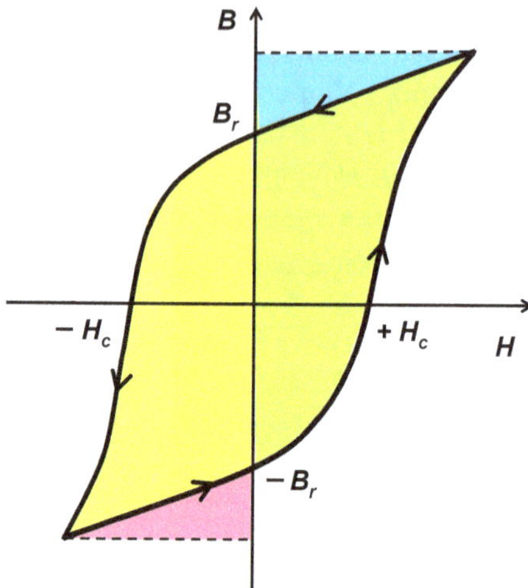

Fig. 6.8 The *B-H* loop upon change of *H* between positive and negative values after the initial magnetization. The area within the loop is the energy loss per cycle. The curve for the initial magnetization is not shown, but is shown in Fig. 6.5.

Magnetization requires energy input. Energy is partly needed to form magnetic dipoles. It is also needed to cause the magnetic domain boundaries to move. Such movement is involved during magnetization. Demagnetization is associated with release of the stored magnetic energy. However, demagnetization may not be complete, due to the difficulty of movement of the magnetic domain boundaries. The difference between the energy input during magnetization and the energy output during demagnetization is the energy loss. The energy loss is largely due to the energy consumed in causing the magnetic domain boundaries to move. When magnetic domain boundaries cannot move easily, hysteresis tends to be large. The more severe is the hysteresis, the greater is the energy loss. The energy lost usually becomes heat. In case of a material that is electrically conductive, the heat can also be generated by the eddy current induced in the material by the changing magnetic field (i.e., resistance heating).

The green + blue region in the right half of Fig. 6.8 (i.e., the part with $H > 0$) is the energy input (per unit volume) to magnetize in the positive B direction. The blue region in the right half of Fig. 6.8 is the energy release (per unit volume) associated with subsequent partial demagnetization to reach B_r. Thus, the energy loss (per unit volume) is the difference between the energy input and the energy release, i.e., the green area inside the right half of the loop. Similarly, the yellow + pink region in the left half of Fig. 6.8 (i.e., the part with $H < 0$) is the energy input (per unit volume) to magnetize in the negative B direction. The pink region in the left half of Fig. 6.8 is the energy release (per unit volume) associated with subsequent partial demagnetization to reach $-B_r$. Thus, the energy loss (per unit volume) is the difference between the energy input and the energy release, i.e., the yellow area inside the left half of the loop. Therefore, the energy loss (per unit volume) due to the process described by the entire loop is the green + yellow area inside the entire loop.

For a ferromagnetic material, the first term on the right side of Eq. (6.13) is negligible compared to the second term, so that B is approximately equal to $\mu_o M$. Therefore, for a ferromagnetic material, the plot of B vs. H is often replaced by a plot of M vs. H, as shown in Fig. 6.9. The M-H loop saturates in M. In contrast, the B-H loop does not saturate in B, due to Eq. (6.13).

The ease of magnetization is described by the energy input required to magnetize during the first application of H. This energy is given by the

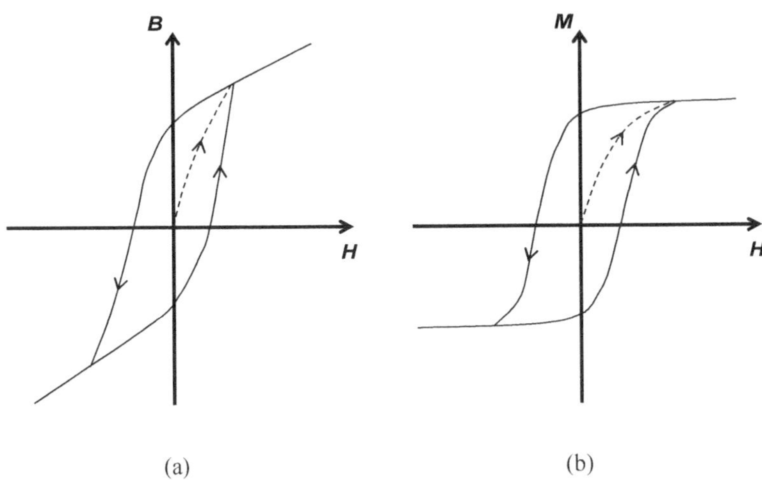

(a)   (b)

Fig. 6.9 (a) *B-H* loop. (b) *M-H* loop.

shaded area in Fig. 6.7. For a given material, this ease depends on the crystallographic direction, due to the interaction of the spin magnetic moment of the electron with the crystal lattice (i.e., the spin-orbit coupling). Depending on the crystallographic direction in the magnetic field, magnetization saturates at different values of *H*. The crystallographic direction that is the easiest for magnetization is known as the easy axis. The crystallographic direction that is the hardest for magnetization is known as the hard axis. This type of anisotropy is known as magnetocrystalline anisotropy. For example, for BCC iron, the [100] direction is the easy axis, the [111] direction is the hard axis and the [110] direction is intermediate in the easy of magnetization. The shaded area in Fig. 6.10 is the energy per unit volume to magnetize iron in the [100] direction. This energy is higher for the [110] and [111] directions. As a result of this anisotropy, the texture (preferred crystallographic orientation) of a polycrystalline material affects the ease of magnetization.

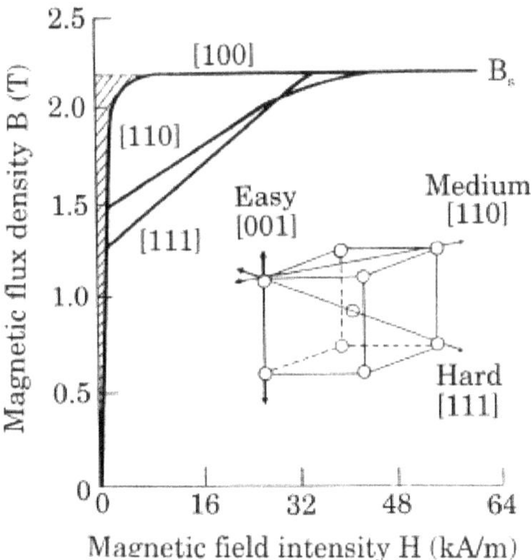

Fig. 6.10 Variation of B with H during first application of H along the [100], [110] and [111] directions of iron. The [100] direction is the easy axis for iron. The shaded area is the energy per unit volume to magnetize iron along the [100] direction. This energy is higher for the [110] and [111] directions.

Some applications require the change in the magnetization from one crystallographic direction to another. The magnetocrystalline anisotropy energy is the energy needed to deflect the magnetization of a single crystal from the easy axis to the hard axis. The energy to remove the magnetization in a particular direction relates to the energy to magnetize in the same direction, since both magnetization and demagnetization involve movement of the magnetic domains.

A ferromagnetic material becomes paramagnetic (Sec. 6.4) upon heating past a certain temperature, which is also known as the Curie temperature. This is a reversible second-order phase transition which involves the change in the magnetic structure (structure associated with the ordering of the directions of the magnetic moments of different atoms or ions) rather than a change in the crystal structure. As the temperature increases, thermal agitation is increasingly severe and the communication (known as the exchange interaction) between atoms becomes more and more hindered. At the Curie temperature, the communication vanishes and the material becomes paramagnetic, with the magnetic moments of different atoms or ions responding individually to the applied magnetic field. In contrast, the different atoms or ions within a domain of a ferromagnetic material respond cooperatively to the applied field. Figure 6.11 shows the decrease of B as the temperature is increased past the Curie temperature. The Curie temperature is 358°C for nickel, 768°C for iron, 1,117°C for cobalt, 747°C for $SmCo_5$ and 312°C for $Nd_2Fe_{12}B$. For a wide temperature of use of a ferromagnetic material, a high Curie temperature is preferred.

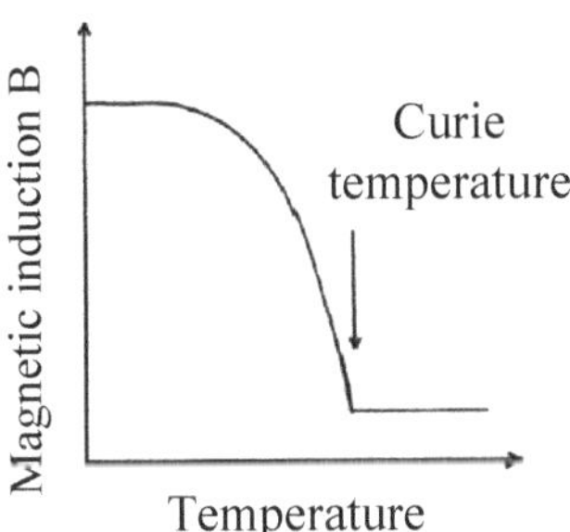

Fig. 6.11 Change of a ferromagnetic material to a paramagnetic material upon heating past the Curie temperature.

The change of the ferromagnetic state to the paramagnetic state upon heating is a phase transition that is used in some devices, since the paramagnetic state is in practice a non-magnetic state. The transition allows temperature-controlled magnetization. An application relates to magnetized soldering tips of soldering irons. The soldering tip is a ferromagnetic material. When the temperature is below the Curie temperature, the tip is magnetic and clings to the heating element. However, above the Curie temperature, the tip is non-magnetic, thus losing contact with the heating element and cooling down.

Another application relates to magneto-optical data storage, i.e., magneto-optical hard disk drives. The storage medium is a ferromagnetic material, the magnetic state of which indicates the information (0 or 1, depending on the magnetic state). It is a type of optical disc drive, as the reading of the information involves using a laser light that is reflected by the surface of the ferromagnetic material. The polarization and reflectivity of the reflected light depend on the magnetic state. The effect of the magnetization on the reflected light is known as the magneto-optical Kerr effect. To erase the information, laser light is used to heat the ferromagnetic material to temperatures above the Curie temperature. An electromagnet is used to write information through magnetization.

Example Problems 1–5, 15 and 16 at the end of the chapter are relevant.

6.4 Paramagnetic behavior

A paramagnetic material has its atoms or ions separately having unpaired electrons and hence non-zero magnetic moments, but the atoms or ions fail to communicate sufficiently (say, because the atoms or ions are relatively far apart) so that the directions of the magnetic moments of different atoms or ions (even adjacent atoms or ions) are independent of one another when $H = 0$. There are no domains. $B = 0$ when $H = 0$. When $H > 0$, more atoms have their magnetic moments in the direction of H, so $B > 0$, but there are still no domains. Because there is no interaction between the atoms, a large H is necessary for aligning the magnetic dipoles. In addition, the magnetization is quickly lost when H is returned to zero. For a paramagnetic material, $\mu_r = \mu/\mu_o$ is slightly greater than 1 (up to 1.01) (i.e., χ_m is positive), so B exceeds the corresponding value for vacuum. As always,

$$B = \mu H.$$

The value of μ_r of a paramagnetic material is much less than that of a ferromagnetic material (Fig. 6.12). The curve of B vs. H is essentially linear for a paramagnetic material, in contrast to the nonlinearity for a ferromagnetic material (Fig. 6.12). More exactly, the curve of M vs. H is linear for a paramagnetic material. Examples of paramagnetic materials include aluminum and titanium.

Since thermal energy opposes alignment of the magnetic moments, the higher the temperature, the lower is μ_r. For a paramagnetic material, the magnetization is inversely proportional to the temperature in K, i.e.,

$$M = CB/T, \tag{6.16}$$

where B is the magnetic flux density, T is the temperature in K and C is a material-specific Curie constant.

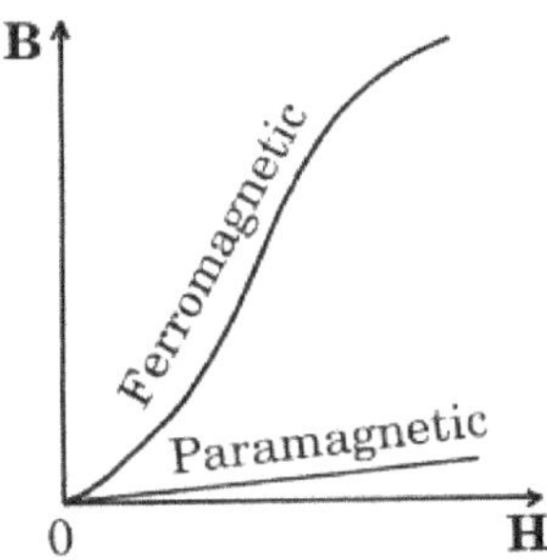

Fig. 6.12 Variation of B with H during first application of H for (a) a ferromagnetic material and (b) a paramagnetic material.

6.5 Ferrimagnetic behavior

In a ferrimagnetic material, the atoms (or ions) having non-zero magnetic moments (referred to as magnetic atoms or ions) within a ferrimagnetic domain communicate with one another via a form of quantum-mechanical coupling called superexchange interaction (as the magnetic atoms or ions are often separated by non-magnetic ones), so that the magnetic moment of every magnetic atom or ion in the domain is in one of two anti-parallel (opposite) directions and the vector sum of the magnetic moments of all these atoms or ions in the domain is non-zero. Just as a ferromagnetic material, a ferrimagnetic material has $B = 0$ before any H is applied and cycling H results in a hysteresis loop in the plot of B versus H.

Ferrimagnetic materials are ceramics materials with ions A magnetically aligned with H and ions B aligned to oppose H. Since the strengths of the magnetic dipoles are different for ions A and B, the net dipole moment is non-zero. Ferrimagnetic oxides such as Fe_3O_4 and $NiFe_2O_4$ are known as ferrites, which are unrelated to the ferrite (α) in the Fe-C phase diagram. In particular, Fe_3O_4 is called magnetite (or lodestone). Ferrites are examples of ceramic magnets (rather than metallic magnets).

The high electrical resistivity of ceramic magnets is attractive for lessening the eddy current (Sec. 4.3), which causes energy loss due to resistance heating. In general, the higher the frequency, the greater is the eddy current. Therefore, a ceramic magnet allows operation at high frequencies.

For Fe_3O_4 to have electrical neutrality, every four O^{2-} ions must be accompanied by one Fe^{2+} ions and two Fe^{3+} ions. This material exhibits the inverse spinel structure. In this crystal structure, the O^{2-} ions occupy the corners and face centers of a cube, the Fe^{2+} ions occupy ¼ of the octahedral interstitial sites (formed by the anion structure), and the Fe^{3+} ions occupy ¼ of the octahedral interstitial sites and 1/8 of the tetrahedral interstitial sites. For every four O^{2-} ions, there must be one Fe^{2+} ion and two Fe^{3+} ions in order to have electrical neutrality and be consistent with the chemical formula Fe_3O_4.

Let us count the number of each type of ions per cube.

$$O^{2-}$$

$$8\left(\frac{1}{8}\right) + 6\left(\frac{1}{2}\right) = 4$$

Corners Face centers

$$Fe^{2+} \qquad 4\left(\frac{1}{4}\right) = 1$$

There are 4 octahedral interstitial sites per cube.

$$Fe^{3+}$$

$$4\left(\frac{1}{4}\right) + 8\left(\frac{1}{8}\right) = 2$$

Octahedral Tetrahedral

There are 8 tetrahedral interstitial sites per cube.

Since a fraction of certain sites are occupied in a cube, different cubes can have different arrangements of the ions even though the ion counts are the same for all the cubes. Because of this, a cube is not a true unit cell. It turns out that the true unit cell consists of eight cubes (Fig. 6.13). The lattice constant of the true unit cell is 8.37 Å. In a true unit cell, there are 4 (8) = 32 O^{2-} ions, 2 (8) = 16 Fe^{3+} ions and 1 (8) = 8 Fe^{2+} ions. The arrangement of the ions in a true unit cell is quite complex (Fig. 6.14).

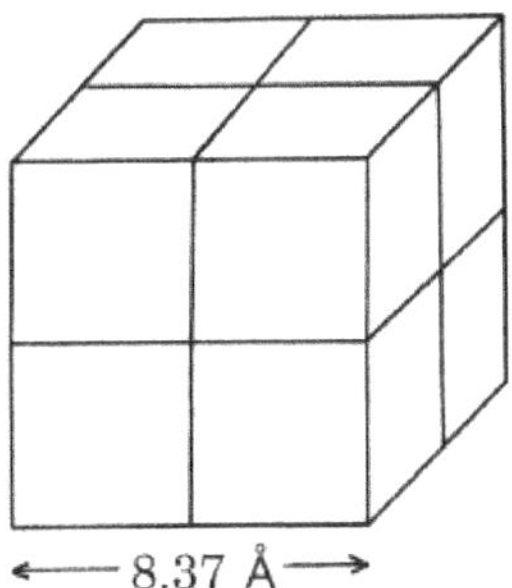

Fig. 6.13 Eight cubes making up a true crystal structural unit cell of dimension 8.37 Å for Fe₃O₄.

The O^{2-} ions have zero magnetic moments. Half of the Fe^{3+} ions (i.e., the Fe^{3+} ions occupying octahedral intersitial sites) and all of the Fe^{2+} ions have their magnetic moments in the same direction, while the remaining Fe^{3+} ions (i.e., those occupying tetrahedral interstitial sites) have their magnetic moments in the opposite direction. In other words, all cations occupying octahedral sites have their magnetic moments in the same direction, while all cations occupying tetrahedral sites have their magnetic moments in the opposite direction.

                    Functional Materials

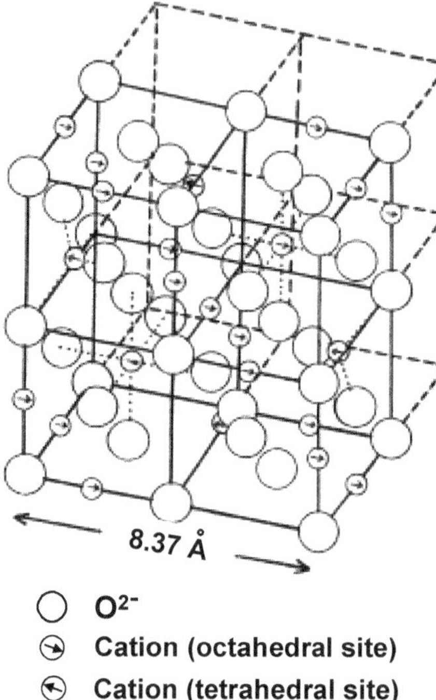

Fig. 6.14 The arrangement of the ions in a true unit cell (Fig. 6.13) of Fe_3O_4, which is ferrimagnetic. Only the front half of the unit cell is shown. The magnetic moment (directions shown by arrows) of the cations in octahedral (six-fold) sites and the cations in tetrahedral (four-fold) sites are opposite in direction. The large circles are O^{2-} ions, which have no magnetic moment.

The magnetic moment of Fe ($3d^6\ 4s^2$) is such that $\beta = 4$; that of Fe^{2+} ($3d^6$) is such that $\beta = 4$; that of Fe^{3+} ($3d^5$) is such that $\beta = 5$.

Let us calculate the magnetic moment per unit cell.

Interstitial site	Spin	Fe^{2+}	Fe^{3+}	β	Magnetic moment (A.m^2)
Octahedral	↑	8		+32	+32 (9.27×10^{-24})
Octahedral	↑		8	+40	+40 (9.27×10^{-24})
Tetrahedral	↓		8	-40	-40 (9.27×10^{-24})
				+32	+32 (9.27×10^{-24})

The saturation magnetization (i.e., magnetic moment per unit volume at saturation)

$$= \frac{\text{Magnetic moment per unit cell}}{\text{Volume of a unit cell}}$$

$$= \frac{32 \left(9.27 \times 10^{-24}\right) \text{ A.m}^2}{\left(8.37 \times 10^{-10} \text{ m}\right)^3}$$

$$= 5 \times 10^5 \text{ A.m}^{-1} \tag{6.17}$$

$NiFe_2O_4$ (called nickel ferrite) consists of Ni^{2+} (instead of Fe^{2+}), Fe^{3+} and O^{2-} ions. For Ni^{2+}, $\beta = 2$. Repeating the above calculation for $NiFe_2O_4$ shows that the magnetic moment per unit cell is +16 (9.27×10^{-24} A.m^2).

6.6 Antiferromagnetic behavior

In an antiferromagnetic material, each atom (or ion) has a non-zero magnetic moment and the atoms within an antiferromagnetic domain communicate with one another so that the magnetic moment of every atom in the domain is in one of two anti-parallel (opposite) directions and the vector sum of the magnetic moments of all the atoms in the domain is zero.

An example of an antiferromagnetic material is NiO, which consists of Ni^{2+} and O^{2-} ions arranged in the rock salt structure (Fig. 6.15). The O^{2-} ions have no magnetic moment, but the Ni^{2+} ions ($\beta = 2$) do. However, half of the Ni^{2+} ions have their magnetic moments one way, while the remaining Ni^{2+} ions have their magnetic moments in the opposite direction. Thus, the material has no magnetization.

Another example of an antiferromagnetic material is MnO, which consists of Mn^{2+} and O^{2-} ions arranged in the rocksalt structure (Fig. 6.16). The magnetic unit cell, which is shown in Fig. 6.16, is 8 times the volume of the crystal structural unit cell. The magnetic moments of the Mn^{2+} ions in adjacent (111) planes are oppositely aligned.

Upon heating past the Neel temperature, the thermal energy is high enough for an antiferromagnetic material to lose its magnetic ordering and the material typically becomes paramagnetic.

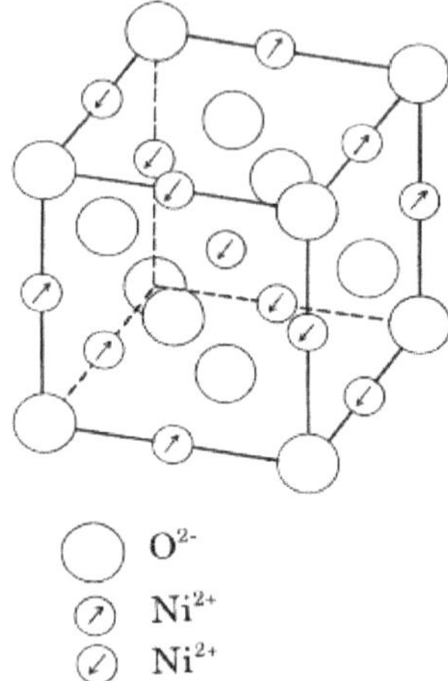

Fig. 6.15 The arrangement of the ions in a crystal structural unit cell of NiO, which is antiferromagnetic. The small circles are Ni^{2+} ions (with magnetic moment in direction shown by arrow); the large circles are O^{2-} ions, which have no magnetic moment.

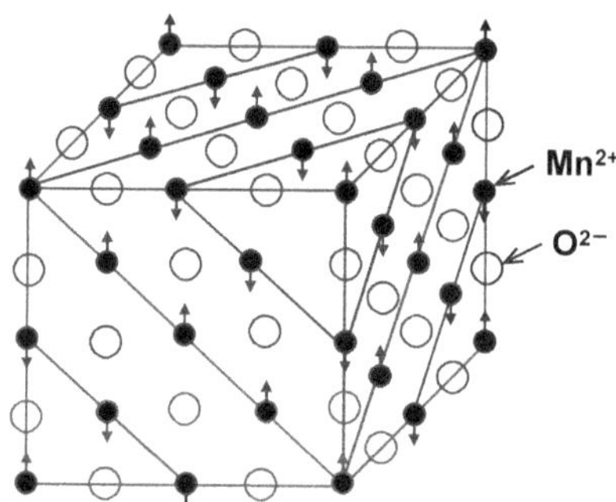

Fig. 6.16 The arrangement of the ions in a magnetic unit cell of MnO, which is antiferromagneticcand exhibits the rocksalt crystal structure. The magnetic unit cell is 8 times the volume of a crystal structural unit cell. The small circles are Mn^{2+} ions (with magnetic moment in direction shown by arrow); the large circles are O^{2-} ions, which have no magnetic moment.

6.7 Hard and soft magnets

Precipitates, grain boundaries, dislocations, etc., tend to anchor domain boundaries, thereby causing a large remanent magnetization and a large

H_c (i.e., a wide hysteresis loop). For example, steel is magnetically harder than iron, due to its multiphase microstructure.

A hard (permanent) magnet is characterized by a high saturation magnetization, a high coercive field (> 10,000 A/m), and hence a large hysteresis loop. In contrast, a soft magnet is characterized by a high saturation magnetization, a small coercive field (< 500 A/m), a small remanence and a narrow hysteresis loop (for minimizing the energy loss) (Fig. 6.17). A material with coercive field between 500 and 10,000 A/m is said to be semi-hard. Note that magnetic hardness is totally different from mechanical hardness.

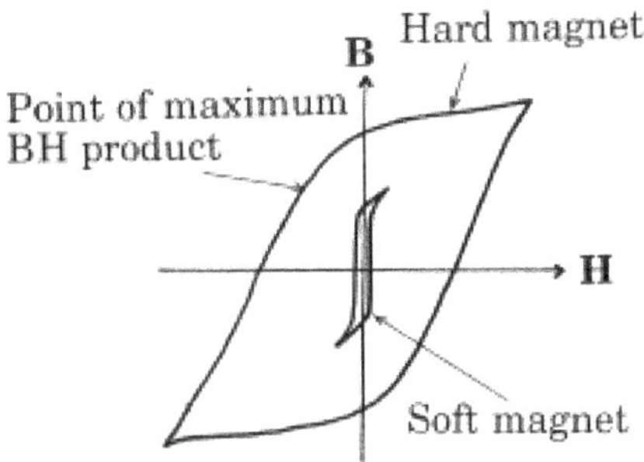

Fig. 6.17 The B-H loop for a hard magnet and a soft magnet.

Hard magnets are needed for magnetic levitation and numerous other applications which require a permanent magnetization that is always in the same direction. Soft magnets are needed for electromagnetic cores, transformer cores (Sec. 4.3.3), motors and generators (Sec. 4.3.4).

For computer memories (e.g., hard disk drive, also called hard drive), which require switching between the two directions of magnetization (as the two directions represent binary information 0 and 1), a small square loop, a low saturation magnetization, a low remanence and a low coercive field are preferred. In this application, an electromagnetic head (i.e., a recording head) is used to store or retrieve information from the disk. In the process of storage, a current in the head magnetizes domains in the disk. In the process of retrieval (i.e., reading), the domains induce a current in the head. A commonly used magnetic material for this application is a magnetic coating in the form of Fe_2O_3 particles. By using smaller particles, the data storage density can be increased. By magnetizing in the direction perpendicular to the plane of the coating (i.e., perpendicular recording) rather than the direction parallel to the plane of the coating, the data storage density can also be increased.

The "power" of a hard magnet is defined as the *B-H* product, i.e., the area of the largest rectangle that can be drawn in the demagnetization quadrant of the hysteresis loop, as shown in Fig. 6.18. The five rectangles drawn in Fig. 6.18 are different in area. It is clear that the rectangles labeled 1, 2, 3 and 5 have smaller areas than rectangle 4. The *B-H* product relates (but is not equal) to the energy per unit volume needed to demagnetize a magnet, i.e., to bring B from B_r to 0. For this product to be large, both the remanence B_r and the coercive field H_c need to be high. A high value of this product is preferred for a hard magnet. The value is 1.6 kJ/m^3 for carbon steel (0.9C-1Mn), 12 kJ/m^3 for Cunife (20 Fe-20 Ni-60 Cu), 36 kJ/m^3 for Alnico V (50Fe-14Ni-25Co-8Al-3Cu), 140 kJ/m^3 for SmCo$_5$, and 220 kJ/m^3 for Nd$_2$Fe$_{14}$B. Among these materials, Nd$_2$Fe$_{14}$B is most attractive, due to its high *B-H* product.

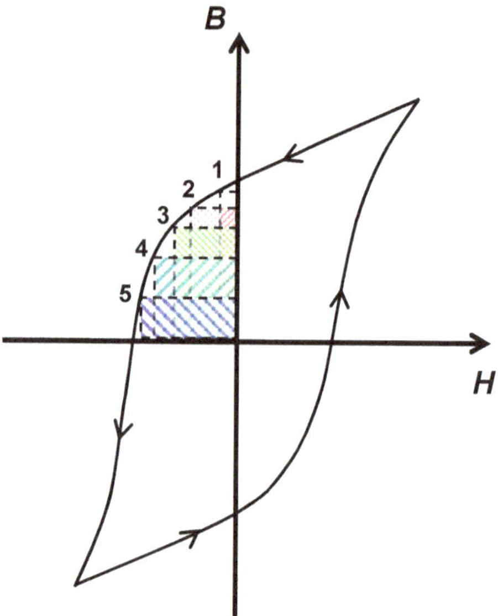

Fig. 6.18 The *B-H* product, which is defined as the area of the largest rectangle that can be drawn in the demagnetization quadrant of the hysteresis loop.

6.8 Magnetic shielding

Magnetic shielding refers to the blocking of magnetic field, so that the field cannot enter the area behind the shield. This is valuable for the protection of electronic equipment, which may malfunction in the presence of a substantial magnetic field. It is also useful for the protection of people, due to perceived health hazards associated with magnetic field exposure. Magnetic fields are generated by AC currents, such as the current in a transmission line. The frequency of the AC current is commonly low, such as 60 Hz, in contrast to the GHz frequency that is associated with EMI shielding (Sec. 4.4). Homes in the vicinity of transmission lines are prone to this perceived health hazard.

Materials for magnetic shielding are soft magnetic materials with very high values of κ (e.g., 50,000–350,000), as needed to trap the magnetic flux lines in the shield. A low hysteresis energy loss is desired, since shielding is not aimed at magnetization. Magnetic shielding materials are commonly metals in a continuous form, such as mesh and foil forms. Magnetic continuity allows the magnetic flux lines to be continuously guided. Electrical continuity allows eddy current flow, which helps dissipate the magnetic energy. A material that is highly effective for magnetic shielding is mumetal, which is a nickel-iron alloy with very high κ. A more commonly used material for magnetic shielding is steel. Because of the low H that is commonly encountered in shielding applications, the initial κ is more relevant than the average κ. In case that the shielding is for high H, the magnetic shielding material should be able to provide a sufficiently high B at saturation.

Due to the magnetic field that is emitted by underground transmission lines and underground electric facilities (e.g., transformers and switches), concrete that is capable of magnetic shielding is desired. Concrete can be rendered magnetic shielding ability by the embedment of a steel mesh. However, the steel mesh cannot be incorporated in the concrete mix. Magnetic particles can be incorporated in a concrete mix, but their connectivity is inadequate. Steel paper clips, as commonly used in offices, are an effective admixture. The clips are discontinuous and are thus suitable for incorporation in a concrete mix, but their tendency to intertwine during mixing results in continuity. Thus, concrete containing 5 vol.% clips is as effective as steel mesh for magnetic shielding at 60 Hz, as shown in Fig. 6.19.

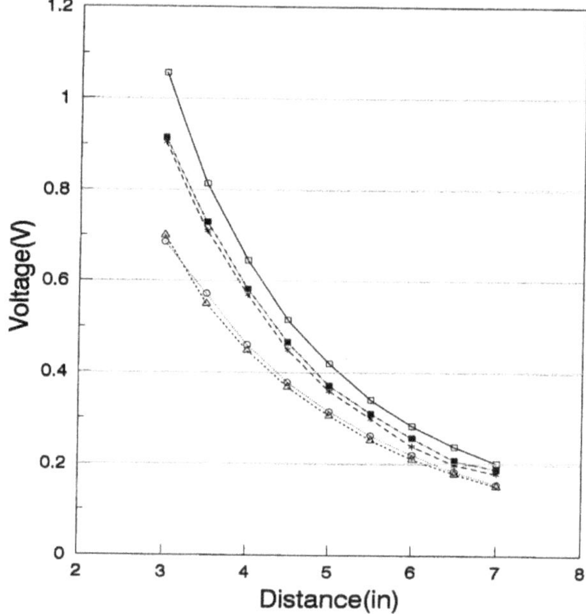

Fig. 6.19 Magnetic shielding effectiveness, as shown by the voltage detected by the receiver as a function of the distance between the transmitter and the receiver. □: No shielding material. Δ: Steel mesh. ○: Concrete with paper clips, cured horizontally. *: Concrete with paper clips, cured vertically. ■: Plain concrete. (Zeng-Qiang Shi and D.D.L. Chung, "Concrete for Magnetic Shielding," Cem. Concr. Res. 25(5), 939-944 (1995))

The testing of the magnetic shielding effectiveness involves placing the source of the magnetic field (i.e., the transmitter) on one side of the vertically positioned shield to be evaluated and placing the receiver (which measures the field that leaks through and around the shield) on the other side, as illustrated in Fig. 6.20. For the data in Fig. 6.19, the transmitted touches the shield and the distance between the near ends of the transmitter and receiver is varied from 3 to 7 in (7.6 to 18 mm) by moving the receiver. The transmitter and receiver are centered along the same axis, which is perpendicular to the shield.

The transmitter is an electromagnet, i.e., it generates a magnetic flux density by a coil that carries a current, such that the flux density is along the axis of the coil. For the data in Fig. 6.19, the coil is made of #12 wire coiled 100 times around a plastic cylinder (7 cm diameter); the wire carries an AC (60 Hz) current of 8.5 A during testing. The magnetic flux density B at the transmitter is given by

$$B = \mu_o NI/(2r), \tag{6.18}$$

where N is the number of turns, I is the current (8.5 A) and r is the average radius of the coil (3.5 cm).

The receiver is also a coil, but no current is applied to this coil. Rather, the time-varying magnetic flux density present at this coil due to the leakage of magnetic field from the transmitter through the shield induces a voltage in the coil, according to Faraday's Law (Eq. (4.7)), which states that the induced voltage relates to the rate of change of the magnetic flux Φ. For the data in Fig. 6.19, the receiver coil comprises 2,000 turns of #40 wire wound concentrically around a 4 inch diameter polyvinyl chloride pipe. Based on Faraday's Law, the magnitude of the output voltage V_{out} of the receiver is related to B at the receiver by the equation

$$V_{out} = \omega BAN, \tag{6.19}$$

where ω is the angular frequency (with the frequency $v = 60$ Hz), A is the area (m^2) and N is the number of turns (2,000) in the receiver. The term ω in Eq. (6.19) comes from the differentiation of the flux Φ (which varies with time as sin ωt) with respect to time t.

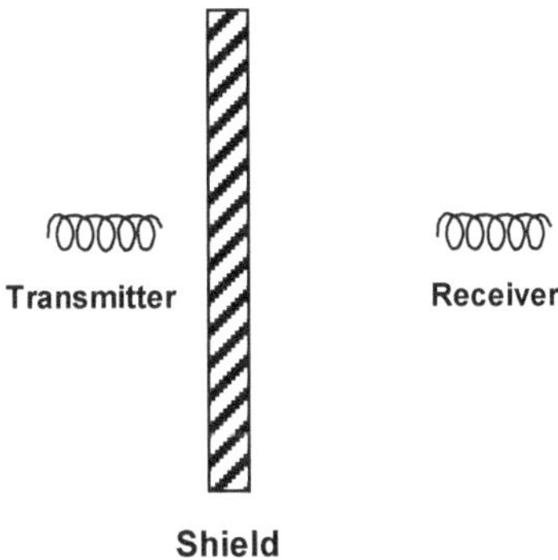

Fig. 6.20 Method of magnetic shielding effectiveness testing, involving a transmitter coil and a receiver coil on the two sides of the shield under evaluation.

Equation (6.18) is valid only at zero distance from the coil. For points at a distance z from the coil, B is given by

$$B = \mu_o I r^2 / [\, 2(r^2 + z^2)^{3/2}\,], \qquad\qquad (6.20)$$

where r (in m) is the radius of the transmitter coil, z is the axial distance (in m) from the coil and I is the current in the transmitter coil.

6.9 Composite material with a magnetic filler and a non-magnetic matrix

A composite material involving a non-magnetic matrix and a magnetic filler is attractive for magnetic property tailoring through composite design. This tailoring involves the composition, size and volume fraction of the magnetic filler. In particular, a higher volume fraction of the magnetic filler increases the saturation magnetization of the composite. In addition, a higher filler content increases the degree of contact between filler units in the composite, thereby enhancing the magnetic continuity. In case that the filler is electrically conductive, a higher filler volume fraction also enhances the electrical continuity, which promotes eddy current. An increase in size of the filler unit (e.g., the diameter of the filler in the form of a fiber) tends to promote a larger grain size in the magnetic filler, thus tending to increase the magnetic domain size. On the other hand, a smaller size of the filler unit increases the amount of filler-matrix interface area per unit volume of the composite, thereby possibly inhibiting domain boundary movement, thus increasing the coercivity. In case that the matrix is a polymer, the composite is further attractive for its molderability and consequent ease of shaping. In addition, due to the low cost of the matrix compared to the filler, the composite route is attractive for cost reduction.

Nickel is a ferromagnetic material. A comparison of the magnetic properties of solid nickel and polymer-matrix composites with various volume fractions (3–19%) of nickel fibers of various diameters (0.4, 2 and 20 μm) is described below. For comparison of materials with different nickel contents, it is scientifically revealing to divide the measured magnetic flux density (i.e., induction) B by the volume fraction of nickel in the material. Figure 6.21 shows the B-H loops for solid nickel and composites with 19 vol.% nickel fiber of three diameters, with the B values having been scaled by dividing by the nickel volume fraction. Figure 6.22 shows corresponding results for composites with 0.4 μm diameter nickel fiber (nanofiber) at various volume fractions. Figure 6.22(a) and Fig. 6.22(b) differ in the range of H. Table 6.1 lists the magnetic properties of solid nickel and the various composite materials, with the B values having been scaled by dividing by the nickel volume fraction. The maximum B, abbreviated B_m, is taken arbitrarily as the scaled B value at $H = 238.7$ kA/m, which is the highest H used in the investigation.

For the same fiber diameter of 0.4 μm, B_m is essentially the same for filler contents of 3, 7 and 13 vol.%, and is slightly smaller for the filler content of 19 vol.%. At the same filler content of 19 vol.%, B_m increases with increasing fiber diameter. This means that the ease of magnetization is quite independent of the nickel volume fraction, but it increases with increasing fiber diameter.

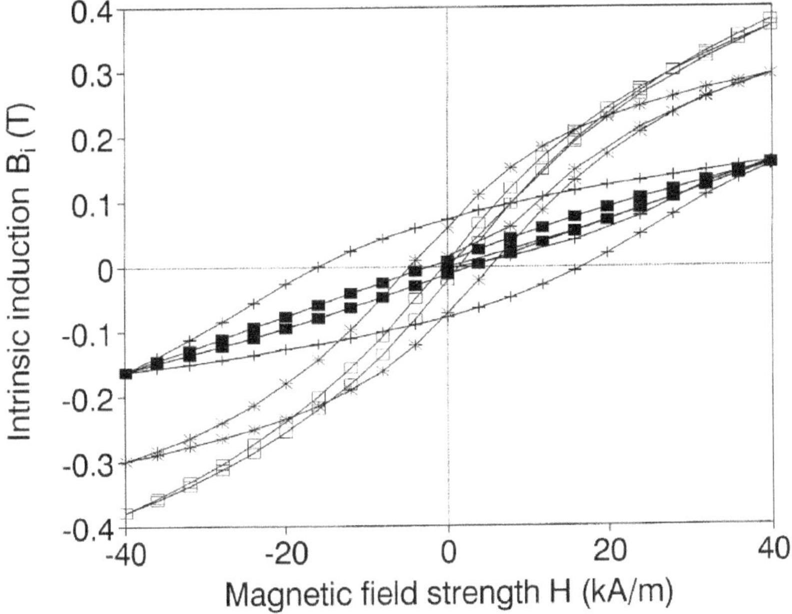

Fig. 6.21 The B-H loops for solid nickel (■) and polymer-matrix composites with 19 vol.% nickel fiber of diameter 0.4 μm (+), 2 μm (*) and 20 μm (□), with the B values having been scaled by dividing by the nickel volume fraction. (Xiaoping Shui and D.D.L. Chung, "Magnetic Properties of Nickel Filament Polymer-Matrix Composites," J. Electron. Mater. 25(6), 930–934 (1996))

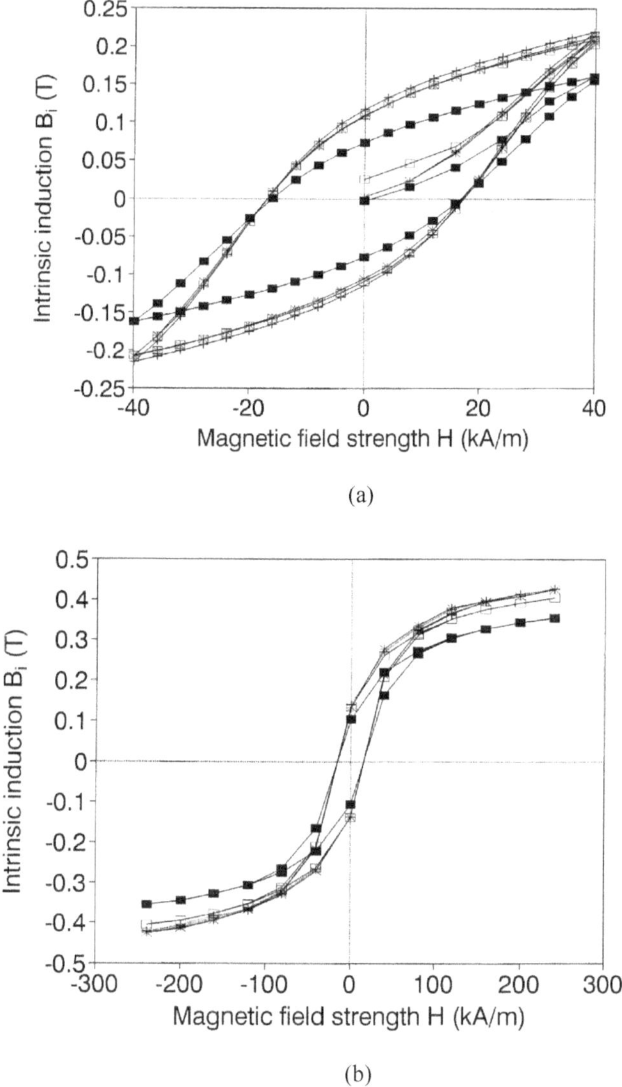

(a)

(b)

Fig. 6.22 The B-H loops for polymer-matrix composites with 0.4 μm diameter nickel fiber at 3 vol.% (□), 7 vol.% (*), 13 vol.% (+) and 19 vol.% (■), with the B values having been scaled by dividing by the nickel volume fraction. (a) H between –40 and +40 kA/m. (b) H between –250 and +250 kA/m. (Xiaoping Shui and D.D.L. Chung, "Magnetic Properties of Nickel Filament Polymer-Matrix Composites," J. Electron. Mater. 25(6), 930–934 (1996).

Table 6.1 Magnetic properties of solid nickel and polymer-matrix composite containing nickel fibers of various diameters at various volume fractions. B_r = remanent B. B_m = maximum B. All B values have been scaled by dividing by the nickel volume fraction. (Xiaoping Shui and D.D.L. Chung, "Magnetic Properties of Nickel Filament Polymer-Matrix Composites," J. Electron. Mater. 25(6), 930–934 (1996))

Filler vol. %	Filler diameter (μm)	B_m (T)	B_r (T)	B_r/B_m	H_c (kA/m)	Hysteresis energy loss (J/m^3)
3	0.4	0.403	0.109	0.27	16.8	10,300
7	0.4	0.426	0.108	0.26	16.8	10,200
13	0.4	0.427	0.116	0.27	16.9	10,815
19	0.4	0.355	0.074	0.21	15.9	7,368
19	2	0.558	0.060	0.10	4.65	4,926
19	20	0.695	6.7×10^{-3}	9.84×10^{-3}	0.45	1,016
Solid nickel		0.616	8.8×10^{-3}	1.43×10^{-2}	2.25	1,283

The ratio B_r/B_m is comparably large for the same filler at different volume fractions (3–19%), but it decreases with increasing fiber diameter at a fixed filler content of 19 vol.%. The coercivity is comparably large for the same fiber at different volume fractions (3–19%), but it decreases sharply with increasing fiber diameter at a fixed filler content of 19 vol.%. These trends of B_r and H_c indicate that the ease of magnetization is independent of the fiber volume fraction and increase sharply with increasing fiber diameter.

The hysteresis energy loss per unit volume of nickel (only the nickel part of the composite) is comparably large for 0.4 μm diameter nickel fiber at 3, 7 and 13 vol.%, but is lower at 19 vol.%. At a fixed filler content of 19 vol.%, the hysteresis energy loss per unit volume of nickel decreases with increasing fiber diameter. The ratio B_r/B_m is also lower at 19 vol.% than the other volume fractions of the same filler. The observed hysteresis energy loss is attributed to the difficulty of domain boundary movement and/or the eddy current loss. The loss is greatly decreased by increasing the fiber diameter, i.e., decreasing the phase boundary concentration (hence facilitating domain boundary movement) and reducing percolation (increasing the electrical resistivity and hence decreasing the eddy current). The loss is decreased by increasing the fiber content from 13 to 19 vol.% for the same filler. This is due to the increased touching of the fibers, so that fiber-fiber interfaces replace some of the filler-matrix interfaces, thereby enhancing the ease of domain boundary movement and/or eddy current.

Correlation of the magnetic behavior and the EMI shielding effectiveness, both measured on the same composites, is revealing. The hysteresis energy loss increases with decreasing fiber diameter, while the EMI shielding effectiveness also increases with decreasing fiber diameter. This means that the magnetic hysteresis energy loss contributes to the absorption of EMI, so that the high EMI shielding effectiveness of the 0.4 μm diameter nickel fiber (Fig. 4.16) is not just due to the skin effect (which makes a fiber with a smaller diameter more effective for shielding).

6.10 Diamagnetic behavior

A diamagnetic material is one with the relative permeability $\mu_r < 1$ (typically about 0.99995) (Fig. 6.23) because the electrons respond to the applied magnetic field by setting up a slight opposing field. This behavior can be a consequence of the electron shells of the atom (or ion) being full so that the electron spins cancel and there is no net magnetic moment associated with the spin. As a result, the magnetic field does not interact with the spins, but rather interacts with the electron orbit (i.e., the shape of the electron cloud) by distorting it. For a diamagnetic material, μ_r is less than 1 (i.e., χ_m is negative), so B is less than the corresponding value for vacuum. The magnetization (which is negative) is linearly related to H.

Examples of diamagnetic materials include copper, gold, silver, bismuth, graphite and alumina. Most organic compounds are slightly diamagnetic. A special class of diamagnetic materials is superconductors, which are strongly diamagnetic. This phenomenon associated with superconductors is known as the Meissner Effect.

Diamagnetism only occurs in the presence of an applied magnetic field. The phenomenon is due to the interaction of the magnetic field with the electrons orbiting the nucleus of an atom. The magnetic field results in a force on the moving electrons, thus affecting the speed of the orbiting electrons. As a consequence, the magnetic moment associated with the electron is changed in a direction against the magnetic field. This opposition of the applied field is responsible for μ_r being less than 1

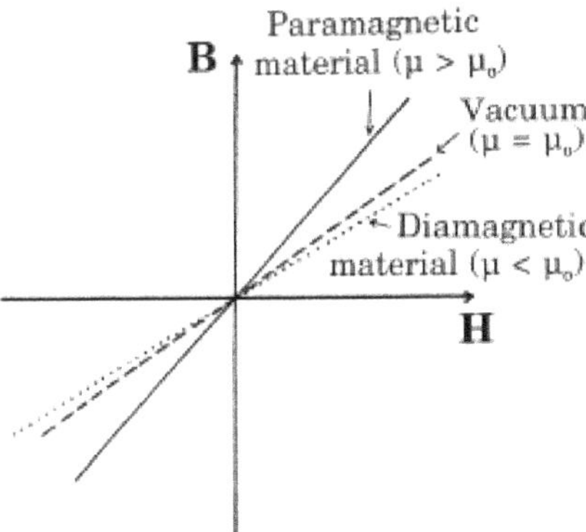

Fig. 6.23 Variation of B with H during first application of H for a paramagnetic material (solid curve), vacuum (dashed curve) and a diamagnetic material (dotted curve).

for a diamagnetic material. Due to the weakness of typical diamagnetism, diamagnetic materials are typically considered non-magnetic.

6.11 Magnetostriction and Villari effect

Magnetostriction refers to the change in dimension upon magnetization of a ferromagnetic material. Thus, it converts magnetic energy to mechanical energy. Magnetostriction provides a mechanism of actuation. In case of the application of a ferromagnetic material as a transformer core, magnetostriction can cause energy loss due to frictional heating. Indeed, an electric hum can be heard near a transformer.

The dimensional change in magnetostriction is reversible and is due to the rotation of the magnetic dipoles (spins) and the resulting change in bond length between atoms (a phenomenon known as spin-orbit coupling, i.e., coupling between the electron spin and the electron orbit). The strain can be expansion (positive magnetostriction) or contraction (negative magnetostriction). It is typically small, of the order of 10^{-6} in magnitude. This strain is denoted by λ and is called the magnetostrictive coefficient, which is defined as the strain upon magnetization from zero to the saturation value. Figure 6.24 shows the strain versus magnetic field H for iron, cobalt and nickel. The strain is negative for all these three materials, except for iron at a low magnetic field. A consequence of magnetostriction is that domain boundaries may be torn at the edges of

 Functional Materials

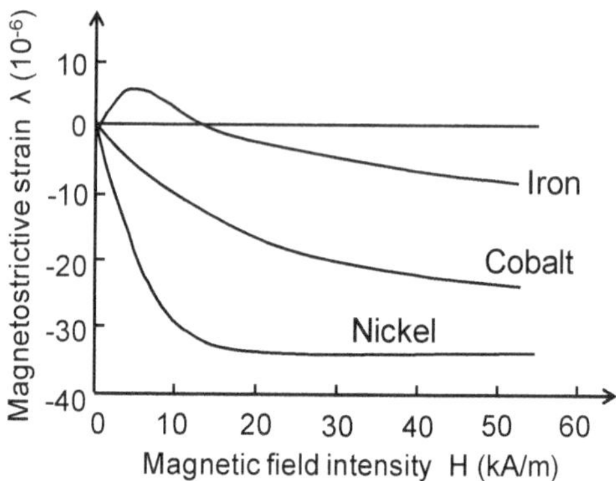

Fig. 6.24 Variation of magnetostrictive strain with the magnetic field H for iron, cobalt and nickel.

the triangular domains (called closure domains, since they allow the loop of magnetic flux path to complete within the solid (rather than in the air around the solid) so as to lower the energy), as illustrated in Fig. 6.25. Since the dimensional change increases with domain size, the problem of Fig. 6.25(a) and (b) can be alleviated by having smaller domains (Fig. 6.25(c)).

The low magnetostrictive strains for the transition metals (elements with 3d subshell, e.g., Fe, Co and Ni, Fig. 6.24) are not suitable for practical application in actuation. The magnetostrictive strains are larger for the rare earth elements (elements with 4f subshell, e.g., samarium (Sm), terbium (Tb) and dysprosium (Dy), because of their large spin-orbit coupling and the highly anisotropic localized nature of the 4f electronic charge distribution. In particular, Tb and Dy are hexagonal and the large magnetostrictive strains are in the basal plane. On the other hand, the localized nature of the 4f electronic charge distribution causes the exchange interaction between atoms to be weak, thereby making the Curie temperature low (<240 K, below room temperature − not practical) compared to those of the transition metals, which have stronger exchange interaction. This shortcoming of the rare earth elements can be alleviated by alloying a rare earth element with a transition metal to form an intermetallic compound. In this way, the Curie temperature is increased. Such compounds have formula AB_2, where A is a rare earth element and B is a transition metal, e.g., $SmFe_2$ (samfenol), $TbFe_2$ (terfenol) and $DyFe_2$. They are cubic in crystal structure. In general, compounds with formula AB_2 (where A and B are any elements) are called Laves phases (e.g., $MgCu_2$, $MgZn_2$ and $MgNi_2$). In $TbFe_2$, saturation strain is 3.6×10^{-3} at room temperature and above, and the Curie temperature is 431°C. The phenomenon is known as giant magnetostriction.

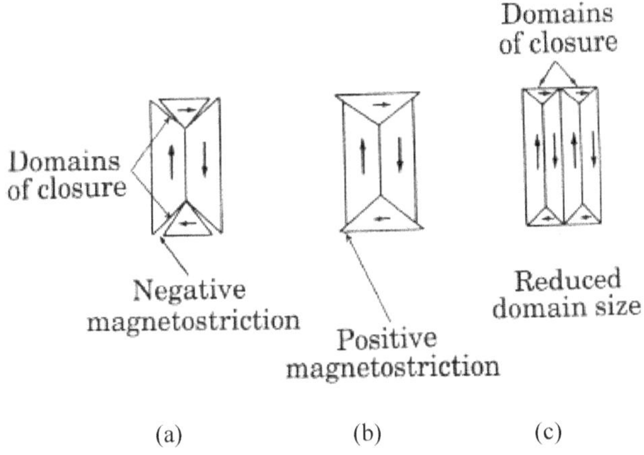

Fig. 6.25 Boundaries of closure domains being torn due to (a) negative magnetostriction and (b) positive magnetostriction. The problem in (a) or (b) is alleviated by reducing the domain size, as shown in (c).

Magnetostriction involves rotation of magnetic dipoles, i.e., domain rotation. The ease of rotation depends on the crystallographic orientation of the dipole. Easy axes are directions that are easy (requiring little energy) to rotate into or out of. Hard axes are directions that are difficult (requiring large energy) to rotate into or out of. The energy required to rotate the magnetization out of the easy direction is called the magnetocrystalline anisotropy constant (K), which is also called the magnetocrystalline anisotropy energy. It is equal in magnitude to the energy required to magnetize in the easy direction. This parameter is usually positive. A negative value of this parameter can occur when there are many equivalent easy axes at an angle of 90° from the easy axis – a situation known as easy-plane anisotropy.

For example, the spins of the iron atoms prefer to be in the [100] direction, so the easy direction for iron is [100] and the hard axis is

[111]; a small magnetic field is enough to cause magnetic saturation in the easy axis, but a larger field is necessary in the hard axis. The [110] direction is intermediate. The shaded area in Fig. 6.10 is the magnetocrystalline anisotropy energy per unit volume for iron. For $TbFe_2$ and $SmFe_2$, the easy axis is [111].

A high magnetocrystalline anisotropy energy inhibits domain rotation, particularly at low temperatures. Furthermore, it causes the domain wall motion to be irreversible, thereby resulting in hysteresis. To decrease the magnetocrystalline anisotropy energy, $TbFe_2$ and $DyFe_2$ (both highly magnetostrictive, with λ positive) are combined, because the magneto-crystalline anisotropy constant K is negative for $TbFe_2$ and positive for $DyFe_2$. Magnetostriction/anisotropy ratio maximization (anisotropy compensation, $K \cong 0$) occurs in $Tb_xDy_{1-x}Fe_2$, where $x = 0.27$, at room temperature. It differs at other temperatures. This ternary alloy is called Terfenol-D, because Tb (ter in Terfenol), Fe (fe in Terfenol) and Dy (D in Terfenol-D) are in the alloy and the Naval Ordnance Laboratory (NOL, nol in Terfenol), now Naval Surface Warfare Center, investigated this material first. This alloy exhibits the largest room temperature mag-netostrictive strain of any commercially available material. Figure 6.26 shows the variation of the magnetostrictive strain with the applied magnetic field for Terfenol-D. Note that the strain is positive for both positive and negative values of the magnetic field intensity.

A related phenomenon is the change of the magnetization with stress. This is associated with the effect of stress on the magnetic permeability μ. The phenomenon, known as the Villari effect or the converse magneto-strictive effect, converts mechanical energy to magnetic energy and can be used for stress sensing. The effect stems from the effect of stress on the bond distance, which affects the exchange interaction. In particular, a sufficiently high stress can cause a particular crystallographic direction to be the easy axis. The Villari effect is used for sensing sound waves, as needed for microphones.

Two other effects are related to magnetostriction. Matteuci effect is the creation of a helical magnetic flux density in a material that is subjected to a torque, which affects the magnetic permeability in an anisotropic fashion. The Wiedemann effect is the twisting of a material when a helical magnetic field is applied.

Example Problem 18 at the end of the chapter is relevant.

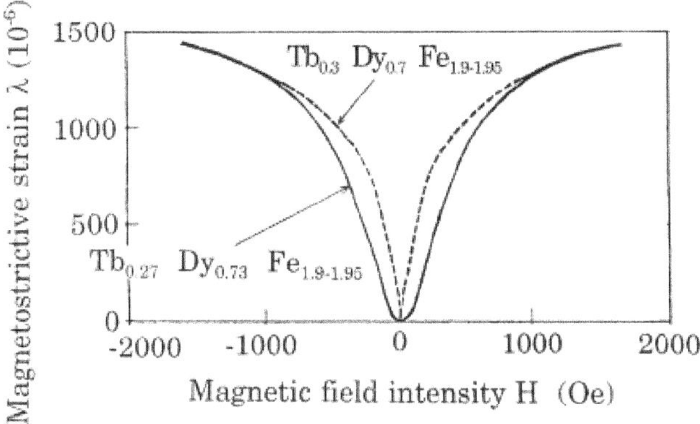

Fig. 6.26 Variation of the magnetostrictive strain with the applied magnetic field H for Terfenol-D.

6.12 Ferromagnetic shape memory effect

The ferromagnetic shape memory effect (abbreviated FSM effect) most commonly refers to the deformation (as much as 10%) of the martensite phase upon exposure to a magnetic field. The field causes alignment of the magnetic moments of regions that are different in the crystallographic orientation due to twinning, so that the regions that have their magnetic moments favorably oriented with respect to the applied magnetic field grow at the expense of other regions, as illustrated in Fig. 6.27. Along with the magnetic moment alignment is the change in dimension, which is known as magnetoelastic deformation. This is a form of magneto-striction. The effect is reversible. Upon removal of the applied magnetic field, the original shape returns. This effect differs from the conventional shape memory effect (Sec. 1.6.4) in that it occurs completely in the martensite phase.

The ferromagnetic shape memory effect is valuable for actuation that is controlled by a magnetic field. In this application, a stress is first applied to the material in order to produce a single-variant state in the absence of an applied magnetic field, as shown in Fig. 6.27(a). This is known as biasing. Then a magnetic field is applied to cause magnetoelastic deformation.

Another form of the ferromagnetic shape memory effect involves the deformation that accompanies the change of austenite to martensite,

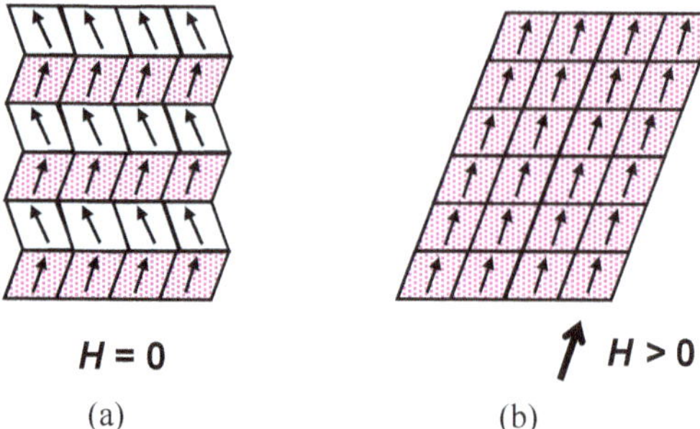

Fig. 6.27 The ferromagnetic shape memory effect. (a) In the absence of an applied magnetic field H. (b) In the presence of an applied magnetic field H. The dotted regions are favorably oriented with respect to the applied magnetic field and grow at the expense of the unshaded regions.

such that this change is induced by an applied magnetic field. This is akin to the stress induced martensitic transformation (also known as pseudoplasticity, as illustrated in Fig. 1.17), but, in the ferromagnetic shape memory effect, the martensitic transformation is induced by a magnetic field rather than stress.

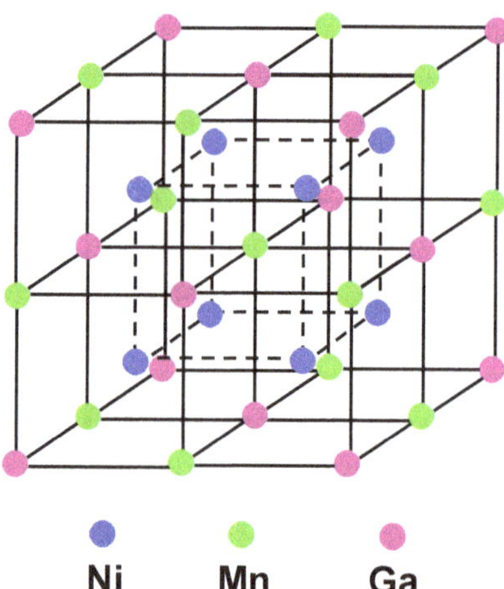

Fig. 6.28 The cubic Heusler crystal structure, as exhibited by the ferromagnetic shape memory alloy Ni_2MnGa.

For the ferromagnetic shape memory effect, the shape memory alloy must be ferromagnetic. The well-known shape memory alloy Ni-Ti is non-magnetic. Ferromagnetic shape memory alloys include Ni_2MnGa, Fe-Pd and Fe_3Pt. The Ni-Mn-Ga alloys, which are among the best ferromagnetic shape memory alloys, typically exhibit maximum strain 100 μm/mm, magnetic field strength for maximum strain 400 kA/m, Curie temperature 103°C, and compressive strength 700 MPa.

The Ni_2MnGa compound is ferromagnetic, with Curie temperature about 85°C. Its austenite phase exhibits the cubic Heusler crystal structure (with the unit cell shown in Fig. 6.28). The martensite obtained upon cooling austenite below M_s (about –10°C, with $M_s - M_f$ less than 3°C) is tetragonal in crystal structure. In this transformation from a cubic structure to a tetragonal structure, contraction occurs in one of the three orthogonal axes of a unit cell, while expansion occurs in the remaining two axes. Any one of the three orthogonal axes of the cubic unit cell can become the axis that contracts during the transformation. Thus, three variants of martensite result, corresponding to the three possible directions of contraction of the cubic unit cell. The three variants tend to coexist in martensite, such that adjacent variants meet at twin boundaries. Thus, the easy axes (the c-axis of the tetragonal unit cell) of adjacent twin bands are essentially perpendicular to one another. The relative proportions of the variants can be controlled by an applied stress or an applied magnetic field. This means that a stress or a magnetic field can be used to bias the material so that a certain variant dominates. Biasing serves to enhance the magnetoelastic deformation.

6.13 Magnetoresistance and magnetic multilayer

The magnetic flux density in a material affects the movement of charged particles in the material, as expected from the Lorenz force (Eq. (6.1)). Magnetoresistance refers to the phenomenon in which the electrical resistance changes with the magnetic field. Mathematically, it is described by the magnetoresistance ratio (abbreviated MR ratio), which is defined as

$$\text{MR ratio} = (R_o - R_H)/R_H, \tag{6.21}$$

where R_o is the resistance in the absence of an applied magnetic field and R_H is the resistance in the presence of an applied magnetic field. The

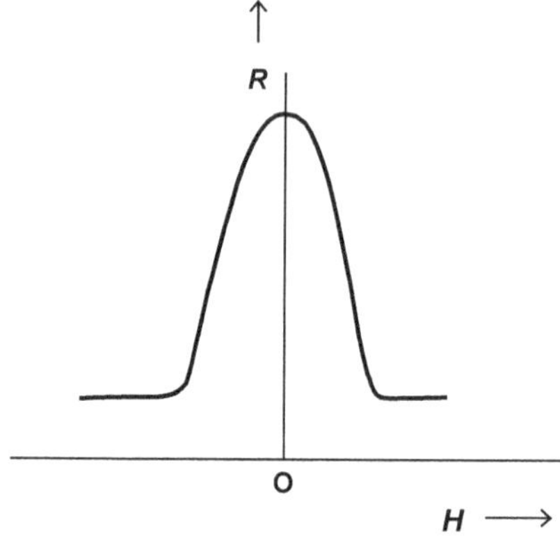

Fig. 6.29 Plot of the electrical resistivity vs. the magnetic field strength *H*, illustrating positive magnetoresistance.

MR ratio is positive if the resistance decreases with increasing magnetic field (as shown in Fig. 6.29). The magnetic field may be parallel or perpendicular to the current used to measure the resistance. The MR ratio in the two directions may be different, even different in sign. The MR ratio tends to be small for a non-magnetic (i.e., diamagnetic, paramagnetic or antiferromagnetic) material, but it is larger for a magnetic material (e.g., a ferromagnetic or ferrimagnetic material). This phenomenon is useful for magnetic field sensing, which is important for magnetic recording (e.g., read heads), data storage (e.g., random access memory) and manufacturing control.

A multilayer is a stack of two or more thin films which are deposited successively (one on top of another) on a substrate. When there are only two layers, the multilayer is known as a bilayer. Each layer is typically less than 2000 Å in thickness and is deposited by sputtering, vacuum evaporation or other vapor methods.

The MR ratio is typically 5% or less for ferromagnetic materials. This weak effect is known as ordinary magnetoresistance (OMR). However, by using a multilayer consisting of ferromagnetic thin films (e.g., iron films of thickness 30 Å each) that are separated by a non-magnetic thin film (e.g., chromium film of thickness 9–18 Å), such that the adjacent (but separated) ferromagnetic thin films exhibit antiferromagnetic coupling (i.e., naturally opposite in the direction of the magnetic moment, as shown in Fig. 6.30(a)), the MR ratio reaches 80% in the direction perpendicular to the multilayer. This strong effect is known as giant magnetoresistance (abbreviated GMR). The non-magnetic layer, which is necessary for the antiferromagnetic coupling, must be elec-trically conductive. Otherwise, the resistance will be high, whether a magnetic field is present or not. Without the separation provided by the non-magnetic layer, the antiferromagnetic coupling of the ferromagnetic layers is not possible. However, an excessive thickness of the non-magnetic layer reduces the antiferromagnetic coupling, thus diminishing the magnetoresistance. For example, the magnetoresistance increases as the thickness of the chromium layer in Fe/Cr/Fe decreases from 18 to 9 Å.

The strong magnetoresistance of the multilayer is due to the scattering of the electrons at the interfaces between the layers of the multilayer. The scattering, which is due to the antiparallel spins of the adjacent ferromagnetic layers, is known as spin scattering. In the absence of an applied magnetic field, the spin scattering associated with the antiferromagnetic coupling causes the resistance in the direction perpendicular to the layers to be high. In the presence of an applied magnetic field, the coupling between adjacent layers changes from antiferromagnetic coupling to ferromagnetic coupling (i.e., the magnetic moment in the adjacent magnetic layers being in the same direction, as shown in Fig. 6.30(b)), thus greatly reducing the spin scattering and decreasing the electrical resistance.

The two magnetic layers are usually made of different materials, with the layer to undergo magnetic moment direction switching (the lower layer in Fig. 6.30) being a soft magnet (e.g., Co) and the other magnetic layer being a hard magnet (e.g., CoPt). A device can consist of numerous alternating magnetic layers, with all adjacent layers being separated by non-magnetic layers. For the discovery of GMR, Fert and Grunberg received the 2007 Nobel Prize in physics. A device based on GMR is an example of spintronics (also known as magnetoelectronics), which refer to solid-state devices that exploit the spin of electronics.

Another application of a magnetic multilayer relates to magnetic data storage, which is known as magnetoresistive random access memory (abbreviated MRAM). The state of antiferromagnetic coupling (Fig. 6.30(a)) and the state of ferromagnetic coupling (Fig. 6.30(b)) provide the two states that correspond to 1 and 0, as needed for data storage. An applied magnetic field is used to switch from the state of antiferromagnetic coupling to the state of ferromagnetic coupling. Reading involves measuring the electrical resistance. A magnetic memory is a grid of cells,

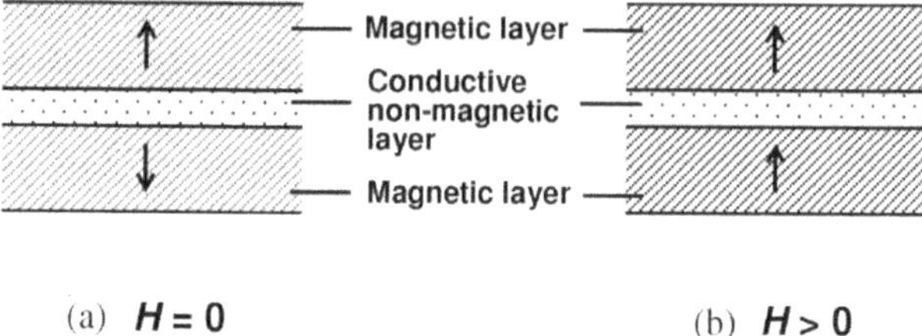

Fig. 6.30 Giant magnetoresistance exhibited by a multilayer consisting of ferromagnetic thin films (shaded regions) that are separated by a non-magnetic but electrically conductive thin film (dotted region). (a) The high resistance state, which occurs in the absence of an applied magnetic field H. (b) The low resistance state, which occurs in the presence of an applied magnetic field *H*.

each of which is a multilayer involving two ferromagnetic layers (one being a hard magnet and the other being a soft magnet) separated by a very thin electrically insulating layer (e.g., MgO of thickness 10 Å). The insulating layer is in contrast to the conductive layer in GMR. The small thickness of the insulating layer allows current flow in the direction perpendicular to the layers by quantum tunneling, which refers to the crossing of a particle (an electron in this case) across a potential barrier (hill) by tunneling through the hill rather than going from one side to the opposite side of the hill via the surface of the hill. The tunneling is possible, due to the tail of the wave-function (probability distribution) of the particle extending from one side of the hill to the opposite side. Tunneling is possible only when the distance across the hill is small – about 10 Å or less. In spite of the insulating character of the separating layer, current can flow by tunneling. This magnetoresistance phenomenon is known as the tunnel magnetoresistance effect (abbreviated TMR, also known as the magnetic tunnel effect). The Fe/MgO/Fe multilayer exhibits MR ratio exceeding 200% at room temperature. Thus, TMR devices have replaced GMR devices in the magnetic storage market.

Yet another application of the multilayer relates to the tailoring of the coercivity and other aspects of the magnetic hysteresis loop. This is because the proportions and compositions of the hard and soft magnetic components can be selected as needed.

Still another application involves a multilayer with ferromagnetic layers separated by electrically insulating layers, which are usually non-magnetic. The purpose is to suppress energy loss due to the eddy current induced by a magnetic field.

Giant magnetoresistance is usually positive (Fig. 6.29). However, a giant negative magnetoresistance occurs in polymer-matrix composites containing iron oxide (ferrite) nanoparticles (~10 nm).

An even larger effect than GMR is colossal magnetoresistance (CMR), which is associated with a MR ratio as high as 100,000%. The origin of CMR is not well understood. CMR is exhibited by certain oxides such as $(La, A)MnO_3$ (A = Ca, Sr or Ba). The crystal structure is perovskite but orthorhombic (with the lattice parameters a $\cong$ b $\neq$ c). Intrinsic to such oxides is ferromagnetic ordering in the a-b plane of the unit cell and antiferromagnetic ordering in the c-axis. The ferromagnetically ordered Mn-O layers of the a-b plane are separated by a nonmagnetic La(A)-O layer.

Example Problem 17 at the end of the chapter is relevant.

6.14 Magnetorheology

Magnetorheology refers to the effect of a magnetic field on the rheology. A magnetorheological fluid (abbreviated MR fluid) is a suspension of micrometer-sized (0.1–10 μm) magnetic particles in a liquid, which is typically an oil. The fluid is positioned between two magnetic pole pieces (corresponding to the north and south poles), as illustrated in Fig. 1.18. The gap between the two pole pieces is typically 0.5–2 mm. In the presence of an applied magnetic field, the particles are magnetized and attract one another, thereby forming columns in the direction of the magnetic field, as illustrated in Fig. 1.18. The columns hinder shear perpendicular to the columns, thus increasing the shear yield stress and the apparent viscosity, so that the fluid resembles a viscoelastic solid. Both MR and ER fluids (Sec. 3.14) are smart fluids, as they are useful for control devices, such as shock absorbers, hydraulic valves and haptic controllers (control of the sense of touch). An MR fluid responds to a magnetic field, whereas an ER fluid responds to an electric field. The effect of a magnetic field on an MR fluid is akin to the effect of an electric field on an ER fluid (Fig. 3.57 and 3.58). An MR fluid is attractive in that the yield stress can be controlled accurately by varying the magnetic field, which is provided by an electromagnet. This means that the ability of the fluid's ability to transmit force can be controlled by a magnetic field.

Table 6.2 shows that MR fluids are advantageous over ER fluids in the high maximum yield stress, wide operable temperature range, low voltage requirement and stability in the presence of impurities. Indeed, MR fluids are dominating over ER fluids in control applications. However, recent improvement in ER fluids has greatly increased the maximum yield stress through the use of urea $((NH_2)_2CO$ organic compound) coated barium titanium oxalate nanoparticles in silicone oil. The high dielectric constant of the particles, the small size of the particles and the urea coating all contribute to causing the high maximum yield stress.

Table 6.2 Properties of typical MR and ER fluids

	MR	ER
Maximum yield stress	50–100 kPa	2–5 kPa
Plastic viscosity	0.2–1.0 Pa.s	0.2–1.0 Pa.s
Maximum field	≈ 250 kA/m	≈ 4 kV/mm
Response time	ms	ms
Density	3–4 g/cm^3	1–2 g/cm^3
Operable temperature range	–50 to 150°C	+10 to 90°C
Power supply	2–25 V 1–2 A 2–50 W	2–5 kV 1–10 mA 2–50 W
Stability	Not affected by most impurities	Cannot tolerate impurities

6.15 Nondestructive evaluation using magnetic particles

Nondestructive evaluation (abbreviated NDE), also known as non-destructive testing (abbreviated NDT), refers to the evaluation of the condition or damage of materials, components or structures by nondestructive methods. Due to the aging of aircraft and the civil infrastructure, NDE is a topic of current importance. In particular, structural composites are used in strategic applications such as aircraft, helicopter rotors, etc. Nondestructive assessment of the structural health is needed in order to prevent hazards.

One of the simplest methods of NDE is magnetic particle inspection, which involves the application of magnetic particles on the surface of the part to be evaluated. The particles tend to gather at the defects present in the part, due to the leakage of magnetic flux lines at defects, as illustrated in Fig. 6.31. The flux leakage is a consequence of the distortion of the flux lines, which prefer to stay in a medium of high magnetic permeability. A defect such as a crack has low magnetic permeability, due to the air present. Thus, both surface and subsurface defects can be detected, though the subsurface defects are limited to those that are near the surface. For best sensitivity, the applied magnetic field should be perpendicular to the length of the defect, as in Fig. 6.31. A defect that has its length parallel to the applied magnetic field cannot be detected, as shown in Fig. 6.32. The magnetic particle inspection method is limited to the evaluation of materials that are magnetic (ferromagnetic or ferromagnetic).

The magnetic particles can be ferromagnetic (e.g., iron) or ferromagnetic (e.g., Fe_3O_4). The particles are often colored to enhance visual inspection. They are typically suspended in a liquid, so as to facilitate the movement of the particles. However, the use of dry power is also possible. The suspension may be sprayed or painted on the part to be inspected. The particle size is typically 20–30 μm.

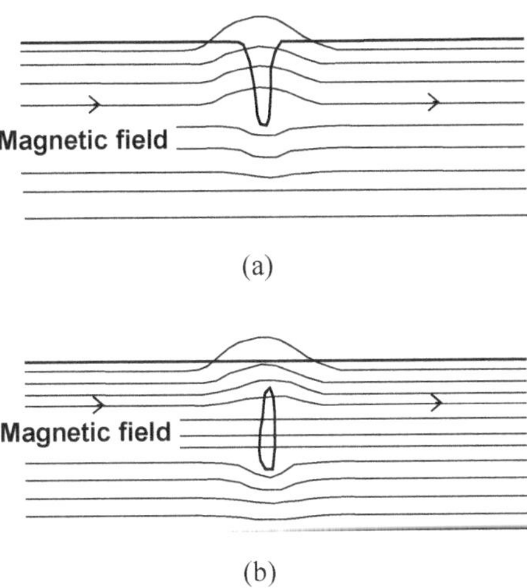

Fig. 6.31 Distortion of the magnetic flux lines at a defect causing flux leakage. The applied magnetic field is perpendicular to the length of the defect. (a) A surface defect. (b) A subsurface defect.

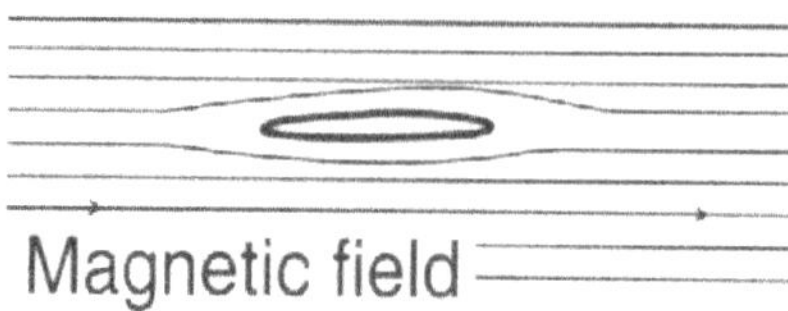

Fig. 6.32 Little distortion of the magnetic flux lines when the applied magnetic field is parallel to the length of the defect.

To generate magnetic flux in the part under evaluation, a common method (called the direct method) involves the passing of a current in the direction along the length of the defect, so that the magnetic flux lines generated by the current are perpendicular to the length of the defect, as illustrated in Fig. 6.33(b). An alternate method (called the indirect method) involves passing current through a coil that surrounds the part under evaluation, so that the generated magnetic flux is along the axis of the coil, as illustrated in Fig. 6.33(a). With the flux being axial, transverse and circumferential defects are perpendicular to the flux and are thus detected. The suitable method depends on the orientation of the defect. In either method, the current can be AC or DC. Due to the skin effect, AC generated flux tends to follow the surface topography of the part and does not penetrate deeply into the part. In contrast, DC generated flux penetrates deeply, though it does not follow the surface topography. A shortcoming of the magnetic particle inspection method is that the failure to detect any defect does not necessarily mean that defects are absent, because the method may not have been applied in an optimal fashion.

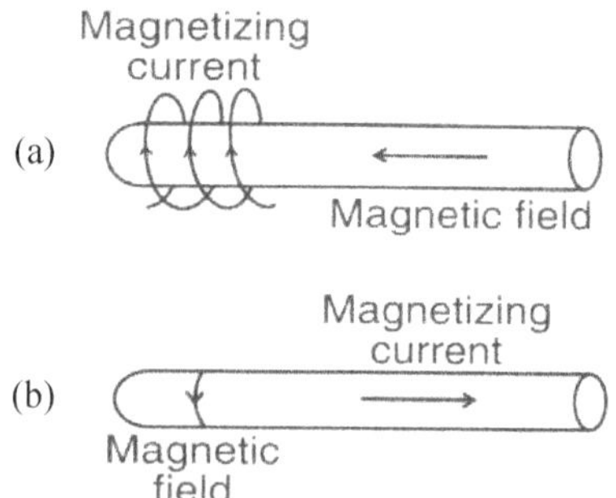

Fig. 6.33 Configurations for generating magnetic flux for the purpose of magnetic particle inspection.

6.16 Composite materials for magnetic applications

6.16.1 *Magnetic composites for nondestructive evaluation*

A method of nondestructive monitoring of the structural health involves adding ferromagnetic particles to a structural composite during composite fabrication. This is known as the particle tagging technique. It allows nondestructive evaluation to be performed, as a crack in the composite will cause distortion of the magnetic flux lines.

A related method for monitoring stress and defects involves adding to a structural composite particles that exhibit the Villari effect (Sec. 6.11). This technique is known as magnetotagging. It is also known as the particle tagging technique. The stress monitoring ability is based on the ability of the Villari particles to exhibit a change in magnetic field when they are mechanically stressed. The defect monitoring ability is based on the distortion of the magnetic flux lines (i.e., magnetic permeability variation) by the defects, as in the case of tagging using ferromagnetic particles.

6.16.2 *Metal-matrix composites for magnetic applications*

Because metals are among the most common ferromagnetic materials, metal-matrix composites are quite common among magnetic composite materials. In general, these composites can be classified into three groups, namely (i) composites with a magnetic (ferromagnetic or ferrimagnetic) metal matrix and a magnetic filler, (ii) composites with a magnetic metal matrix and a non-magnetic (anti-ferromagnetic, paramagnetic or diamagnetic) filler and (iii) composites with a non-magnetic metal matrix and a magnetic filler. In any group, ferromagnetic metals are dominant among the magnetic metals.

Composites in group (i) include those containing an α-Fe (a soft magnetic material) and $R_2Fe_{14}B$ (a hard magnetic material, with R = a rare earth element such as Nd). The soft magnetic grains cause a large spontaneous magnetization, while the hard magnetic grains induce a large coercive field. Other examples of composites in group (i) are those containing an nickel-iron alloy and nickel zinc ferrite, those containing cobalt and EuS, and those containing Sb and MnSb.

Composites in group (ii) include those containing a ferromagnetic metal (e.g., Co, Ni and Fe) and an antiferromagnetic phase (e.g., CoO,

NiO and FeS). The large antiferromagnetic-ferromagnetic interface causes enhancement of the room temperature coercivity of the ferromagnetic phase.

Composites in group (iii) include those with aluminum, Zn-22Al, Cu-Zn-Al, $Nb_{0.33}Cr_{0.67}$, silver, copper and other metals as matrices. The fillers include iron fibers, barium ferrite powder, strontium powder, Fe-Cr flakes, $SmCo_5$, iron nitride, iron oxide, CoFe and ferrite. In spite of the non-magnetic character of the matrix, these composites are attractive in the electrical conductivity and formability resulting from the metal matrix. The formability is particularly high for composites with a superplastic metal matrix, such as Zn-22Al.

6.16.3 *Polymer-matrix composites for magnetic applications*

Although polymers are usually not magnetic, their processability makes the fabrication of polymer-matrix composites relatively simple. These composites contain a non-magnetic polymer matrix and a magnetic filler, which is commonly in powder form. They provide a monolithic and low-cost form of magnetic material. Examples of fillers are ferrite, Fe_2O_3, iron, nickel and cobalt-nickel. Both conventional and specialized magnetic applications benefit from magnetic polymer-matrix composites. The specialized applications are described below.

Due to the interaction of a magnetic material with electromagnetic radiation, magnetic polymer-matrix composites are used for electromagnetic interference (EMI) shielding. Composites with elastomeric matrices (e.g., silicone) are attractive for magnetostrictive actuation, heat-shrink applications, magnetoresistive switching and piezoinductive current generation. Some of these applications (e.g., magnetoresistance and piezoinduction) benefit from a certain degree of electrical conductivity, so the conductivity of the ferromagnetic particles and, less commonly, that of the polymer matrix, are sometimes exploited. Examples of conducting polymer matrices are polypyrrole and polyaniline. Nevertheless, non-conducting polymer matrices, such as polyethylene, wax and others, are most common, due to their low cost, wide availability and processability.

For reducing the energy loss due to the eddy current flow in a magnetic core of a current transformer, low electrical conductivity is desired for the core material. To attain a low conductivity, polymer-matrix composites with a non-conducting matrix are used. Due to the requirement for AC operation of the transformer, the filler is a soft magnetic powder.

Ferromagnetic and electrically conductive particles are used as a filler in composites which require anisotropic conduction, because the particles can be aligned during the composite fabrication by using a magnetic field.

In case that the ferromagnetic particles are asymmetric, the particles can be oriented by the flow of the polymer during injection molding. The orientation results in enhanced magnetic permeability when the direction of orientation and the magnetic field are parallel to one another. Akin to asymmetric particles are flakes, particularly nanoflakes (submicron thickness and aspect ratio 10–100), which are potentially useful for microwave effective media.

Ferromagnetic particle polymer-matrix composites in the form of films formed by spin coating on glass or semiconductor substrates are potentially useful as magneto-optical media for optical devices and integrated optics. A related type of composite is a polymer (e.g., silicone) substrate implanted with iron and cobalt ions.

6.16.4 *Ceramic-matrix composites for magnetic applications*

For transformer cores and related applications, polymer-matrix composites are attractive, due to the low conductivity of most polymers. In contrast, metal-matrix composites have high conductivity. However, polymer-matrix composites cannot withstand high temperatures. Therefore, ceramic-matrix composites, which typically have low conductivity and high temperature resistance, are desirable. However, ceramic-matrix composites suffer from the high cost of fabrication.

The most common ceramic matrix is aluminum oxide (alumina, or Al_2O_3), which can be obtained from alumina gels (AlOOH) by calcination and reduction. The ferromagnetic fillers used in alumina-matrix composites include iron particles, nickel particles and cobalt particles. The metal particles can be obtained by reduction of the corresponding metal oxide particles, e.g., Fe from Fe_2O_3, and Ni from NiO, or reduction of the corresponding metal salts, e.g., $Ni(NO_3)_2$, to the metal oxides, e.g., NiO, prior to further reduction to the metal, e.g., Ni.

One of the next most common ceramic matrices is zirconium oxide (zirconia, or ZrO_2), which is commonly doped with Y_2O_3 (yttria). The fillers used include ferrimagnetic ferrite particles such as barium hexaferrite ($BaFe_{12}O_{19}$) and ferromagnetic nickel particles. Other ceramic matrices used include aluminum nitride, boron nitride and silicon nitride.

Ceramic-matrix composites in the form of films can be prepared by reactive sputtering of appropriate alloys or compounds (such as nickel aluminide) to form a metal nitride (such as nickel nitride), followed by heat treatment in vacuum to obtain the metal from the metal nitride. The heat treatment can be performed locally by means of a focused laser beam in order to generate microscopic features that are useful for magnetic data storage. Instead of reactive sputtering, composite films can be deposited by vacuum codeposition, e.g., codeposition of Fe and MgO to form an MgO-matrix Fe particle composite film.

Materials that absorb microwave are needed for avoiding the detection of aircraft, missiles and ships by radar. This is known as Stealth technology or low observability. The interaction of a magnetic material with electromagnetic radiation provides a mechanism for absorption. Therefore, aluminosilicate ($Al_2O_3.2SiO_2$) and related low-cost ceramic matrices are used to form magnetic structural composites.

Glass-matrix composites containing magnetic particles are attractive for magnetooptical applications. Examples are a sodium borosilicate glass matrix composite containing ferrimagnetic garnet particles, a silica (SiO_2) matrix composite containing Fe_3O_4 particles, and an aluminum borate matrix composite containing Fe particles.

In the hyperthermia treatment of cancers, bone repairing materials in the form of bioactive glass-ceramics containing ferrimagnetic particles are useful as thermoseeds.

Lead zirconate titanate (PZT) is both ferroelectric and ferromagnetic, so its use, especially in a ferrite matrix, presents electric dipoles and magnetic dipoles to the electromagnetic radiation, thereby attaining high electromagnetic absorption ability.

A superconductor is characterized by the ability to exclude magnetic flux. This is known as the Meissner Effect. A composite with a ferromagnetic matrix and a superconductor particle filler is potentially useful for novel magnetic devices, because the ferromagnetic matrix affects the flux exclusion of the superconductor filler.

Review questions

1. What is meant by a Bohr magneton?
   Answer: A Bohr magnton is the magnetic moment of one unpaired electron.

2. What is meant by the *B-H* product?
   Answer: The *B-H* product is the product of B and H, as typically obtained in the demagnetization quadrant of the *B-H* curve. It relates to the magnetic energy per unit volume. Although it is often referred as the power, it is energy rather than power.

3. Give an example of a ferromagnetic material.
   Answer: Iron.

4. Give an example of a ferrimagnetic material.
   Answer: Fe_3O_4.

5. Give an example of an antiferromagnetic material.
   Answer: NiO.

6. What is the difference between a hard magnetic material and a soft magnetic material?
   Answer: A hard magnetic material is hard to magnetize or demagnetize, whereas a soft magnetic material is easy to magnetize or demagnetize.

7. What is an advantage for a magnetic material to be electrically non-conductive?
   Answer: Nonconductivity avoids eddy current, which causes energy loss.

8. What is the advantage of using a nanometer-sized magnetic filler instead of a micrometer-sized magnetic filler in a composite?
   Answer: A smaller size of the filler unit increases the amount of filler-matrix interface area per unit volume of the composite, thereby possibly inhibiting domain boundary movement, thus increasing the coercivity.

9. Name an application of the Villari effect.
   Answer: The Villari effect (or the converse magnetostrictive effect) converts mechanical energy to magnetic energy and can be used for stress sensing.

10. What is the difference between the ferromagnetic shape memory effect and the conventional shape memory effect?
    Answer: This effect differs from the conventional shape memory effect in that the deformation occurs completely in the martensite phase.

11. What are the main disadvantages of the magnetic particle inspection method?
    Answer: (a) Both surface and subsurface defects can be detected, though the subsurface defects are limited to those that are near the surface. (b) A defect that has its length parallel to the applied magnetic field cannot be detected. (c) The method is limited to the evaluation of materials that are magnetic (ferromagnetic or ferromagnetic).

12. Name an application for a multilayer involving alternately magnetic and non-magnetic layers.
    Answer: The main application relates to magnetic field sensing, which is important for magnetic recording (e.g., read heads), data storage (e.g., random access memory) and manufacturing control.

13. Why are ferromagnetic particles added to a structural composite?
    Answer: The addition of ferromagnetic particles to a structural composite enables nondestructive monitoring of the structural health of the structural material.

14. What are the advantages of a magnetorheological fluid compared to an electrorheological fluid?
    Answer: MR fluids are advantageous over ER fluids in the high maximum yield stress, wide operable temperature range, low voltage requirement and stability in the presence of impurities.

15. How many atoms of each type are there per unit cell of the cubic Heusler structure?
    Answer: As shown in Fig. 6.28, a unit cell contains 8 Ni atoms, 4 Mn atoms and 4 Ga atoms.

Example problems

Problem 1 (Solenoid)

A solenoid is a coil of wire acting as a magnet when it carries and electric current. Calculate the magnetic flux density B inside a solenoid which is 2 m long and has 2000 turns. The solenoid is in air and carries a current of 1600 A. Hints:

The magnetic field strength H is given by nI, where n is the number of coils per unit length of the solenoid and I is the current.

H is abbreviation for Henry $= \Omega.s$; 1 Henry corresponds to the situation in which a current of 1 Ampere flowing through a coil produces a magnetic flux of 1 Weber per turn of the coil.

> The unit of the magnetic flux density B is T (abbreviation for Tesla). $1\ T = 1\ Wb/m^2$
> $Wb = V.s$

The permeability of free space $\mu_o = 4\pi \times 10^{-7}$ H/m. The value is essentially the same for air instead of vacuum.

Solution:

The number of coils per unit length $= n = 2000/(2\ m) = 1000\ m^{-1}$

$H = nI = (1000\ m^{-1})(1600\ A) = 1.6 \times 10^{6}\ A/m$

$B = \mu_o H$, where the permeability of free space $\mu_o = 4\pi \times 10^{-7}$ H/m.

$B = (4\pi \times 10^{-7}\ H/m)(1.6 \times 10^{6}\ A/m) = 2\ H.A/m^2 = 2\ T$

Note: $H.A/m^2 = (\Omega.s)A/m^2 = V.s/m^2 = T$

Problem 2 (Magnetic flux density)

What is the magnetic flux density B for a magnetic field strength H of 80 kA/m in (a) vacuum, (b) a material with relative permeability $\mu_r = 5000$? (c) What is the magnetization M in (b)?

                    Functional Materials

Solution:

(a) In vacuum,

$$B = \mu_o H$$

$$= (4\pi \times 10^{-7} \text{ H/m}) (80 \text{ kA/m})$$

$$= 1.0 \times 10^{-4} \text{ H.kA/m}^2$$

$$= 1.0 \times 10^{-1} \text{ H.A/m}^2$$

Since the unit H = wb/A,

$$\text{H.A/m}^2 = \frac{\text{wb}}{\text{A}} \frac{\text{A}}{\text{m}^2} = \text{wb/m}^2 .$$

Hence,

$$B = 1.0 \times 10^{-1} \text{ wb/m}^2$$

(b) In a material with $\mu_r = 5000$,

$$B = \mu H$$

$$= \mu_o \, \mu_r \, H$$

$$= (4\pi \times 10^{-7} \text{ H/m}) (5000) (80 \text{ kA/m})$$

$$= 500 \text{ wb/m}^2$$

(c) $B = \mu_o (H + M) = \mu_o H + \mu_o M$

Hence,

$$M = (B - \mu_o H)/\mu_o = (B/\mu_o) - H$$

$$= \frac{500 \text{ wb/m}^2}{4\pi \times 10^{-7} \text{ H/m}} - 80 \text{ kA/m}$$

$$= 4 \times 10^8 \text{ wb/H.m} - 8 \times 10^4 \text{ A/m}$$

Since the unit H = wb/A,

$$\frac{\text{wb}}{\text{H.m}} = \frac{\text{wb}}{(\text{wb/A}).m} = \text{A/m}.$$

Hence,

$$M = 4 \times 10^8 \text{ A/m} - 8 \times 10^4 \text{ A/m} = 4 \times 10^8 \text{ A/m}$$

Note that $\mu_o H$ is negligible, so that $B \sim \mu_o M$. This is because μ_r is large.

Problem 3 (Magnetic flux density)

What is the magnetic field strength H needed to obtain a magnetic flux density B of 2.01 T (T being the abbreviation for Tesla) in a material with relative magnetic permeability $\mu_r = 127$?
Hints:
H is abbreviation for Henry = Ω.s; 1 Henry corresponds to the situation in which a current of 1 Ampere flowing through a coil produces a magnetic flux of 1 Weber per turn of the coil.
    1 T = 1 Wb/m^2
    Wb = V.s
The permeability of free space $\mu_o = 4\pi \times 10^{-7}$ H/m.

Solution:

$$B = \mu_o \mu_r H,$$

$$H = B/(\mu_o \mu_r) = (2.01 \text{ T})/[(4\pi \times 10^{-7} \text{ H/m}) (127)]$$

$$= 1.26 \times 10^{-3} \text{ T.m/H} = 1.26 \times 10^4 \text{ A/m}.$$

Note: T.m/H $= [(V.s)/m^2][m/(\Omega.s)] = V/(m.\Omega) = A/m$

Problem 4 (Magnetization)

What is the magnetization M corresponding to a magnetic flux density B of 2.01 T (T being the abbreviation for Tesla) in a material with relative magnetic permeability $\mu_r = 127$?
Hints:
$$B = \mu_o H + \mu_o M$$

$B = \mu H$, where μ is the magnetic permeability of the material.

$\mu = \mu_o \mu_r$, where μ_r is the relative magnetic permeability of the material.

The permeability of free space $\mu_o = 4\pi \times 10^{-7}$ H/m.

Solution:

$$B = \mu H$$

$$H = B/\mu = B/(\mu_o \mu_r) = (2.01 \text{ T})/[(4\pi \times 10^{-7} \text{ H/m}) (127)] = 1.26 \times 10^4$$

$$\text{T.m/H} = 1.26 \times 10^4 \text{ A/m}$$

Note: T.m/H $= [(V.s)/m^2][m/(\Omega.s)] = V/(m.\Omega) = A/m$

$$B = \mu_o H + \mu_o M$$

$$M = (B/\mu_o) - H = (2.01 \text{ T})/(4\pi \times 10^{-7} \text{ H/m}) - (1.26 \times 10^4 \text{ A/m})$$

$$= (1.6 \times 10^6 \text{ A/m}) - (1.26 \times 10^4 \text{ A/m}) = 1.6 \times 10^6 \text{ A/m}$$

Problem 5 (Magnetization and magnetic susceptibility)

(a) What is the magnetization of a material that exhibits a relative permeability (μ/μ_0) of 46? Express the magnetization in terms of the magnetic field H.
Hint: $B = \mu_o (H + M) = \mu H$

(b) $M = \chi H$, where M is the magnetization, H is the magnetic field and χ is the magnetic susceptibility. What is the magnetic susceptibility for a material that exhibits a relative permeability (μ/μ_o) of 46?

Solution:

(a) $H + M = (\mu/\mu_o) H = 46\ H$

$M = (46 - 1)\ H = 45\ H$

(b) $\chi = M/H = 45$

Problem 6 (Lorentz force)

Calculate the Lorentz force resulting from a magnetic flux density B of 0.79 T and a current I of 1.47 A, such that B and I are perpendicular to one another. The wire carrying the current has diameter 1.74 mm. The wire material has 1.68×10^{22} electrons per cm^3. The drift of these electrons is responsible for the current.
Hint:
From Eq. (2.2) in Chapter 2, the current I is given by

$$I = qNvA,$$

where N is the number of electrons for unit volume, A is the cross-sectional area, v is the drift velocity and q is the magnitude of the electronic charge.

Solution:

$$v = I/(qNA)$$

$$A = \pi \, [(1.74 \times 10^{-3} \text{ m})/2]^2$$

$$F = qvB = qIB/(qNA) = IB/(NA)$$

$$= (1.47 \text{ A})(0.79 \text{ T})/\{(1.68 \times 10^{28} \text{ m}^{-3})\,[\pi\,(1.74 \times 10^{-3} \text{ m}/2)^2]\}$$

$$= 2.9 \times 10^{-23} \text{ N}$$

Note: A.T m^3/m^2 =A (V.s/m^2) m = C.V/m = J/m = N

Problem 7 (Mass spectrometry)

In relation to mass spectrometry, derive an expression for the mass-to-charge ratio (m/q) of an ion as a function of the magnetic induction B, the radius r of the ion trajectory, and the ion velocity v. Hint: Lorentz force = centripetal force.

Solution:

$$qv \times B = mv^2/r$$

$$m/q = Br/v$$

Problem 8 (Magnetic moment)

What is the magnetic moment in 1 cm^3 of ferrite (Fe_3O_4) that has been fully magnetized?

Solution:

The saturation magnetization of ferrite is 5×10^5 A.m^{-1}.

$$\text{Magnetic moment} = (\text{saturation magnetization}) \, (\text{volume})$$

$$= (5 \times 10^5 \text{ A.m}^{-1}) \, (1 \times 10^{-6} \text{ m}^3)$$

$$= 0.5 \text{ A.m}^2$$

Problem 9 (Magnetic moment)

(a) What is the number of Bohr magnetons associated with an atom of the element Eu (europium)? The electronic configuration of Eu is [Xe] $4f^7 6s^2$.

(b) What is the number of Bohr magnetons associated with an F$^-$ (fluoride) ion?

Solution:

(a) $4f^7$ gives 7 unpaired electrons, since there are 7 orbitals in the f subshell. Thus, there are 7 Bohr magnetons per Eu atom.

(b) F$^-$ has valence electronic configuration $2s^2 2p^6$. There is no unpaired electron. The number of Bohr magnetons is zero.

Problem 10 (Magnetic moment)

Magnesium oxide (MgO) consists of Mg^{2+} and O^{2-} ions. It exhibits the rocksalt (NaCl) crystal structure, just like NiO (Fig. 6.15). Estimate the magnetic moment per unit cell of MgO by considering the contribution due to the spin of the electrons. Explain your answer.

Solution:

Both Mg^{2+} and O^{2-} ions have full shell electronic configurations, so there are no unpaired electrons for either ion. Hence, the magnetic moment is estimated to be zero.

Note that Mg is not a transition metal, whereas Ni is a transition metal (with d electrons).

Problem 11 (Magnetic moment)

Metallic iron is ferromagnetic and is body-centered cubic (BCC) in crystal structure, with lattice parameter (length of an edge of the cubic unit cell) 2.8665 Å. Estimate the maximum magnetic moment per unit volume (i.e., saturation magnetization) by considering the contribution due to the spin of the electrons.

Solution:

From Sec. 6.2, the magnetic moment per Fe atom is 4 (9.27×10^{-24} A.m^2). There are 2 atoms per BCC unit cell. Thus,

$$\text{magnetic moment per unit volume}$$

$$= \frac{\text{Magnetic moment per unit cell}}{\text{Volume of a unit cell}}$$

$$= \frac{(2)(4)(9.27 \times 10^{-24}\,\text{A.m}^2)}{(2.8665 \times 10^{-10}\,\text{m})^3}$$

$$= 3.15 \times 10^6 \,\text{A.m}^{-1}$$

Problem 12 (Bohr magnetons)

How many Bohr magnetons of magnetic moment are associated with (a) a samarium (Sm) atom, (b) a terbium (Tb) atom?

Solution:

(a) The electronic configuration of Sm is … $4f^6 5d^0 6s^2$.

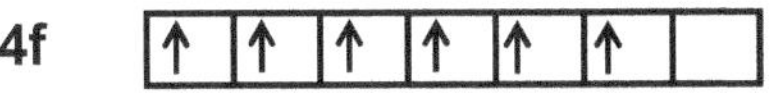

There are six unpaired electrons, so $\beta = 6$.

(b) The electronic configuration of Tb is … $4f^9 5d^0 6s^2$.

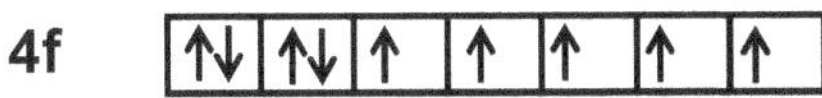

There are five unpaired electrons, so $\beta = 5$.

Problem 13 (Magnetic moment)

Neodymium is an element with electronic configuration [Xe] $4f^4\ 6s^2$. What is the magnetic moment in unit of the Bohr Magneton for an atom of neodymium? Indicate how you obtain your answer.

Solution:

4 Bohr magnetons.

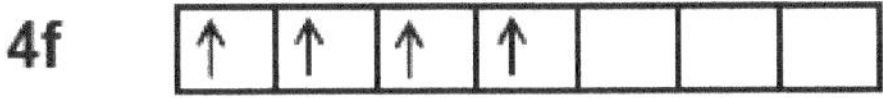

Problem 14 (Magnetic moment)

What is the magnetic moment in 15 cm^3 of ferrite (Fe_3O_4) that has been fully magnetized?
The saturation magnetization of ferrite is 5×10^5 A.m^{-1}.

Solution:

Magnetic moment = (saturation magnetization) (volume)

$$= (5 \times 10^5 \text{ A.m}^{-1})\,(15 \times 10^{-6} \text{ m}^3) = 7.5 \text{ A.m}^2$$

Problem 15 (Saturated magnetization)

Calculate the saturated magnetization of BCC iron, which is ferromagnetic. The lattice constant (length of an edge of the conventional cubic unit cell) is 2.8665 Å.
Hint: Magnetization = magnetic moment per unit volume.

Solution:

BCC iron has 2 atoms per cubic unit cell. Each atom has 4 Bohr magnetons. Hence, each unit cell has $4 \times 2 = 8$ Bohr magnetons.

3d | ↑↓ | ↑ | ↑ | ↑ | ↑ |

The magnetic moment corresponding to 1 Bohr magneton is 9.27×10^{-24} A.m^2. Thus,

Magnetic moment per unit cell = $8\,(9.27 \times 10^{-24} \text{ A.m}^2)$

Saturated magnetization

= Magnetic moment per unit cell/volume of a unit cell

= $[8\,(9.27 \times 10^{-24} \text{ A.m}^2)]/(2.8665 \times 10^{-10} \text{ m})^3$

= 3.15×10^6 A/m

Problem 16 (Magnetic permeability and magnetic energy)

A ferromagnetic material of volume 2579 mm^3 has relative magnetic permeability 5787. A magnetic field strength H of 58.7 A/m is applied to this material. The permeability of free space $\mu_o = 4\pi \times 10^{-7}$ H/m.

(a) Calculate the magnetic permeability μ of this material.

(b) Calculate the magnetic flux density B.

(c) Calculate the magnetization M.

(d) Calculate the magnetic moment in this volume of material.

(e) Estimate the magnetic energy per unit volume of this material.

Hints:
$$H = Wb/A$$

$$T = Wb/m^2$$

$$J = Wb.A$$

$$B = \mu_o \, (H + M)$$

Solution:

(a) $\mu = 5787 \, (4\pi \times 10^{-7} \text{ H/m}) = 7.27 \times 10^{-3}$ H/m

(b) $B = \mu H = (7.27 \times 10^{-3} \text{ H/m}) \, (58.7 \text{ A/m}) = 0.427$ T

(c) $B = \mu_o \, (H + M)$

$H + M = B/\mu_o$

$M = (B/\mu_o) - H = [(4.27 \times 10^{-1} \text{ T})/(4\pi \times 10^{-7} \text{ H/m})] - (58.7 \text{ A/m})$

$= 3.398 \times 10^5 \text{ A/m} - 58.7 \text{ A/m}$

$= 3.40 \times 10^5$ A/m

(d) Magnetic moment $= M$ (volume)

$$= (3.40 \times 10^5 \text{ A/m}) (2.579 \times 10^{-6} \text{ m}^3)$$

$$= 0.877 \text{ A.m}^2$$

(e) Magnetic energy per unit volume $\sim$ BH

$$= (0.427 \text{ T}) (58.7 \text{ A/m})$$

$$= 25.1 \text{ J/m}^3$$

Problem 17 (Magnetoresistance)

A magnetoresistive material has its electrical resistivity decreased by 45% upon application of a magnetic field. What is the magnetoresistance ratio?

Solution:

From Eq. (6.21),

$$\text{MR ratio} = (R_o - R_H)/R_H$$

$$= (R_o - 0.55 \ R_o)/(0.55 \ R_o)$$

$$= (1 - 0.55)/0.55 = 0.45/0.55$$

$$= 0.82$$

Problem 18 (Magnetostriction)

A magnetoresistive material has its electrical resistivity decreased by 55% upon application of a magnetic field. What is the magnetoresistance ratio?

Solution:

$$\text{MR ratio} = (R_o - R_H)/R_H$$

$$= (R_o - 0.45\ R_o)/(0.45\ R_o)$$

$$= (1 - 0.45)/0.45$$

$$= 0.55/0.45$$

$$= 1.22$$

References

1. https://www.khanacademy.org/science/physics/magnetic-forces-and-magnetic-fields/magnetic-flux-faradays-law/a/what-is-magnetic-flux

2. https://eng.libretexts.org/Bookshelves/Electrical_Engineering/Electro-Optics/Book%3A_Electromagnetics_I_(Ellingson)/02%3A_Electric_and_Magnetic_Fields/2.05%3A_Magnetic_Flux_Density

3. https://www.youtube.com/watch?v=5-gVHE2VY0o

4. https://www.askiitians.com/iit-jee-magnetism/magnetic-properties-of-materials/

5. https://www.toppr.com/guides/physics/magnetism-and-matter/magnetic-properties-of-materials/

6. https://chem.libretexts.org/Courses/Saint_Marys_College_Notre_Dame_IN/CHEM_431%3A_Inorganic_Chemistry_(Haas)/CHEM_431_Readings/14%3A_Magnetism/14.02%3A_Magnetic_Properties_of_Materials